International Conference on Emerging Electronic Solutions for IoT (ICEESI 2017)

MATEC Web of Conferences Volume 140 (2017)

Penang, Malaysia
9 - 10 October 2017

Editors:

Ping Jack Soh
Zahereel Iswar Abdul Khalib
Thennarasan Sabapathy
Muzammil Jusoh

ISBN: 978-1-5108-5351-5

Printed from e-media with permission by:

Curran Associates, Inc.
57 Morehouse Lane
Red Hook, NY 12571

CURRAN ASSOCIATES INC. proceedings.com

Some format issues inherent in the e-media version may also appear in this print version.

Printed by Curran Associates, Inc. (2018)

For additional information, please contact EDP Sciences – Web of Conferences
at the address below.

EDP Sciences – Web of Conferences
17, Avenue du Hoggar
Parc d'Activité de Courtabœuf
BP 112
F-91944 Les Ulis Cedex A
France

Phone: +33 (0) 1 69 18 75 75
Fax: +33 (0) 1 69 28 84 91

contact-edps@webofconferences.org

Additional copies of this publication are available from:

Curran Associates, Inc.
57 Morehouse Lane
Red Hook, NY 12571 USA
Phone: 845-758-0400
Fax: 845-758-2633
Email: curran@proceedings.com
Web: www.proceedings.com

TABLE OF CONTENTS

Preface

The International Conference on Emerging Electronic Solutions for IoT (ICEESI 2017) is the first joint conference between School of Computer & Communication Engineering, Universiti Malaysia Perlis (UniMAP) and the Department of Computer Science & Engineering of Dhaka University of Engineering & Technology (DUET). The themes of the conference have been chosen in view of the revolution and rapid developments in the field of electronics and communication design. This conference focuses on future technologies, including Information and Communication Technology (ICT), green computing/ communication technology, multimedia technology, mobile and wireless communication technologies, wireless sensor networks, system-on-chip (SOC) technology and embedded computing technology.

ICESSI 2017 organized in Penang, Malaysia, on 9th and 10th October 2017 brought together professionals, researchers, academicians and industries to share and generate interest in cutting-edge applications, discuss the challenges and issues in the area of Internet of Things (IoT). This conference provided an avenue for productive dialogue between researchers in the same fields and different backgrounds which also provide an invaluable opportunity for networking among participants.

 DOI: 10.1051/matecconf/201714000002

Welcoming Address

Selamat Datang and Welcome to all participants of ICEESI2017

On behalf of the Organizing Committee, it is my great pleasure to welcome all local and international delegates to International Conference on Emerging Electronic Solution for IoT (ICESSI) 2017 and to the beautiful island known as "Pearl of the Orient". This event is jointly organized by SCCE UniMAP, Malaysia and CSE DUET, Bangladesh.

The Technical Program Committee undertook a challenging task in evaluating the large number of submitted papers, considering the merits of each through detailed reviews discussing, and selecting the best papers. These papers of highest calibre are planned into the two days program consisting of two keynote speakers and five (5) parallel technical sessions/tracks. I hope your experience with ICEESI2017 will be a fruitful and long lasting one. With your support and participation, it is envisioned that ICEESI2017 will continue its success for a long time in the future.

I would like to gratefully acknowledge a wide spectrum of generous sponsors and technical support, without whom we may struggle to host such a prestigious conference. They include the Malaysian Communication and Multimedia Commission (MCMC), Tekmark and Keysight Technologies, and UniMAP Holdings. My greatest gratitude goes to all the committee members who have been working very hard to ensure everything falls into place during this special occasion.

Finally, I would like to apologize for any weaknesses that have been experienced throughout organizing this conference. I sincerely hope that all the participants will enjoy a memorable and a unique overall experience during ICEESI 2017. Wish u a cheerful and great stay in Penang and Malaysia.

Thank you.

Dr. Muzammil Jusoh, C. Eng
ICEESI 2017 Conference Chair

Statement of Peer review

In submitting conference proceedings to MATEC Web of Conferences, I certify to the Publisher that I adhere to the **Policy on Publishing Integrity** of the journal in order to safeguard good scientific practice in publishing.

1. All articles have been subjected to peer review administered by the proceedings editors.
2. Reviews have been conducted by expert referees, who have been requested to provide unbiased and constructive comments aimed, whenever possible, at improving the work.
3. Proceedings editors have taken all reasonable steps to ensure the quality of the materials they publish and their decision to accept or reject a paper for publication has been based only on the merits of the work and the relevance to the journal.

Title, date and place of the conference

The 2017 International Conference on Emerging Electronic Solutions for IoT (ICEESI 2017), October 9 to 10, 2017 Penang, Malaysia

Proceedings editor(s):

Dr. Ping Jack Soh
Dr. Zahereel Iswar Abdul Khalib
Dr. Thennarasan Sabapathy
Dr. Muzammil Jusoh

Date and editor's signature 22 November 2017

DOI: 10.1051/matecconf/201714001001

Performance Evaluation of Downlink Design Based on Free Space Optical System

A. A. Anis[1,*], *C.B.M.Rashidi*[1], *A. K. Rahman*[2], *S.A.Aljunid*[1] and *Yaakob S*[3]

[1]Advance Communication Engineering, Centre of Excellence School of Computer & Communication Eng, Universit Malaysia Perlis, (UniMAP) Kangar, 01000, Kangar, Perlis
[2]Deparment of Electrical and Electronic Eng, Universiti Malaysia Sarawak, (UNIMAS), Sarawak
[3]Deparment of Computer and Communication Systems Eng, Faculty of Eng, Universiti Putra Malaysia (UPM), Serdang, Selangor

Abstract. In this paper, the performance of downlink design utilizing the free space optic (FSO) is presented. The permissible BER of 10^{-58} is achieved when the system utilized 40 cm aperture diameter. The highest system sensitivity will lead to less error and enable more transmission output. The result revealed, at a bit rate 155Mbps and effective received power -27.5 dBm, FSO downlink system at bit rate 155 Mbps improved by 13% as compared to 1.5 Gbps and 2.5 Gbps in single transmission line. It is truly shown that, the FSO system utilized 155 Mbp/s with 40 cm aperture diameter shown good performance of a reliable FSO downlink system.

1 Introduction

Free space optical (FSO) has to be included in the set of answers for qualifying the bandwidth requirements of the modern internet. FSO system is an incredible alternative to overcome the bottleneck broadband connectivity problem. Recently, FSO is very active with numerous fundamental and technological problems to enhance its performance in a wide situations range. The swift continuity in the increasing of data rate transmission by FSO communication system in both long- and short-range applications implementing the rapid development of the system performance. Figure 1 shows a basic scheme of the FSO communication system with optical wireless and diverse user connectivity.

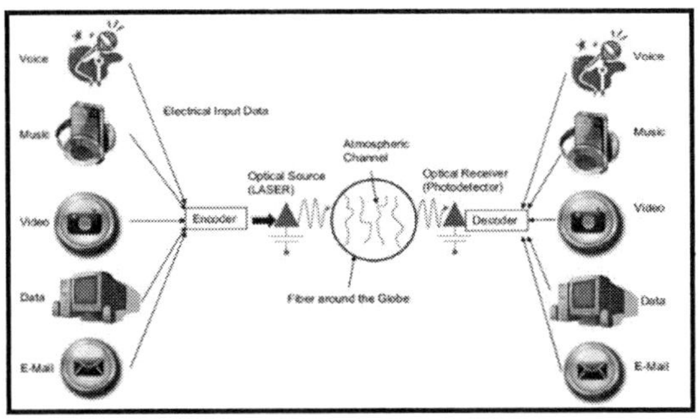

Fig. 1. A basic scheme of the FSO communication system with optical wireless and diverse user connectivity [1]

Various FSO link conditions must fulfil multiple system needs. These involve various propagation channel characteristics, different technique of transmission, and detection methods. The advantage of indoor applications of FSO communications such as secure communications and sensor networking over short ranges is drawing attention globally. It is made as a non-interfering replacement to RF communications [1]. Speeds usually up to 2.5 Gbps are achievable and ranges up to a few km, depends on technology and conditions of the atmosphere [2]. In a FSO link the transmitters transmit high power light. The power light however appears as low power light at the receiver after transferring through the air.. The link margin of the connection shows the light amount received by an optical terminal over the lowest value crucial to maintain the link active. Fig. 2 displays the basic communication system which is consisting of a transmitter, a receiver, and a communication channel.

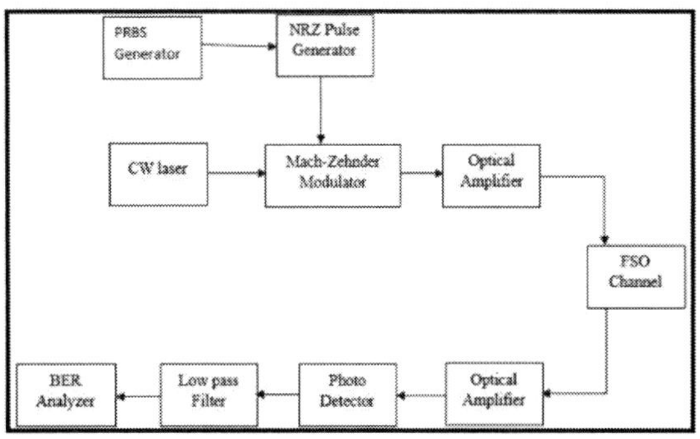

Fig. 2. The basic communication system [3]

There are some parameters related to the performance degradation of a FSO link. They are the

* Corresponding author: anis_afina@ymail.com

DOI: 10.1051/matecconf/201714001001

link range between optical transceiver, line of sight, alignment of optical transceiver, atmospheric conditions, water vapour and hot air, strong electromagnetic interference, and optical signal wavelength [4]. An expendable microwave link is inevitably crucial, as the laser link can stop working under various changing conditions. Numerous investigations and applications of FSO have been reported [5-6]. A laser beam faces power loss due to various factors as it propagates through the air. One of them is absorption and scattering loss as defined in Beer's Law [7]. Absorption happens due to the photons of the beam bumping with haze particles. This phenomenon can reduce the link margin, distance and the availability of the link [8]. Scattering is the dispersion of a radiation beam due to physical interactions.

2 Bit Error Rate Simulation Analysis

The structural design of FSO can be categorized into three types of connections. They Point-to-Point connections, Point-to-Multipoint connections, and Multipoint-to-Multipoint connections. Point-to-Point connections consist of one transmitter and one receiver. This type of connection provides greater bandwidth although their availability is lower. The second connection is Point-to-Multipoint connections. They consist of one transmitter and multiple receivers. This second design provides low cost connections. Its advantage is that it facilitates node addition of the network. Even so, this type of connection provides poor bandwidth level compared to the former. The third design of FSO connection is Multipoint- to- Multipoint connections. This connections utilize multiple receivers and transmitters. They are also called as Mesh Network which provides redundancy and better reliability with easy node addition. Nonetheless, these connections have a significant disadvantage which shorter is distance limitation than other FSO designs. The FSO networks are shown in Figure 3, Figure 4 and Figure 5.

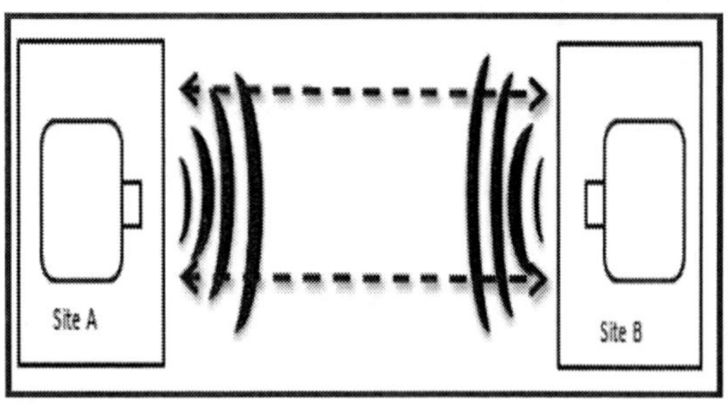

Fig. 3. Point-to-Point Connection

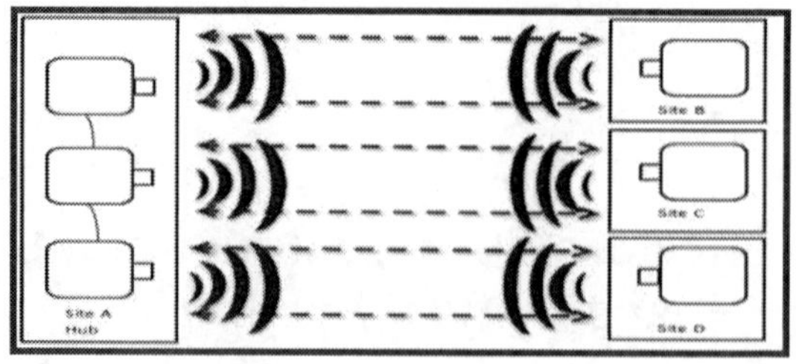

Fig. 4. Point-to-Multipoint Connection

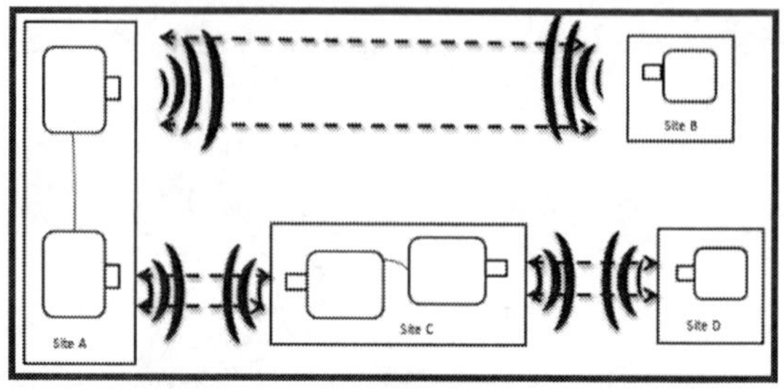

Fig. 5. Multipoint-to-Multipoint Connection

FSO systems provide numerous advantages to the users. Examples are being low start-up and economical, swift deployment, and high bandwidths due to the technology optical nature[9]. The optical bandwidth can reach up to 1 Gbps or more. It is primarily to assist wide bandwidth range over a long distance communication [10]. Although at high bandwidth, long distance cost efficient communication the environmental factors such as weather linked factors continue as major challenges to minimize and may limit the data rate and link availability via LOS (Line of Sight) [11]. Recent studies suggest ways to solve LOS line of sight problems with minimal bit-error-rate (BER) performance in FSO communication system under worst interference conditions [12]. But the increasing of laser power can resolve this problem up to some point [13]. As the transmitter reaching maximum power limit, the bandwidth or transmission distance of the physical links which are under severe weather conditions such as heavy haze has to be reduced. This will cause some alterations in the structures of FSO networks. Thus, the effectiveness and performance of Free Space Optical network will be affected.

3 Bit Error Rate Simulation Analysis

Bit error rate (BER) is one of the main elements of FSO communication system. BER can be estimated using equation 1

$$BER \approx \frac{Ne}{Nb}$$

where,
Ne=received error bits number
Nb=transmitted bits for a period

Generally, the BER values for a digital transmission system are usually specified and influenced on the Signal-to-Noise-Ratio (SNR) of the optical receiver.
This feature determined the probability of errors when one bit of data is being transmitted. The internal noise of the optical receiver affects the value of BER [14]. The value of BER can be calculated using the following formula.

$$BER = \frac{1}{2} erfcQ$$

However in this paper, the results for tht BER are obtained from simulation using OptiSys™ simulation tool. The data collected are analysed in the form of graph and evaluated in the section below.

4 Results and Discussion

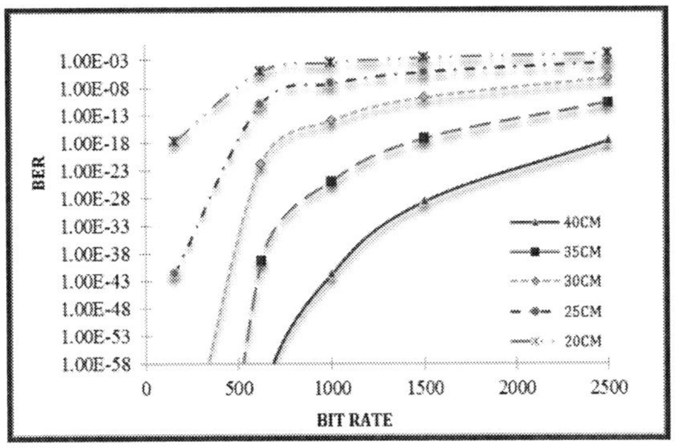

Fig. 6. Graph of bit rate against bit error rate (BER) with different receiver aperture diameter

Figure 6 demonstrate the relationship between bit error rate (BER) and bit rate using apertures diameter which ranges from 20 to 40 cm. From the graph, it can be seen that the bit error rate increases when there is bit rate increment. It is observed that the receiver aperture diameter of 40 cm has achieved the lowest bit error rate throughout the investigation (from bit rate 155 mbps to 2.5 gbps) compared to aperture diameter of 20, 25, 30, and 35 cm respectively. The FSO with receiver aperture diameter 40 cm has the lowest BER of 10^{-58} at 155 mbps. Then it finishes at BER 10^{-18} with bit rate of 2.5 Gbps. For receiver aperture diameter of 20 cm, the BER is 10^{-18} at the bit rate 155 mbps until BER 10^{-2} at bit rate of 2.5 Gbps. From the figure, it can be concluded that by increasing the receiver aperture diameter, the receiver's sensitivity is also increased. Higher sensitivity will allow more outputs and reduce the error received by the system. This will directly improve the system at high performance.

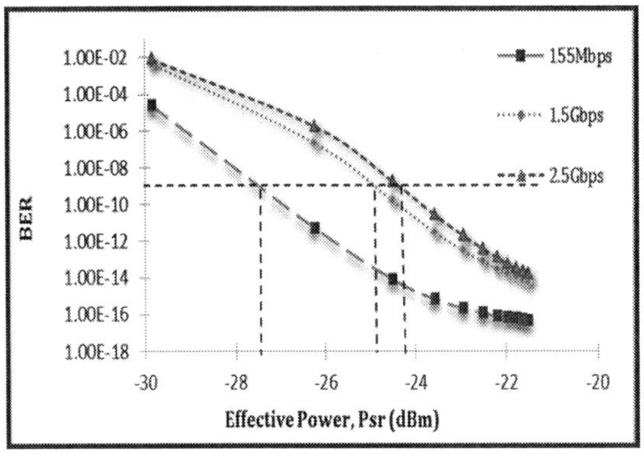

Fig. 7. Graph of bit error rate (BER) versus effective power with different bit rate

Figure 7 shows plot a BER versus effective power for bit rate 155 Mbps, 1.5 Gbps and 2.5 Gbps, respectively by considering all at range 1km and receiver aperture diameter at 20cm. the value caused of effective power P_{sr} varies from -30 dBm to -20 dBm. It can be seen that, the bit rate 155Mbps have a better performance contrast with the bit rate 1.5 Gbps and 2.5 Gbp, respectively, when the effective receives power P_{sr} for bit rate 1.5 Gbps and 2.5 Gbps are larger (i.e. when P_{sr} >-25 dBm). The effective receives power for bit rate 155 Mbps is -27.5 dBm at BER= 10^{-9} better than bit rate 1.5 Gbps and 2.5 Gbps are -24.8 dBm and -24.4 dBm respectively. The effective receiver power for bit rate 155 Mbps is the lower receiver power that can be reached to the receiver as compared with the bit rate 1.5 Gbps and 2.5 Gbps. The bit rate 155 Mbps proved that high bit rate can enhance the transmission link by 13% compared to bit rate 1.5 Gbps and 2.5 Gbps are 2% impeovement.

4 Conclusions

The downlink FSO communication system with 155 Mbps bit rate at BER 10^{-9} is best suited in comparison with a bit rate 1.5 Gbps and 2.5 Gbps, respectively. The result had shown the bit rate 155 Mbps with low BER 10^{-9} can obtain optical receive power P_{sr} -27.5 dBm. In addition, by using large aperture diameter, consequently, will improve system sensitivity.

References

1. Majumdar, A.K.: *Fundamentals of Free-Space Optical (FSO) Communication System.(2015)*
2. Pacheco de Carvalho, J.A.R., Marques, N., Veiga, H., Ribeiro Pacheco, C.F., Reis, A.D.: Extended Field Performance Evaluation of a Gbps FSO Link. ICETE. (2012) 439-446
3. Chauhan, N.R., Vala, M.K.: System Design and Performance Analysis on the Free Space Optics (FSO) System in Atmospheric Turbulence. IRJET, **Vol. 4**. (2017) 1789-1793

4. Rockwell, D.A., Mecherle, G.S.: Wavelength Selection for Optical Wireless Communication Systems. SPIE, **Vol. 4530**. (2001) 27-35
5. D'Amico, M., Leva, A., Micheli, B.: Free Space Optics Communication Systems. First Results From a Pilot Field Trial in the Surrounding Area of Milan, Italy. IEEE. (2003) 305–307
6. Löschnigg, M., Mandl, P., Leitgeb, E.: Long-term Performance Observation of a Free Space Optics Link. 10th International Conference on Telecommunications. (2009) 305-310
7. Kaushal, H., Kaddoum, G.: Free Space Optical Communication: Challenges and Mitigation Techniques. (2015)
8. Alkholidi, A., Altowij, A.: Effect of Clear Atmospheric Turbulence on Quality of Free Space Optical Communications in Western Asia. (2012) 41-74
9. Srinivasan, R., Vijayaraj, A.: Mobile Communication Implementation Techniques to Improve Last Mile High Speed FSO Communication. (2011) 55-63
10. Kedar, D., Arnon, S.: Urban optical wireless communication networks: the main challenges and possible solutions. (2004)
11. Liu, Q., Qiao, C., Mitchell, G., Stanton, S.: Optical wireless communication networks for first- and last-mile broadband access. Journal of Optical Networking. (2005) 807–828
12. Arnon, S., Kopeika, N.S.: Adaptive optical transmitter and receiver for space communication through thin clouds. Applied Optics. (1997) 199-1987
13. CBM Rashidi, SA Aljunid, F Ghani, HA Fadhil, MS Anuar, AR Arief, Optik-International Journal for Light and Electron Optics 125 (17), 4889-4894
14. Vorontsov, M.A., Dudorov, V.V., Zyryanova, M.O.,Kolosov, V.V., Filimonov, G.A.: Bit error rate in free-space optical communication systems with a partially coherent transmitting beam. Atmospheric and Oceanic Optics, **Vol. 26**. (2013) 185-189.

MATEC Web of Conferences **140**, 01002 (2017) DOI: 10.1051/matecconf/201714001002
ICEESI 2017

Performance Evaluation of a Novel Optimization Sequential Algorithm (SeQ) Code for FTTH Network

*C.A.S.*Fazlina[1,*], *C.B.M.*Rashidi[1], *A.K.*Rahman[1], *S.A.*Aljunid[1]

[1]Advanced Communication Engineering, Centre of Excellence- School of Computer & Communication Eng, Universiti Malaysia Perlis, (UniMAP) Kangar, 01000, Kangar, Perlis
[2]Department of Electrical and Electronic Eng, Universiti Malaysia Sarawak, (UNIMAS), Sarawak

Abstract. The SeQ codes has advantages, such as variable cross-correlation property at any given number of users and weights, as well as effectively suppressed the impacts of phase induced intensity noise (PIIN) and multiple access interference (MAI) cancellation property. The result revealed, at system performance analysis of BER = 10^{-09}, the SeQ code capable to achieved 1 Gbps up to 60 km.

1 Introduction

Multiple Access are used to allow a large number of mobile users to share the allocated spectrum in the most efficient manner. The types of multiple access are TDMA (Time Division Multiple Access), FDMA (Frequency Division Multiple Access) and CDMA (Code Division Multiple Access) [1]. In CDMA, the same bandwidth is occupied by all the users. However, they are all assigned separate codes, which differentiates them from each other's [2]. Optical CDMA system provide users both simultaneous and asynchronous access to networks with high security. CDMA is a strong candidate for creating effective multiple methods for the multiple methods for the optical subcarrier access network because of its asynchronous access and code multiplexing [3]. Optical network provide higher capacity and reduced costs for new applications such as internet, video, multimedia and advanced digital services. The OCDMA network system allow to each subscriber a specific code word. The code word permits to the transmitter to modulate its data sequences. In order to satisfy faster and more reliable optical communication system requirements and optimize the huge optical bandwidth sharing [4]. The main advantages of OCDMA compared with others techniques are the high capacity for higher connectivity, have more flexible bandwidth usage, higher granularity and scalability within optical networks, capable improved crosstalk performance, asynchronous access, and have potential for improved system security as number of users on the network has specific address code [5]. In OCDMA system, a bit of "1" is to be transmitted as an assigned code and a bit of "0" does not transmitted anything [5]. In OCDMA, most important consideration are code design. Multiple Access Interference (MAI) is the interference from other users transmitting at the same time, which will limit the effective error probability with the presence of noise in the overall system [6].

*Corresponding author: eng_fazz@yahoo.com

In OCDMA, PIIN will increase between codes sequences because of inappropriate cross users and cannot be improved by increasing the transmitted power [7]. There have several codes have proposed for OCDMA system, such as Flexible Cross-Correlation (FCC), Modified Double Weight (MDW), Dynamic Cyclic Shift (DCS), Modified Frequency Hopping (MFH) and Hadamard code. This paper present, algorithm called Sequential Algorithm (SeQ) code has been developed. This proposed code was developed based on superior performance of Flexible Cross Correlation proposed by [7] where, SeQ code utilizing even and odd properties in order to enhance the limitations occurs in the existing codes.

2 Sequential Algorithm (SeQ) Code Design Approach

The SeQ code design is concerned about mathematical preliminaries to design codes for any given number of users and weights to have minimum code length. Since, optical codes may be represented as vectors; linear algebra has been used for the design of these codes. The codes are treated as rows of a matrix of finite dimension; hence, the methods of design and analysis of matrix algebra is best suited in these cases. For the codes be unique, the code matrix is required to have a finite value of its rank for a given set of users. The in phase auto-correlation functions and the cross-correlation functions of the codes can easily be defined in term of the dot and cross product of the rows of the matrix. The auto-correlation of each codeword $a_i = (a, a_2, a_3 .. a_N)$ and the cross-correlation between any two distinct codeword $A_i = (a_1, a, a .. a_N)$ and $B_j = (b_1, b_2, b_3 .. b_N)$, respectively. Optical code are family of K (for K users) binary [0, 1] sequences of length N, code weight W (the number of "1" in each codeword) and the maximum cross-correlation, λ_a. The auto and cross-correlation functions of these sequences are defined by [7];

$$\lambda_a(\tau) = \sum_{i=1}^{N} a_i a_{i+\tau} = W \; for \; \tau = 0 \quad (1)$$

$$\lambda_{ab}(\tau) = \sum_{i=1}^{N} a_i b_{i+\tau} \leq 1 \; for \; \tau = 0 \quad (2)$$

 DOI: 10.1051/matecconf/201714001002

Hence, major bottleneck in the successful implementation of all optical networks is basically when all the users try to transmit their data simultaneously. It is noteworthy when designing coding sequences such that, it must cause least overlapping between data chips. Now, let $A = \{a_n\}$ and $B = \{b_n\}$ be the sequences of length N such that the vector form can be written as [8];

$$\left.\begin{aligned} \{a_n\} &= '0' \; or \; '1', i = 0, \ldots N-1 \\ \{b_n\} &= '0' \; or \; '1', i = 0, \ldots N-1 \end{aligned}\right\} \tag{3}$$

Since a_n is a {0, 1} binary sequence, the maximum value of $\lambda_a(\tau)$ in Eq (1) is for $\tau = 0$ and is equal to W, the code weight of the sequence can be expressed as;

$$\lambda_a(0) = W \tag{4}$$

The codes described by Eq (3) can also be represented in vector form as;

$$\left.\begin{aligned} A &= \{a_i\} = for \; i = 0, 1, \ldots N-1 \\ B &= \{b_i\} = for \; i = 0, 1 \ldots N-1 \end{aligned}\right\} \tag{5}$$

where, X and Y are vectors of length N with elements as defined by Eq (5). In terms of the vectors X and Y, Eqs (1) and (2) can be written as [9];

$$\left.\begin{aligned} \lambda_a(0) &= AA^T = W \\ \lambda_{ab}(0) &= AB^T \end{aligned}\right\} \tag{6}$$

where, A^T and B^T denote the transpose of vectors A and B, respectively.

Table 1: Codeword of SeQ code for W=3, K=4 and $\lambda_c \leq 1$

Basic	Parity
(1 2 4)	1 1 0 1 0 0 0
(2 3 5)	0 1 1 0 1 0 0
(3 4 6)	0 0 1 1 0 1 0
(4 5 7)	0 0 0 1 1 0 1

Notice that the code weight of each row more than two, and the relation between N and K for SeQ code can be express as:

$$N = K + 2W - 3 \tag{7}$$

2.1 Signal to Noise Ratio Performance Analysis

The Signal Noise Ratio (SNR) is defined as the average signal to noise power, $SNR = \left[\frac{I^2}{\sigma^2}\right]$, where σ^2 is the average power of noise which is given by [9],

$$\sigma^2 = \langle i^2_{shot} \rangle + \langle i^2_{PIIN} \rangle + \langle i^2_{thermal} \rangle \tag{8}$$

$\mathcal{R}$ represents as the responsively of the photo-detectors. Hence the photo current I can be expressed as;

$$I = \Re\left[\frac{P_{sr}[W]}{N}\right] \tag{9}$$

The power of *Shot* noise can be written as [7];

$$\langle i^2_{Shot} \rangle = 2eB\mathcal{R}\left[\frac{P_{sr}}{N}\right][3W + 1] \tag{10}$$

The Phase Induced Intensity Noise (PIIN) noise will dominate the broadband sources. The PIIN noise at the receiver output is given by;

$$\langle i^2_{PIIN} \rangle = B(I_1^2 \tau_{c1} + I_1^2 \tau_{c2}) = I^2 * \tau_c * B \tag{11}$$

Furthermore, the variance of the PIIN noise at the receiver can be expressed as;

$$\langle i^2_{PIIN} \rangle = \frac{B\mathcal{R}^2 P_{sr}^2 KW}{N^2 \Delta_V}[3W + 1] \tag{12}$$

Thermal noise is given as [8];

$$\langle i^2_{Thermal} \rangle = \frac{4K_b T_n B}{R_L} \tag{13}$$

From Eqs. (9), (10), and (12), the SNR for the proposed SeQ code in the ODCMA coding system is derived as [6];

$$SNR = \frac{\left[\frac{\Re P_{sr} W}{N}\right]^2}{\left[\frac{2eB\Re P_{sr}}{N}\right][3W+1] + B\Re\left[\frac{P_{sr}^2 KW}{N^2 \Delta V}\right][3W+1] + \frac{4K_b T_n B}{R_L}} \tag{14}$$

Since, the corresponding BER can be obtained as follows [6];

$$BER = \frac{1}{2} efrc\left(\sqrt{\frac{SNR}{8}}\right) \tag{15}$$

Finally, Eqs. (14) and (15) will be used for the theoretical calculation for an evaluation of the proposed OCDMA coding system utilizing SeQ code. The performance of the SeQ code will be compared with the existing OCDMA codes, respectively. The theoretical parameters are shown in Table 2.

Table 2: Typical parameters for theoretical calculations.

Parameter	**Value**
Electron's charge	$e = 1.60217646 \times 10^{-19}$ coulombs
PD quantum	$\eta = 0.75$
Electrical bandwidth	B = 80 MHz
Boltzmann constant	$K_b = 1.38 \times 10^{-23}$ W/K/Hz
Receiver noise	$T_n = 300K$
Receiver load resistor	$R_L = 1030\ \Omega$
Data transmission rate	$R_b = 155$ Mbps
Broadband line width	$\Delta\lambda = 3.75$ THz

 DOI: 10.1051/matecconf/201714001002

3 Result and Discussion

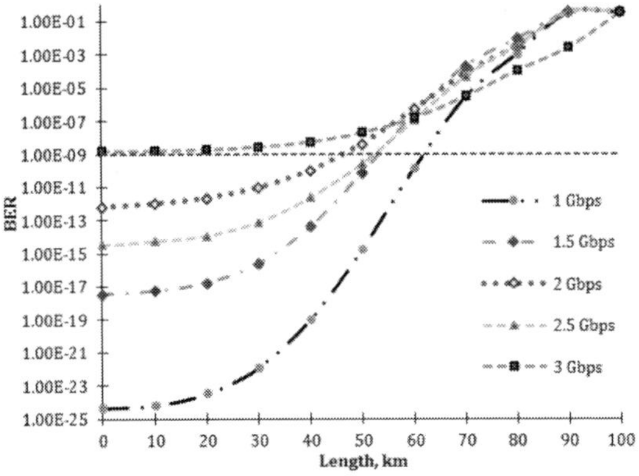

Fig. 1. Performance BER versus Fiber Length (km) for SeQ code (W=4) at Different Bit Rates.

Fig. 1 shows the performance of BER versus fiber length (km) for SeQ code (W=4) at variance bit rate of 1 Gbps, 1.5 Gbps, 2 Gbps, 2.5 Gbps, and 3 Gbps, respectively. From the graph, 1 Gbps bit rate shows the excellent performance of BER compared with 1.5 Gbps, 2 Gbps, 2.5 Gbps, and 3 Gbps. Even the 3 Gbps the worst performance yet still at BER = 10^{-09}. However, at permissible BER = $10^{-09,}$ 1 Gbps bit rate capable to support the performance of BER up to 60 km. Contrast with 1. 5 Gbps, 2 Gbps, 2.5 Gbps and 3 Gbps bit rate only can perform at 55 km, 50 km, 45 km and 30 km, respectively. From these observation, SeQ (W=4) at 1 Gbps can successfully eliminate and suppressed the effects of PIIN and MAI for the OCDMA coding system.

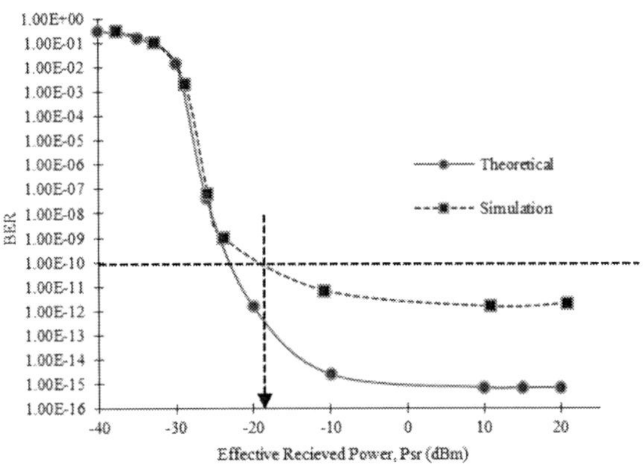

Fig. 2. Relationship Performance of BER versus effective power theoretical and simulation of SeQ code (W=4).

Fig. 2 illustrates the comparison between theoretical and simulation for BER performance versus various effective received power, P_{sr} for SeQ code. From the graph plots, the theoretical results on BER performance with effective receive power P_{sr} are close to the simulation results when the effective received power, P_{sr} are increased. At error floor BER= 10^{-09}, the effective received power, P_{sr} for theoretical and simulation are -24 dBm. Nevertheless, the performance of BER increasingly degrade after the effective received power, P_{sr} are increased. However, the theoretical result supposed to be excellent compared with the simulation result. It can be observed that, when the effective received power, P_{sr} are increased the performance BER are constant. For theoretical analysis result the BER performance is 10^{-15} whereas, for simulation analysis the BER performance is 10^{-11}, respectively. This performance are not parallel due to very tough to achieve a fix BER in simulation rather that theoretical result. In addition, there are addition losses in the components used in simulation which are not included in the theoretical formula Eqs. (14) and (15).

Conclusion

The new SeQ code address code capable to enhance the impact of noises influences has been presented. The SeQ code had shown superior performance where at BER = 10^{-09} can achieve bit rate of 1 Gbps up to 60 km as compared to 15 Gbps, 2 Gbps, 2.5 Gbps and 3 Gbps, respectively. The system successfully degrade the BER, due to PIIN and MAI can be suppressed by using SeQ code. From the result, 1 Gbps shows the excellent performance compared with existing bit rate. For OCDMA system, 1 Gbps was the suitable enough speed for FTTH and LAN network. Besides that, as an evidence that SeQ have enhanced the system compared with existing code, the simulation result compared with theoretical result for validation that SeQ code capable to enhanced the system. From the result obtain, theoretical result can achieve good performance compared with theoretical result, quite difficult to achieve exactly effective receive power due to come component can't be included in theoretical result.

Acknowledgments. This project was supported by the Government of Malaysia under the Fundamental Research Grant Scheme at Universiti Malaysia Perlis, Malaysia under the research grant # 9003-00456.

References

1. C. B. M. Rashidi, S. A. Aljunid, F. Ghani, H. A. Fadhil, M. S. Anuar , "Cardinality Enhancement using Flexible Cross Correlation (FCC) Code for Spectral Amplitude Coding Optical Code Division Multiple Access Systems," vol. **8**, no. 12, (2012).
2. S. A. Aljunid, F. Ghani, H. A. Fadhil, and M. S. Anuar, "Cardinality Enhancement Using Flexible Cross Correlation (FCC) Code for Spectral Amplitude Coding Optical Code Division Multiple Access Systems," vol. **8**, no. 12, pp. 5614–5626, (2012).
3. C. B. M. Rashidi, S. A. Aljunid, M. S. Anuar "Study of Direct Detection Technique for Zero Cross Correlation Code in OCDMA," *International Conference on Computer and Communication*, (2010)
4. G. D. F. Guimarães, "Analysis of the Performance of Optical Code-Division Multiple-Access System (OCDMA), Operating with Gold Codes under Nonlinear Effects," J. Electromagn. Anal. Appl., vol. **5**, no. 02, pp. 49–57, (2013).
5. S. A. Aljunid, F. Ghani, H. A. Fadhil, and M. S. Anuar, "Cardinality Enhancement Using Flexible Cross Correlation (FCC) Code for Spectral Amplitude Coding Optical Code

DOI: 10.1051/matecconf/201714001002

Division Multiple Access Systems," vol. **8**, no. 12, pp. 5614–5626, (2012).

6. C. B. M. Rashidi, S. A. Aljunid, F. Ghani, H. A. Fadhil, M. S. Anuar, and A. R. Arief, "Cardinality enrichment of flexible cross correlation (FCC) code for SAC-OCDMA system by alleviation interference scheme (AIS)," Optik, vol. **125**, no. 17. pp. 4889–4894, (2014).
7. T. H. Abd, S. A. Aljunid, H. A. Fadhil, I. F. Radhi, and N. M. Saad, "Performance Evaluation of Hybrid SCM/SAC-OCDMA Transmissions System using Dynamic Cyclic Shift Code," J. Opt. Commun., vol. **33**, no. 4, pp. 279–283, (2012).
8. P. Yang, S. Xiao, H. Feng, L. Yi, H. Chen, J. Shi, and W. Hu, "A Modification of Hadamard Codes and Its Implementation in SAC-OCDMA PON," vol. **1**, no. July, pp. 311–312, (2012).
9. C. B. M. Rashidi, S. A. Aljunid, M. S. Anuar, and A. k. Rahman, "Capacity Enrichment OCDMA Based On Algorithm of Novel Flexible Cross Correlation (FCC) Address Code," no. 115, pp. 583–588, (2015).

MATEC Web of Conferences **140**, 01003 (2017) DOI: 10.1051/matecconf/201714001003
ICEESI 2017

An Efficient Multi-sensing and GSM Equipped Fire Monitoring System

Md. Fahad Bin Zamal[1*], *Shehrin Sayed*[2], *Touhid Bhuiyan*[1], *Mostafijur Rahman*[3]

[1]Department of Software Engineering, Daffodil International University (DIU), Bangladesh
[2]Department of Electrical & Electronic Engineering, Bangladesh University of Engineering and Technology (BUET), Bangladesh
[3]School of Computer and Communication Engineering, Universiti Malaysia Perlis (UNIMAP), Malaysia

Abstract. The principal goal of fire monitoring system is to react promptly to a fire and not to misleading particulate signatures produced by nuisance sources. In this paper we proposed a system that not only able to detect and prevent fire at early stages but also capable of interact with surround environment. Recent researches lead us to detect fire by light and heat sensor, image processing, smoke detection mechanism but failed to integrate those in one The advancement on fire detection technologies has been significant over the last few decade due to rapid progress in communication technologies, advances in sensing devices and greater understanding of fire physics. But lack of intelligence among fire monitoring system often failed to make an impact on fire incidents. Our proposed fire monitoring system is incorporated in such a way that can communicate with environment by itself through the help of GSM Modem. Here we introduce an intelligent and advance fire monitoring system that can communicate by itself with fire station and can detect fire at its early stage and extinguish it in the shortest time subject to a few effective factors.

1 Introduction

Human lives, infrastructure and ecological system is always been threaten by fire and still facing challenges to reduce forged alarms and unable to provide support in evacuation. Rapid progresses can make the communication technologies, advances in sensing devices and establish greater understanding on fire but still we are lagging behind to accumulate those knowledge and technology together. Expensive, complicated and sensitive devices are being used but lack of intelligence is present there.

Most of the fire detection systems use single sensor and operate without any controlling mechanism. Those devices either use temperature sensing or smoke detection concept to identify the presence of fire [1]. But if we use multi sensor detectors, it increases the efficiency in early fire detection as well as it can monitor multiple aspects of a wide variety of fires. Distributed temperature monitoring system is currently used on fire protection system in cities [2]. Another mechanism being used in recent years which is distributed fiber optic temperature measurement system but it has some drawbacks of excessive cable and very high cost [3]. Wireless sensor network in other hand will be a good solution as it is small, easy to maintain, cheap and suitable for large-scale application [4].

This paper studies the design standards and principles of fire detection mechanism, communication mechanism and multiple sensing mechanisms. Multiple sensing mechanisms enhance the accuracy to detect fire incidents and inclusion of stepper motor will provide mitigation of fire in early stage. Microcontroller (ATmega32/Arduino) associated with temperature sensor (LM35) and smoke detector (LDR) will operate to accurately identify the presence of fire. Introduction of GSM modem in this system will make the device capable of communicate with fire stations and localities. The system consists Temperature Detection Module (TDM) where LM-35 sensor being used, Smoke Detection Module (SDM) create by Light Dependent Resister (LDR), Automatic Action Module (AAM) consists stepper motor, and Notification Module (NM) operate with GSM modem.

2 System Design and Architecture

2.1 System Overview

Our proposed system contains four (4) modules. Those are:

A. Temperature Detection Module (TDM)

B. Smoke Detection Module (SDM)

C. Automatic Action Module (AAM) and

D. Notification Module (NM)

In temperature detection module (TDM) it will always check some parameters before generating a fire alarm. Missing any condition will not trigger the fire alarm which will make the system more accurate and

* Corresponding author: fahad.swe@diu.edu.bd

precise. The temperature detection modules first detect temperature that is above the critical temperature and secondly it will check the rate of temperature rises along with the time or not. Here critical temperature (Tc) will be set manually in the microcontroller and it will always aware of the instantaneous value of the temperature (T). If T > Tc, then the first condition will be satisfied and it will then concentrate on checking whether temperature is rising or not. As an example if temperature for a certain time is T1. After Δt time the temperature raises to T2. Then, rate of temperature rise, $r = T2{-}T1\ /\ \Delta t$.If r is very high then the second condition is also satisfied.

Smoke detection module (SDM) will finally check the presence of smoke by the help of smoke detector sensor. If there is smoke, the smoke detector circuit gives a digital '1' to the microcontroller. Otherwise it gives digital '0'. If these three conditions are fulfilled, then microcontroller turns on the alarm. The Automatic Action Module (AAM) automatically turns on the stepper-motor which is attached to a mechanical system providing water/gas to minimize the effect. In Notification Module (NM) incorporate with GSM modem will send a caution message to nearby residences and also firefighting department with an exact location of the incident.

2.2 System Architecture

Figure 1 shows the overall circuit diagram of our proposed fire monitoring system consists Temperature Detection Module (TDM), Smoke Detection Module (SDM), Automatic Action Module (AAM) and Notification Module (NM).

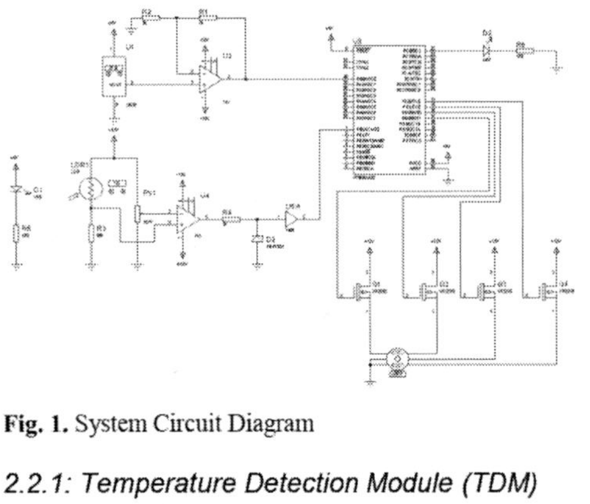

Fig. 1. System Circuit Diagram

2.2.1: Temperature Detection Module (TDM)

In our proposed Temperature Detection Module (TDM) we use LM35 sensor for temperature detection mechanism. In case of LM35 which is integrated circuit temperature sensor has an advantage over other linear temperature sensors as there is no requirement of subtract a large voltage from the output. It does not require any external calibration to provide accuracy. An another added advantage is its lower cost. Because of low output impedance, precise calibration, single power supplies, linear output and low cost makes LM35 sensor a better choice over other temperature detection sensors.

It produces output voltages proportional to °C (10mV/°C), so the nominal output voltage is 250mV at 25°C and 1.000V at 100°C. The LM35 is more accurate (±1°C from -55°C to +150°C vs. ±3°C from -20°C to +100°C) temperature measuring sensor. Figure 2 shows the overall architecture of Temperature Detection Module (TDM).

In Figure 2 section 1(Red labeled) senses the temperature and converts it to an electrical signal in proportion to the temperature. If the temperature is T degree Celsius, then the output of the LM35 is given in equation 1.

$$Vout = T \times 10\ mV \tag{1}$$

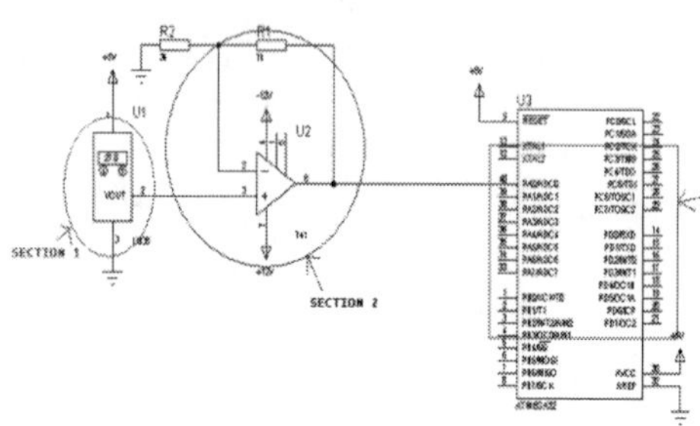

Fig. 2. Temperature Detection Module (TDM)

For example, if the temperature is 30°C then output of LM35 is 300mV or 0.3V. The working range of LM35 is 0°C to 150°C. So the range of its output is from 0V to 1500mV or 1.5V. Figure 3 shows the temperature measuring mechanism by LM35.

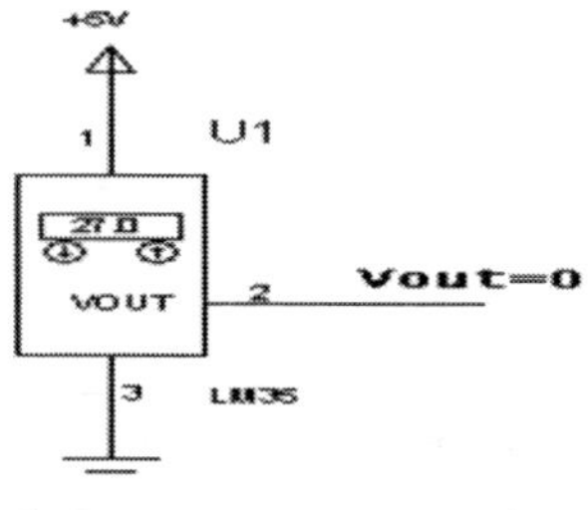

Fig. 3. Temperature Measurement by LM35

The input range of ADC is 0~5 V and the output range of LM35 is 0~1.5 V. So, the output of LM35 must be amplified to make it suitable for the ADC input. It means we have to convert the 0 to 1.5 V range into 0 to 5 V range. In figure 2 section 2 (Blue labeled) we use a non-inverting amplifier which is capable to do this operation. By calculating we found that we have to use a non-inverting amplifier of gain = 5 / 1.5 = 3.3. In our proposed system, we adjust the two resistors R1 and R2 in such a way that it will give a gain of 3.3. For getting desire gain we have to select the resistance of R2 in 33KΩ and R1 in 10KΩ.So setting the value of R1 and R2 in that value (10KΩ,33KΩ) we able to get the desired gain and the non-inverting amplifier doing the rest and gives us the desired range for ADC.

DOI: 10.1051/matecconf/201714001003

In Figure 4 section 3 (Green labeled) ADC of microcontroller coverts an analog signal to a digital signal. It takes an analog signal within the range of 0V to 5V. The converted value is stored in a 10bit register in the microcontroller. If the analog input is 0V then the value stored in the register is 0 and if the analog input is 5V the value stored in the register will be 1023. If the analog input is between 0V and 5V then the stored value may be any number between 0 and 1023. For example, if the analog input is 3.2V, then the stored value in the register is

$$Register = 1023 / 5 \times 3.2 = 655$$

The microcontroller is programmed such that it reads the data from the register and calculates the room temperature. Microcontroller continuously reads the value stored in the register and calculates the temperature which is sensed by the LM35.Suppose, at some point the value stored in the register is 512. Then the analog input at ADC (ADC0 pin of microcontroller) is:

$$\text{ADC} = 5 / 1023 \times 512 = 2.502 \text{ V}$$

This is the output of the non-inverting amplifier. Since the gain of the amplifier is 3.3, the input of the amplifier should be

$$Output / Gain = 2.502 / 3.3 = 0.75\text{V}.$$

Again, the input of the amplifier is the output of LM35. So, output of the LM35 is 0.75V or 750mV.

$$\text{Temperature} = 750mV / 10mV = 75°\text{C}.$$

2.2.2: Smoke Detection Module (SDM)

Figure 4 shows the schematic view of our proposed Smoke Detection Module (SDM). In this module LDR (Light Dependent Resistors) is used to examine the presence of smoke. The LDR is selected because of its exciting features, such as: when the light level is low the resistance is high (this prevents current flow into the circuit), when light level is high resistance falls and current flows into the circuit.

In this module a comparator Op-Amp is integrated, which is used to compare two voltages and switches, and its output to indicate which voltage is larger. Here it compares the voltage value we adjust in the Potentiometer and the voltage we get when the LDR is starting to conduct. So when comparator compare both voltage value and found that voltage value in the LDR section is larger than it switches a + V_s in the circuit. This + V_s then turn on the smoke detection indicator (LED). In this process we get noticed that our smoke detector found the presence of smoke. The output voltage can be generated from the Equation (2)

$$V_{out} = \begin{cases} V_{S+} & V_1 > V_2 \\ V_{S-} & V_1 < V_2 \end{cases} \quad (2)$$

(Where V_s is the supply voltage and the op-amp is powered by + V_s and − V_s.)

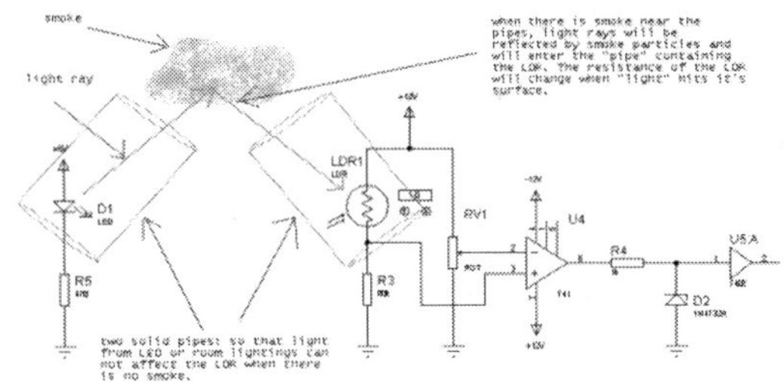

Fig. 4. Schematic view of Smoke Detection Module

Figure 5 shows the operation of zener diode, comparator and potentiometer in proposed Smoke detection module (SDM). In our proposed SDM we used a 5.1V zener diode as we need a 5V input for the micro controller. This 5.1V zener diode will exhibit a voltage drop of 5.1 V if reverse bias voltage applied across it is more than its zener voltage. From the comparator when we get a +Vs voltage then zener diode makes sure a 5.1 V output as it exceed the zener voltage. As zener diode has low impedance excessive voltage may obliterate the diode and the whole process. So here we use some resistance for saving the zener diode from excessive voltage. Now we get almost suitable voltage value for the micro controller but not the exact one. For getting the exact one we have to use a Digital Buffer IC.

A Digital Buffer is a single input device that performs any type of logical operation on its input signal as its output exactly matches that of its input signal [7]. It is a "Non-inverting" device and so will give us the Boolean expression. For the suitable value (5 V and 0 V) we used a Digital Buffer IC (7407) which is capable for doing these operations. As a simple function this Buffer IC converts the 5.1 V to the closest value of 5 V & -0.6 V to the closest value of 0 V and also represents it into the logical form of operation. The micro controller gets a logical '0' which represent the 'Absence of Smoke' and gets a logical '1' which represents the 'presence of Smoke'.

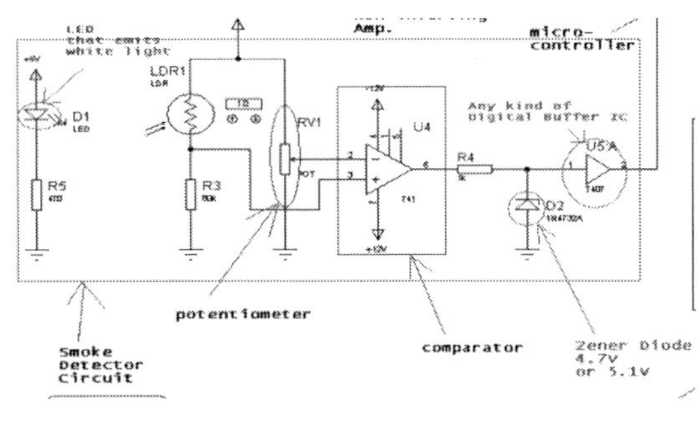

Fig. 5. A closer look into Smoke Detection Module (SDM)

2.2.3: Automatic Action Module (AAM)

In our proposed system Automatic Action Module (AAM) is capable of generating water/gas into the fire to minimize the fire in early stages. We used steeper motor for this purpose. A stepper motor converts electrical pulses into specific rotational movements [8]. The movement created by each pulse is precise and repeatable, which is why stepper motors are so effective for positioning applications [9].

Stepper motors have multiple "toothed" electromagnets arranged around a central gear-shaped piece of iron. The electromagnets are energized by an external control circuit, such as a microcontroller. Stepper motor consists of four coils. At a time when a single coil is excited, the coil creates a magnetic field and the conductor aligns with that coil [10].

Figure 6 shows the mechanism of stepper motor in our proposed system. To excite a coil, we have to put one end of the coil at +Vs (the source voltage) and the other end at ground (0 Volt). In this module, the common wire is kept at +Vs and the other terminals (A, B, C, D) are connected to the ground one at a time. This means when the A terminal is connected to the ground other terminals are open circuited. To implement this idea we used four MOSFETs which acts as a voltage controlled switch. When the gate voltage is 5V, the switch closes and shorts D and S. Otherwise there is an open circuit between D and S. Micro controller serially provides a series of pulses from the pins PD0, PD1, PD2 and PD3.

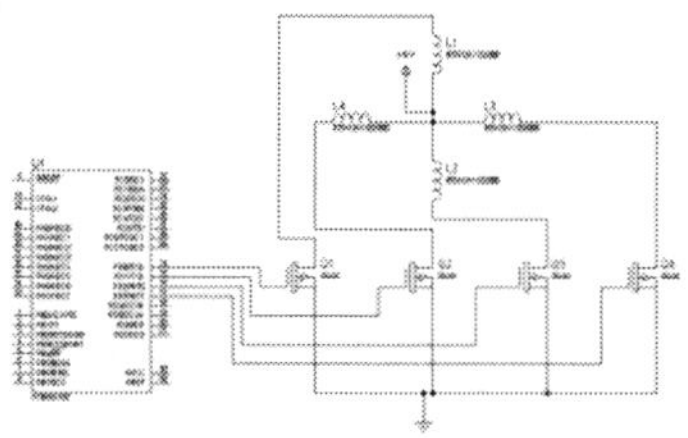

Fig. 6. Operation of Stepper motor

Table 1 show the steeper motor output in AAM. For Type1, the first MOSFET gets a 5V at the gate and the other MOSFETs get 0V. By doing this, the first coil gets excited and the conductor aligns with the first coil. Similarly Type 2, Type 3 and Type 4 and MOSFET 2, MOSFET 3 and MOSFET 4 are on which makes the second, third and fourth coils get excited. Controlling the stepper motor enable the system to provide water/gas in the incident to prevent the fire being spreading.

Table 1: Stepper motor output

	A	B	C	D
Type 1	1	0	0	0
Type 2	0	1	0	0
Type 3	0	0	1	0
Type 4	0	0	0	1

2.2.4: Notification Module (NM)

In our proposed system the Notification Module (NM) is responsible of generating short message service (SMS) to notify the nearby residents and Fire Fighting team about the incidents. It will also provide the exact location of the incidents by the help of GSM modem. An external GSM modem is connected to the microcontroller through a serial cable or a USB cable [11]. It operates at either 900 MHz or 1800 MHz frequency band and it can support data transfer of up to 9.6 Kbit/s [12]. In order to operate a GSM modem is also requires a SIM card by which it establishes and manages the connection. The GSM modem comes with a serial interface which the modem can be controlled using AT command interface [13].

GSM modem support a common set of standard AT commands which integrated with the multi controller. After successful connection the program is written in microcontroller to perform the specific task it is up to [14]. The modem produces warning text SMS to the mobile phone of the authority after it receives a command. Getting the SMS will help the authority to act fast and will help to minimize the fire. Depending on the fire the SMS sending process will be differ. When the incident of fire is limited the system will send SMS only those people of that authority but when the situation getting worse then the system will broadcast the SMS to all. For GSM modem a special subscription to an SMS service provider is not required which makes it more cost effective and efficient [15].

In our proposed system it Notification Module (NM) (Figure 7) will consists two types of sensor nodes mobile nodes and Fire station nodes. Data after aggregated in Microprocessor will be relayed to the base stations where the further processing takes place. Base station (BS) relays the information to the operating personnel who takes necessary actions based on the information received. To ensure the location accuracy data from sensors must be combined and dispatched to base station in such a way those nearby sensors should temper their data with each other. To locate the location the device transmits the data to the central server with GPS co-ordinates using map applications in android mobile through SMS [16] Flow chart of Notification Module (NM) shows how microcontroller integrates with TDM and SDM to generate the operation of AAM and intexgrate GSM/GPRS module to send SMS through BS.

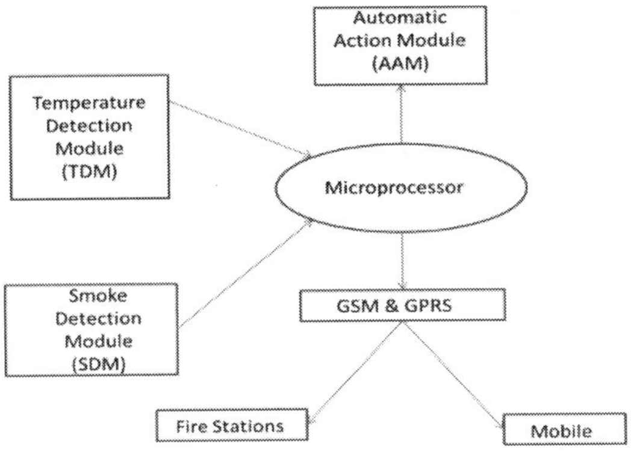

Fig. 7. Flow chart of *Notification Module (NM)*

3 Results

Figure 8, 9 and 10 shows our implemented Temperature Detection Module (TDM), Automatic Action Module (AAM), Smoke detection Module (SDM) and Notification Module (NM) for an Efficient Multi-sensing and GSM Equipped Fire Monitoring System.

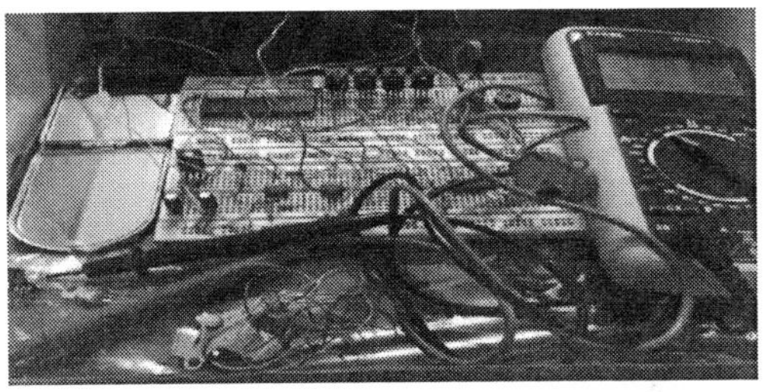

Fig. 8. Temperature Detection Module (TDM) and Automatic Action Module (AAM)

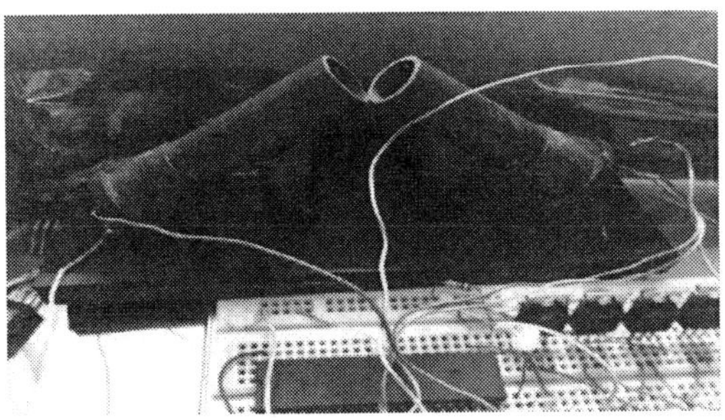

Fig. 9. Smoke detection Module (SDM)

Fig. 10. Notification Module (NM)

4 Conclusions

A principal goal of fire monitoring system is to react promptly to a fire and not to misleading particulate signatures produced by nuisance sources. System that only concentrates on smoke detection mechanism could not discriminate between fire and non-fire sources of smoke. Early detection and quick response is the key to minimize damage caused by fires [17]. By using multi-sensor detectors we can analyze various aspects of a wide selection of signatures produced by flaming fires, non-flaming fires, and nuisance sources [18]

The use of micro controller will bring more intelligence in the system as it now able to distinguish different aspect and can make the best decision out of that. By using GSM modem we not only able to locate the incident spot but also able to warn local authority for quick response [19]. A better understanding on those factors that influence the performance of a fire monitoring system can make a big difference. The outcome from this project can add value to various aspects in manufacturers, designers, authorities, programmer to create further improvement in this sector. As a future work the presence of microcontroller and GSM device will enable the proposed system to generate real time data to monitor the system upfront.

References

[1] L. Liu, R. Sun, Y. Sun, S. Al-Sarawi, "A smart bushfire monitoring and detection system using GSM technology," International Journal of Computer Aided Engineering and Technology, 2010, 2(2/3), pp. 218-232.

[2] K. V. Teja and S. Angadi, "Fire Detection and Notification System in Trains," International Journal of Innovative Research in Science, Engineering and Technology (IJIRSET), vol. 2(4), April 2013, 1149-1152.

[3] Y. Yao, J. Yang, C. Huang and W. Zhu, "Fire Monitoring System Based on Multi-Sensor Information Fusion," 2010 2nd International Symposium on Information Engineering and Electronic Commerce, Ternopil, 2010, pp. 1-3.

[4] K Gaikwad, A. Quadri, S. Akshada, Z. Reshma; "Fire Monitoring and Control System," in International

DOI: 10.1051/matecconf/201714001003

Research Journal of Engineering and Technology (IRJET), 3(5), 2016.

[5] W. Ping-ping, S. Feng-jie, C. Weixin. "Discussion on temperature monitoring system for power cable joints". Telecommunication for Electric Power System, 27(160), 2006, pp. 59-61.

[6] W. Chao, P. Zhen-cun. "Power Cable Junction Monitoring System for Fault Pre-Warning," Electrical Power Automation Equipment, 21(5), 2001, pp. 25-28.

[7] L. Jun, Q. Chengshan, S. Peng, W. Zhiwei. "Power Cable Connector Monitoring System Based on Wireless Sensor Network Monitoring System," Chinese Journal of Electron Devices, 37, 2014, pp. 766-771.

[8] S. Vitusutrapoj "Advance Microcontroller 8051 Series" Seed Education Co,

[9] Douglas V., "Hall Microprocessors and Interfacing (Programming & Hardware)", 1991, Glencoe McGraw-Hill.

[10] J. E. Traister, (2nd edition) "Security/Fire Alarm Systems: Design, Installation, and Maintenance", Columbus, OH: McGraw- Hill, 1995

[11] R. Shams, S. Hossain, S. Priyom, N. Fatema, S. R. Shakil, & M. K. Rhaman, "An automated firefighting system," 2015 12th International Conference on Fuzzy Systems and Knowledge Discovery, FSKD 2015, 2327-2331.

[12] J. J. Chen, M. X. Lu, T. H. Wu and Y. S. Hwang, "Sub-1-V Fast-Response Hysteresis-Controlled CMOS Buck Converter Using Adaptive Ramp Techniques," in IEEE Transactions on Very Large Scale Integration (VLSI) Systems, 9(21), pp. 1608-1618, Sept. 2013.

[13] Z. Liu and A. K. Kim, "Review of Recent Developments in Fire Detection Technologies," Journal of Fire Protection Engineering, 2(13) pp. 129-151, May 1, 2003.

[14] N. Bouguila and K. Daoudi, "A Statistical Approach for Binary Vectors Modeling and Clustering," Lecture Notes in Computer Science book series (LNCS, volume 5476), pp. 184–195, 2009.

[15] M. Reddy and K.Rao. "Fire Accident Detection and Prevention monitoring System using Wireless Sensor Network enabled Android Application" Indian Journal of Science and Technology, 9(17), 2016.

[16] H. Wang. "Discriminating Fire Detection via Support Vector Machines", Lecture Notes in Control and Information Sciences, 2006

[17] G. Healey, D. Slater, T. Lin, B. Drda and A. D. Goedeke, "A system for real-time fire detection," Proceedings of IEEE Conference on Computer Vision and Pattern Recognition, New York, NY, 1993, pp. 605-606.

[18] A. Goswami, T. Bezboruah, K. C. Sarma, "Design of an embedded system for monitoring and controlling temperature and light," IJEER. 2009: 1(1), pp. 27–36.

[19] S. Dhananjeyan, K. Mohana Sundaram, A. Kalaiyarasi, P. G. Kuppusamy, "Design and Development of Blind Navigation System using GSM and RFID Technology," Indian Journal of Science & Technology, 9(2), PP. 1-5, 2016.

MATEC Web of Conferences **140**, 01004 (2017)
ICEESI 2017
DOI: 10.1051/matecconf/201714001004

Performance Verification on UWB Antennas for Breast Cancer Detection

V.Vijayasarveswari[1], *M.Jusoh*[1], *T.Sabapathy*[1], *Rafikha Aliana*[1], *S.Khatun*[2], *Z.A.Ahmad*[1], *Mohamed Nasrun Osman*[1]

[1]Embedded Network and Advance Computing Research Cluster (ENAC), School of Computer and Communication Engineering, Universiti Malaysia Perlis, Perlis.
[2]Faculty of Electrical & Electronic Engineering, Universiti Malaysia Pahang, Pekan, Pahang.

Abstract. Breast cancer is a common disease among women and death figure is continuing to increase. Early breast cancer detection is very important. Ultra wide-band (UWB) is the promising candidate for short communication applications. This paper presents the performance of different types of UWB antennas for breast cancer detection. Two types of antennas are used i.e: UWB pyramidal antenna and UWB horn antenna. These antennas are used to transmit and receive the UWB signal. The collected signals are fed into developed neural network module to measure the performance efficiency of each antenna. The average detection efficiency is 88.46% and 87.55% for UWB pyramidal antenna and UWB horn antenna respectively. These antennas can be used to detect breast cancer in the early stage and save precious lives.

1 Introduction

Breast cancer cases are increasing year by year. 100,000 cancer cases are reported every year in Malaysia. The five main cancers are breast cancer (14.5%), internal cancer (12.1%), lung cancer (11.8%), cervical cancer (5.7%) and throat cancer (5.4%) [1]. Breast cancer is a leading case in Malaysia. However, 89% of women survived after breast cancer detected in early stage [2]. The available breast cancer detection methods are mammography, magnetic resonance imaging (MRI) and ultrasound.

Mammography is a technology that compresses the breast between two plates to get an X-ray image of the breast. Magnetic resonance imaging (MRI) scan is done by constructing images of the internal breast structure. It is using high magnetic field and radio energy wave. Ultrasound is using sound wave to image the breast structure [3-5]. Despite of the negative effect on human health, these types of methods are able to detect breast cancer but not in early stage [6-8]. If a cancer is detected, usually the patient has to undergo biopsy for further investigation. A wrong detection can make patient undergo some painful diagnose process. This traditional equipment are only available in some of the clinics and hospitals because they are very expensive [9,10]. Moreover, this equipment should be handled by an expert operator. So, the patient needs to go to the specific hospital for their check-up. These types of screening systems are not advisable to use regularly due to their side effects. Besides, the detected cancer cannot be determined whether it is benign or malignant by the available equipment except biopsy [5,8]. Determining of benign or malignant is important for further treatment decision.

To overcome the above shortcomings, researchers have been proposed new breast cancer detection system using microwave based UWB imaging technology. Radar based imaging method is widely used for breast cancer detection compared to microwave tomography [9,11,12]. There are a lot of ways to process the received signal. Most of the researchers used VNA to view the received signal. The antenna is attached to the VNA to obtain output and the image of the breast [13,14]. In other hand, researcher has used real time oscilloscope to simulate the received signal from pulse generator [15]. Some researchers have proposed to use artificial intelligence (AI) to analyse the received UWB signal for early tumor detection [4,12,16]. While antenna will transmit and receive the signal. This signal contains the signature of the cancer. Different type of antennas included radar antenna [17], Vivaldia antenna [18,19], horn antenna [20,21], slot on-body antenna [22] and wideband half oval patch antenna, stacked patch antenna and wide-slot antenna [23] have been designed and used to detect breast cancer. Here, the performance of two types of antennas (wearable UWB pyramidal antenna and UWB horn antenna) are analysed to detect the breast cancer.

* Corresponding author: M.Jusoh (muzammil@unimap.edu.my)

DOI: 10.1051/matecconf/201714001004

2 Methods and Material

Figure 1 shows the summary of the methodology taken throughout the experiment. Firstly the forward scattered UWB signals are collected using the two types of antennas (UWB pyramidal and UWB horn antenna). Then, the collected UWB signals are converted into digital form (1632 data points). The data samples are combined together and fed into the developed k-fold based feed forward backpropagation neural network (KFFBPNN) module. KFFBPNN module is trained, validated and tested until it is optimized. The developed KFFBPNN module is used to test using untrained data samples.

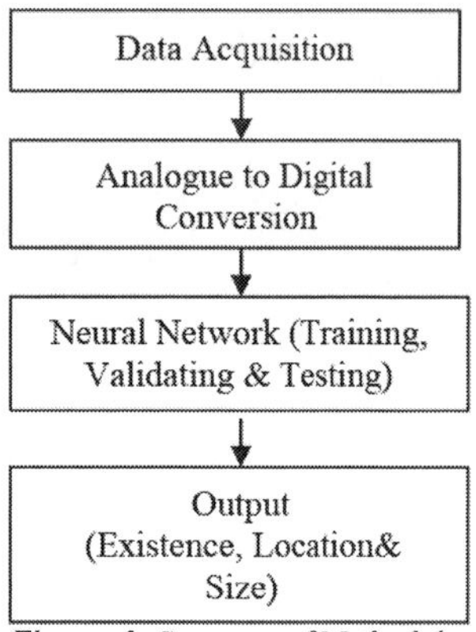

Figure 1: Summary of Methodology

2.1. Antenna Design

Two types of antennas are used here: UWB pyramidal antenna and UWB horn antenna. Both antennas are designed in Computer System Technology (CST) Microwave Studio in frequency range of 3.1GHz – 10GHz. Both UWB pyramidal antenna and UWB horn antenna is fabricated using textile (Zelt textile for patch and Felt textile for substrate) as shown in Figure 2.

(a) Pyramidal Antenna (b) Horn Antenna

Figure 2: UWB Antenna

2.2 Data Acquisition

Heterogeneous breast phantom and tumor are developed with existing chemical materials as shown in Figure 3 [25]. The phantom consists of fat, gland and skin. Tumor is developed with different sizes (3mm and 5mm). The materials are included p-toluic acid, n-propanol, deionized water, 200 bloom gelatine, formaldehyde, safflower oil and ultra-ivory detergent. The amount of each material is different for fat, gland, skin and tumor respectively.

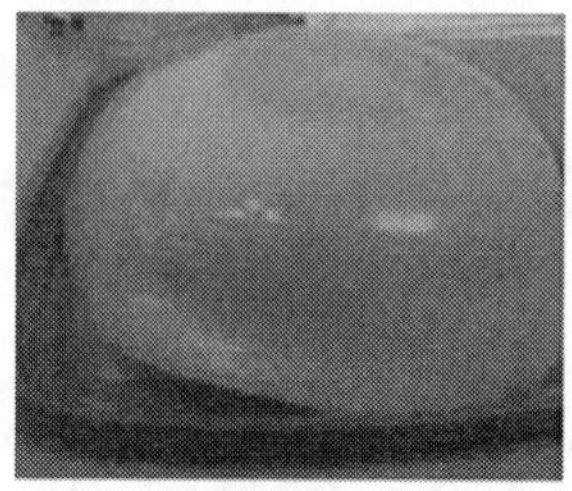

Figure 3: Heterogeneous Breast Phantom

The breast phantom is placed in the middle of a pair of antenna as shown in Figure 4. Antenna will transmit UWB signal and another antenna will receive the UWB signal. This UWB signal contains the information of tumor's location and size. The received UWB signals are saved in PC for further investigation. UWB signals are initially in analogue and converted to digital through Analogue to Digital Conversion.

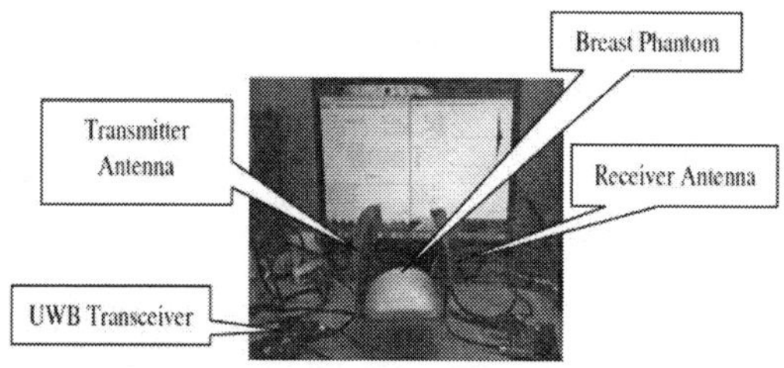

Figure 4: Experiment Set-up

2.3 Neural Network

The 82 data samples are collected for UWB pyramidal antenna (41 data samples) and UWB horn antenna (41 data samples). 41 data samples were divided into two groups. First group consisted of 30 data samples for training, validating and testing while second group consisted of 11 data samples for real time testing. Each data sample consisted 1632 data points. Four data points are left after extracted the important points from 1632 data points. Four data points included mean, median, maximum number and minimum number.

K-fold based feed forward backpropagation neural network (c) module is developed in Matlab software. Hidden neuron, hidden layer and training function are

DOI: 10.1051/matecconf/201714001004

the most important parameter during developing the KFFBPNN module. These parameters are changed depending on input data and target data. First group data samples (30 data samples) are fed into the KFFBPNN module to train, validate and test the network. The network is trained continuously until it is optimized but network should not memorize the output. The optimized network parameters are shown in Table 1.

Table 1: Network Parameters

Neural Network Parameters	
Number of nodes in input layer	4
Number of nodes in first hidden layer	20
Number of nodes in second hidden layer	1
Number of nodes in output layer	4
Training function	Trainlm
Learning rate	0.009
Momentum constant	0.6
Maximum number of epochs	1000
Minimum performance gradient	1e-25

3 Results and Discussion

After the training session, the developed KFFBPNN module is tested using untrained 11 data samples from second group. This process is essential to determine the efficiency of developed module. Table 2 and Table 3 showed the detection accuracy for each output data and detection performance efficiency for UWB pyramidal antenna and UWB horn antenna respectively. Here to mention, negative value indicated the absence of the tumor while positive value indicated the presence of the value.

Table 2: Detection Performance Efficiency for UWB Pyramidal Antenna

Actual Target (mm)				**NN output (mm)**				**Detection Performance Efficiency (%)**			
x	y	z	size	x	y	z	size	x	y	z	size
-1	-1	-1	-1	-1.16	-0.86	-0.33	-1.00	100	100	100	100
32.5	32.5	40	3	36.04	16.35	35.79	13.62	90.28	50.15	89.50	45.33
32.5	32.5	40	5	31.69	19.62	67.02	50.00	97.54	60.30	32.50	100
62.5	32.5	30	3	59.76	40.43	59.59	29.98	95.52	80.45	50.34	99.67
32.5	62.5	30	3	48.78	62.49	58.65	29.99	66.60	99.84	51.19	99.67
32.5	1.0	30	3	31.00	1.96	28.51	29.98	95.38	96.92	95.00	99.67
62.5	32.5	30	5	67.18	32.31	28.73	50.00	93.14	99.38	95.67	100
1.0	32.5	30	3	49.85	29.01	49.59	23.92	48.50	89.23	60.61	79.67
32.5	1.0	30	5	42.11	0.88	30.27	50.01	77.20	90.00	99.34	99.80
32.5	62.5	30	5	30.20	62.57	29.20	44.83	92.92	99.84	97.33	89.60
1.0	32.5	30	5	2.69	33.59	25.94	53.30	80.00	96.73	86.33	93.98

Table 3: Detection Performance Efficiency for UWB Horn Antenna

Actual Target (mm)				**NN output (mm)**				**Detection Performance Efficiency (%)**			
x	y	z	size	x	y	z	size	x	y	z	size
-1	-1	-1	-1	-0.53	-0.85	-1.24	-0.66	100	100	100	100
32.5	32.5	40	3	45.17	27.19	8.23	3.62	61.02	83.65	20.59	79.49
32.5	32.5	40	5	35.16	34.76	39.99	4.66	91.81	93.05	99.99	93.19
62.5	32.5	30	3	54.74	34.21	29.98	2.99	87.52	94.75	99.97	99.97
32.5	62.5	30	3	36.17	67.84	29.99	1.83	88.72	91.46	99.99	61.01
32.5	1.0	30	3	42.88	0.52	29.82	4.12	68.05	52.10	99.42	62.65
62.5	32.5	30	5	67.17	24.19	30.00	5.09	92.53	74.42	100	98.22
1.0	32.5	30	3	5.07	44.12	8.34	2.99	49.70	64.23	27.84	99.68
32.5	1.0	30	5	69.86	43.64	29.60	4.53	46.56	22.94	9.87	90.51
32.5	62.5	30	5	28.06	59.86	29.98	5.01	86.35	95.78	99.95	99.82
1.0	32.5	30	5	4.19	64.87	30.16	5.08	94.88	50.15	99.45	98.37

Table 4 showed the average detection efficiency for UWB pyramidal antenna and UWB horn antenna. UWB pyramidal antenna (88.46%) performed better than UWB horn antenna (87.55%). The antenna is fabricated using the same materials but in different size. UWB pyramidal antenna is fabricated in larger size compared to the UWB horn antenna. UWB pyramidal antenna is able to close the phantom completely and able to transmit and receive all the UWB signals without loss any signals.

Table 4: Average Detection Efficiency

Antenna Type			UWB Pyramidal Antenna	UWB Horn Antenna
Performance Efficiency (%)	**Existence**		100	100
	Location	x	85.19	79.43
		y	87.53	81.09
		z	77.98	96.17
	Size		91.58	93.50
Average Detection Efficiency (%)			88.46	87.55

3 Conclusions

Two types of antennas are used to analyse its performance for breast cancer detection. The designed antennas are fabricated using textile materials so that it can be wear by the patient during the screening process. 100% detection efficiency is achieved in terms of existence of the tumor. The developed system and wearable antenna able to detect breast cancer in the early stage without any difficulties. This will help to save precious lives.

Acknowledgement

This work is supported by Ministry of Higher Education, Malaysia, Grant FRGS – 9003-00418

References

1. The Star Online. Retrieved from http://www.thestar.com.my/news/nation/2016/04/03/about-100000-malaysians-suffer-from-cancer-each-year/ (Date accessed: 6 April 2017)
2. C.H. Yip, P.N. Bhoo, S.H. Teo, Med J Malaysia, **69**, 8 (2014).
3. T. Henriksson, M. KlemM, D. Gibbins, J. Leendertz, T. Horseman, A.W. Preece, R. Benjamin, I.J. Craddock. *Antennas and Propagation Conference (LAPC)*, 1 (2011).
4. K.J. Reza, S. Khatun, M. Jamlos, M. Fakir, M.N. Morshed, Journal of Medical Imaging and Health Informatics, **5**, 1246 (2015).
5. M. Løberg, M.L. Lousdal, M. Bretthauer, M. Kalager, Breast Cancer Research, **17**, 63 (2015).
6. M. Sajjadieh, F. Foroozan, A. Asif, *13th IEEE International Multitopic Conference,* 1 (2009).
7. H. Sylvia, Heywang-Köbrunner, A. Hacker, S. Sedlacek, Breast Cancer (Basel), **6**, 199 (2011).
8. A. Jalalian, S. Mashohor, R. Mahmud, B. Karasfi, M.I. Saripan, A.R. Ramli, Excli Journal, **16**, 113 (2017).
9. V.Vijayasarveswari, S. Khatun, M.Jusoh, M.M. Fakir, Journal of Scientific Research and Development, **3**, 59 (2016).
10. S.D. Placido, C.D. Angelis, M. Giuliano, C. Pizzi, R. Ruocco, V. Perrone, D. Bruzzese, G. Tommasielli, M.D. Laurentiis, S. Cammarota, G. Arpino, British Journal of Cancer, **116**, 821 (2017).
11. K.J. Reza, S. Khatun, M.F. Jamlos, M.M. Fakir, Ikram-e-Khuda, Z. Ishwar, International Journal of Engineering and Technology, **5** (2013).
12. S.A. Alshehri, S. Khatun, Progress In Electromagnetics Research C, **7**, 79 (2009).
13. S.H.M. Salleh, M.A. Othman, N. Ali, H.A. Sulaiman, M.H. Misran, M.Z.A.A. Aziz, ARPN Journal of Engineering and Applied Sciences, **10**, 723 (2015).
14. M. Klemm, J.A. Leendertz, D. Gibbins, I.J. Craddock, A. Preece, R. Benjamin, Antennas and Wireless Propagation Letters, IEEE, **8**, 1349 (2009).
15. J.C.Y. Lai, C.B. Soh, E. Gunawan, K.S. Low, Progress In Electromagnetics Research M, **16**, 19 (2011).
16. S.A. Alshehri, S. Khatun, A.B. Jantan, R.S.A.R. Abdullah, R. Mahmood, Z. Awang, Progress In Electromagnetics Research, **111**, 447 (2011).
17. A.W. Preece, I. Craddock, M. Shere, L. Jones, H.L. Winton, Journal of Medical Imaging, **3**, (2016).
18. Y. Medina, M. Augusto, A.V. Paz, *2016 IEEE Andescon*, 1(2016).
19. C.J. Shannon, E.C. Fear, M. Okoniewski, Electronics Letters, **41**, 388 (2005).
20. R.K. Amineh, A. Trehan, N.K. Nikolova, Progress In Electromagnetics Research B, **13**, 59 (2009).
21. K. L. Narayanan, V. Ayyempillai, International Journal Of Innovative Research In Science, Engineering And Technology, **2**, 508 (2013).
22. X. Xia, S. Hang, W.Z. Jie, W. Liang, Chin. Phys. B, **23**, 1 (2014).
23. A. Sam, A.A. Jone.A, International Journal Of Computer Trends & Technology, **1**, 281 (2013).
24. E. Porter, J. Fakhoury, R. Oprisor, M. Coates, M. Popovic, Antennas and Propagation (EuCAP), 1 (2010).

Energy efficient routing protocols for wireless sensor networks: comparison and future directions

Murukesan Loganathan[1*], *Thennarasan* Sabapathy[1], *Nur Hafizah* Ghazali[1], *R Badlishah* Ahmad[2] and *Mohamed Nasrun* Osman[1]

[1]Bioelectromagnetic (BioEM) Research Group, School of Computer and Communication Engineering, University Malaysia Perlis, Pauh Putra, Perlis, Malaysia. [2]Faculty of Informatics and Computing, Universiti Sultan Zainal Abidin (UniSZA), 22200 Besut, Terengganu.

Abstract – Wireless sensor network consists of nodes with limited resources. Hence, it is important to design protocols or algorithms which increases energy efficiency in order to improve the network lifetime. In this paper, techniques used in the network layer (routing) of the internet protocol stack to achieve energy efficiency are reviewed. Usually, the routing protocols are classified into four main schemes: (1) Network Structure, (2) Communication Model, (3) Topology Based, and (4) Reliable Routing. In this work, only network structure based routing protocols are reviewed due to the page constraint. Besides, this type of protocols are much popular among the researchers since they are fairly simple to implement and produce good results as presented in this paper. Also, the pros and cons of each protocols are presented. Finally, the paper concludes with possible further research directions.

1. Introduction

Wireless sensor networks (WSN) are result of recent advances in micro-electro-mechanical system (MEMS) technology which enables the manufacturing of small and low cost sensors technically and economically feasible. Sensor nodes were initially designed and developed for the use of military and defense applications. In fact, they were used during Vietnam War to detect enemy soldiers in dense forest [1]. However, those sensors had serious limitations such as high energy consumption, bulky size and very basic network capability. Fast forward few decades and now sensor nodes are much smaller in size but with high computational capability as compared to the earlier nodes. Figure 1 shows basic building blocks of a typical sensor node.

The microcontroller is designed to consume ultra-low power since the attached battery is usually non-replaceable once the sensor nodes are deployed. In addition, the size of the battery also limited since sensor nodes are required be small [2]. It also has limited storage capacity. Moreover, IEEE 802.15.4 compatible radio is used since the protocol is designed for devices with limited energy.

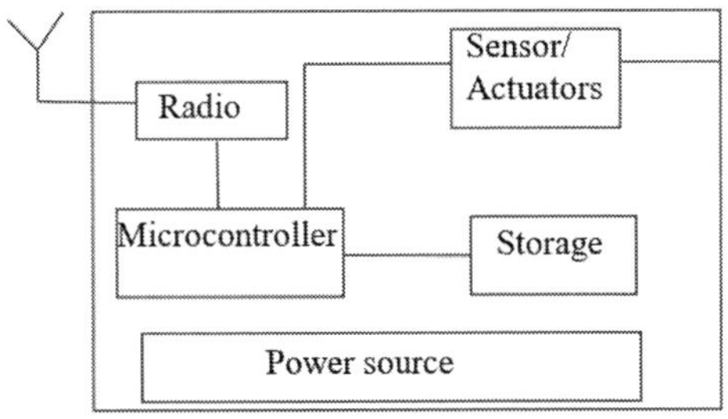

Fig. 1. Building blocks of a typical sensor node.

It is evident from the design itself that the sensor nodes are expected to be energy efficient in order to maximize the lifetime. Hence, it is only natural that there are plethora of energy efficient routing protocols in the literature to support data delivery to the destination with minimal energy consumption. However, these routing protocols are often application- and network architecture-specific since WSNs have been used in many diverse applications. Based on the environment that sensor nodes are being deployed, there are multiple types of WSNs such as wireless body area network (WBAN), wireless multimedia sensor network (WMSN) and etc.

Nevertheless, there are some problems which are common for any type of WSN. For instance, although

*Corresponding author = murukesan.loganathan23@gmail.com

DOI: 10.1051/matecconf/201714001005

current generation sensor nodes are designed to be energy efficient, lifetime of sensor nodes is still one of the main concerns [2]. Also, lifetime of sensor nodes is dependent on both hardware and software implementations. Even though the components that assemble the sensor node save energy, inefficient protocols may lead to early death of the nodes. Energy efficiency can be achieved in WSN through implementation of efficient algorithms and protocols at various network stack level [3].

As shown in the figure 2, a sensor node protocol stack consists of physical layer, data link layer, network layer, transport layer, application layer, power management plane, mobility management plane and task management plane [4]. Physical layer is responsible for frequency selection, carrier frequency generation, signal detection, modulation, and data encryption. The data link layer is responsible for multiplexing of data streams, date frame detection, and medium access and error control. Besides, the network layer is responsible for routing while transport layer provides transmission protocol and congestion control. Also, application layer includes the main application and several management functionalities. Finally, the power, mobility, and task management planes monitor the power, movement, and task distribution among the sensor nodes. These planes help the sensor nodes to coordinate the sensing task and lower the overall power consumption.

The main contribution of this paper is to provide a survey (not exhaustive) from 2011 to 2016 on the energy efficient routing protocols for WSNs. This is so that new researchers can readily know about the current trends in the field. Even though there are several surveys in the literature as shown in Table 1, this work approaches routing protocols from an energy efficiency perspective. Also, we discuss the pros and cons of each protocols by making comparison between them using related metrics (scalability, mobility and robustness).

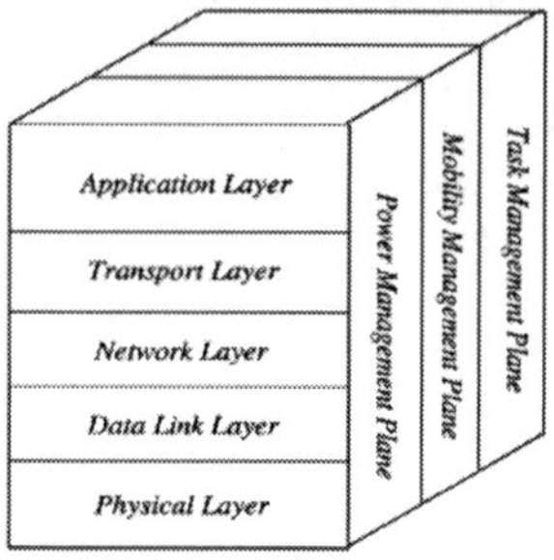

Fig. 2. Network protocol stack of sensor nodes [4].

Besides, we also suggest few future research directions for researchers based on the identified limitations in the present routing protocols.

The main contribution of this paper is to provide a survey (not exhaustive) from 2011 to 2016 on the energy efficient routing protocols for WSNs. This is so that new researchers can readily know about the current trends in the field. Even though there are several surveys in the literature as shown in Table 1, this work approaches routing protocols from an energy efficiency perspective. Also, we discuss the pros and cons of each protocols by making comparison between them using related metrics (scalability, mobility and robustness). Besides, we also suggest few future research directions for researchers based on the identified limitations in the present routing protocols.

This paper is organized as follow: In section 2, classification of routing protocols is explained. In section 3, network structure based routing protocols are reviewed. In section 4, possible future research directions are presented. Finally, this paper is concluded in section 5.

2. Classification of routing protocols

In this paper, we employed the routing protocol classification (Figure 3) as proposed by Pantazis et al [1]. It should be noted that some of the protocols described in following section, may fall under one or more of the routing categories shown in the figure 3, but only the protocols which has network structure scheme as its foundation is discussed here.

Table 1. Previous surveys.

Authors	Characteristic	Time Period
[5]	WSN routing	1988-2004
[6]	Multipath routing	1993-2011
[7]	QoS routing	1990-2012
[1]	WSN routing	2004-2011
[8]	WBAN survey	2000-2013
[9]	Real-time QoS routing	2000-2015

DOI: 10.1051/matecconf/201714001005

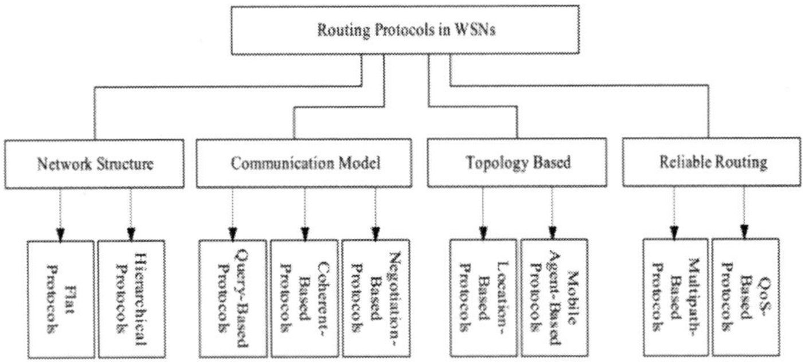

Fig. 3. Classification of WSN routing protocols [1].

Also, tables 2 and 3 summarizes the advantages and disadvantages of each protocol described in this paper. In addition, some metrics for each protocol are presented. These metrics are as follow:

- Scalability – The scalability of a protocol refers to the ability of the protocol to maintain its performance for both small and large networks.
- Mobility – This metric denotes the ability of the protocol to manage mobile nodes. Low mobility means the protocol either does not consider the mobility all together or very limited number of mobile nodes in very slow speed are allowed. Good mobility is when the protocol supports large number of mobile nodes with moderate speed.
- Route Metric – This metric denotes the form of the routing selected to send data from source to destination in such a way that the path taken to send the data is most efficient. Besides, the criteria to choose a path varies across each protocols as shown in these tables.
- Robust – This metric denotes the ability of the protocol to cope up with sudden changes in the network such as topological and link quality changes. The protocol is not robust if it does not consider both the topological and link quality changes. If either one is considered, then it is a moderately robust protocol. It is rated good when both the topological and link changes were considered in the protocol.

3. Routing protocols based on network structure

3.1 Flat protocols

The first category of routing protocols is the multi-hop flat routing protocols. In flat networks, each node typically plays the same role and sensor nodes collaborate together to perform the sensing task. In the rest of this subsection, these protocols are summarized and their limitations are highlighted.

Amiri et al. had studied the feasibility of using fuzzy logic and ant colony optimization (FACOR) for an energy efficient routing in flat WSN architecture [10]. Basically, the ants were used to find best candidate routes to base station while fuzzy value is used by the ants to determine best next hop node. Route that have least number of hops and highest fuzzy value is selected as optimal path. The FACOR protocol was compared with ad-hoc on-demand distance vector (AODV) protocol from the literature in terms of energy efficiency, complexity, end-to-end delay and reliability. Proposed protocol proven to be 40% more energy efficient as compared to AODV in same scenario. However, the setup time for FACOR is higher than as for the AODV protocol. Besides, the ant colony would take longer time to stabilize in large scale WSNs. In addition, this protocol is not suitable for WSNs with heterogeneous nodes and it is not fault-tolerant.

Ghadimi et al. had proposed opportunistic routing in wireless sensor networks (ORW) routing protocol targeting applications with low duty cycles [11]. The protocol maintains a so called expected duty cycle (EDC) metric for all the neighboring nodes. Node with highest EDC is selected to forward the packet to the sink. Unlike other protocols, ORW makes use of both the reliable and unreliable links for data forwarding. The proposed ORW was compared with collection tree protocol (CTP) protocol in terms of delay, reliability, transmission count and radio duty cycle. Using test bed implementation, ORW was found to reduce duty cycles on average by 50% and delays by 30% to 90% while achieving reliability and transmission counts similar to the CTP protocol. However, ORW has several limitations as follow. First, it is not applicable for network with high throughput requirement. Also, the protocol does not support mesh routing. In addition, mobility factor was overlooked.

A routing protocol based on routing by energy and link quality (REL) for internet of things applications is proposed in [12]. The protocol maintains a metric called link quality index (LQI) for each nodes as shown in figure 4. Route which has lesser number of weak links with minimum hop counts and more residual energy is selected as optimal path. The proposed routing protocol was compared with link quality-based lexical routing metric (LABILE) and AODV protocols in terms of energy efficiency, PDR, and latency. In large scale network simulation with high node density (200 nodes), REL increases network lifetime by up to 26.6%, latency by 17.9% and PDR by 12%, as compared to both AODV and LABILE. However, only uniform distribution of nodes is considered in REL. In real life, nodes are rarely

 DOI: 10.1051/matecconf/201714001005

distributed in uniform manner. Since inter-packet-interval of 2s is used, this protocol is ill-suited for real-time multimedia content applications which are delay sensitive such as streaming and monitoring.

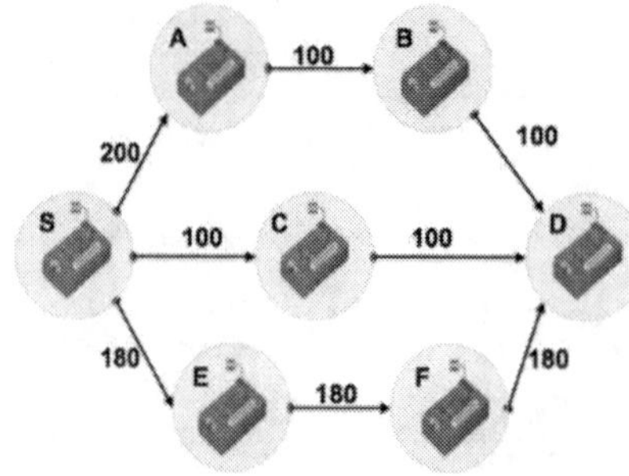

Fig. 4. Nodes with calculated LQI in REL protocol [12].

Jain and Reddy had proposed an orthogonal directional proactive-reactive routing protocol (OD-PRRP) using ant colony optimization and fuzzy logic [13]. The protocol uses directed fuzzy vertex and edge graph of WSN and ant colony to find optimal path to the sink node. Also, nodes are classified into two which are the normal and proactive nodes. The proactive nodes are used for load balancing purposes along with its normal operation. The OD-PRRP was compared with energy aware routing (EARQ), enhanced AODV (EAODV) and energy efficient ant-based routing (EEABR) protocols from literature. Besides, it was evaluated using mobile nodes to evaluate its performance in dynamic environment. OD-PRRP is 57.9%, 114.3% and 160.9% better in energy efficiency as compared to EAODV, EEABR and EARQ, respectively. But, the protocol was not evaluated in heterogeneous networks. Also, the reported performance figure cannot be verified using current generation sensor nodes since they all use omni-directional antenna in contrast to directional antenna as used in this simulation.

In table 2, comparison between flat routing schemes using various metrics such as scalability, mobility, real-time support and robustness of the algorithms is given.

Table 2. Comparison of Flat Routing Schemes

Scheme	Pros	Cons	Scalability	Mobility	Route Metric	Robust	Real-Time
FACOR	Energy efficient, reliable and have good end-to-end delay.	No mobility and fault-tolerance	Good	Not Considered	Minimum hop with highest fuzzy value	Low	Good
ORW	Decreases delay and duty-cycle without sacrificing energy efficiency and reliability.	Not applicable for network with high throughput requirement.	Good	Good	Expected duty-cycle (EDC)	Good	Good
REL	Energy efficiency and load balancing using end-to-end link quality for IOT applications.	Mobility factor was overlooked.	Good	Not Considered	Threshold (link quality, residual energy and hop counts)	Moderate	Good
OD-PRRP	OD-PRRP reduces control overhead without compromising cost and energy efficiency	Directional antenna is not currently available on motes.	Good	Good	Orthogonal direction	Good	Good

3.2 Hierarchical protocols

In hierarchical routing, nodes are grouped into clusters and every cluster has a cluster head. In the rest of this subsection, hierarchical protocols are summarized and their limitations are highlighted.

Shah et al. proposed scaling hierarchical clustering and energy aware routing (SHEAR) protocol with the aim to improve network lifetime by using power efficient and scalable clustering together with an energy aware path selection scheme [14]. This protocol is based on the scaling hierarchical power efficient routing (SHPER) by [15]. In this protocol, WidestShortestPath algorithm is used to find all the paths in the network which have highest residual energy after the CH election process. From these set of paths, a path which minimizes energy consumption is selected for routing. The proposed SHEAR protocol was compared with SHPER from the literature in terms of number of transmission, energy consumption and the number of nodes alive over time. SHEAR was found to increase the network lifetime by 23% as compared to the SHPER protocol. However, the protocol is not resilient to topological changes. Also, the cluster formation and CH election for every transmission round is not energy efficient.

In [16], authors had proposed the low loss energy aware protocol (LLEAP) to address the lack of load balancing and topology updating mechanism in the original energy aware protocol (EAP). In this protocol, after the CH election and cluster formation state, CHs form a collection tree rooted at the sink node. The protocol also incorporates some modifications into the main protocol so that the data packets are not routed to a dead node. The proposed LLEAP was compared with EAP protocol from the literature in terms of throughput, delay and energy efficiency. LLEAP was found to improve energy efficiency by 7.8% as compared to the EAP protocol. The improvement in network lifetime is not significant. Besides, it was achieved by trading-off the latency metric.

Kong et al. proposed a novel instantaneous clustering protocol (ICP) that groups sensor nodes into single-hop clusters in a parallel manner [17]. In this protocol, CHs are locally determined by the pre-assigned probability and its present status. Besides, to save energy, ICP get rids of acknowledgment mechanism and only CHs are allowed to contend for transmission period. This period is elaborately derived to guarantee the connectivity. The proposed protocol was evaluated using both test bed and simulation. It was compared with LEACH, energy constrained minimum dominating set based efficient clustering (ECDS), balanced clustering algorithm with distributed self-organization for wireless sensor networks (DSBCA) and fault-tolerant energy efficient clustering (FT-EEC) protocols from the literature in terms of setup time and energy efficiency. Performance results demonstrate that ICP significantly outperforms existing clustering methods by reducing up to 55%-time consumption and 89% message overheads for energy-saving.

In [18], authors had proposed an evolutionary algorithm based clustering and routing protocol (ERP) for heterogeneous wireless sensor networks in order to optimize the stability period and network lifetime. This was achieved by formulating a new fitness function that incorporates two clustering aspects, namely, cohesion and separation error. Readers are encouraged to refer page 4 of original manuscript for a detailed description on used evolutionary algorithm parameters (mutation, crossover, probabilities of mutation etc.). Routing of data packets is done in multi-hop fashion after the clusters were formed. Proposed ERP protocol was compared with stable election protocol (SEP), LEACH and hybrid cluster routing (HCR) protocols from the literature in terms of stability period and network lifetime. The ERP protocol produced 42%, 62% and 3% lifetime improvement as compared to LEACH, SEP, and HCR protocols, respectively. However, the stability period of ERP is slightly lower than SEP but better than LEACH and HCR protocols. Besides, the protocol can be further optimized by introducing optimal route selection policy and load balancing. Nevertheless, the complexity of the EA algorithms makes it a tough candidate for current generation sensor nodes which are computationally challenged [19].

Building on ERP, Attea and Khalil proposed stable aware evolutionary routing protocol (SAERP) protocol which increases the stability period (time for first node to die) of WSN while reducing the instability period (time from the death of first node until the last node) using an evolutionary algorithm (EA) based approach [20]. The proposed SAERP was compared with LEACH and SEP protocols from the literature in terms of stability, instability periods and throughput. SAERP prolongs the stability period for more than hundreds of rounds as compared to both the SEP and LEACH protocols. Besides, it has very steep instability period which shows its excellent load balancing capability. In addition, SAERP provides more throughput per energy dissipation as compared to the SEP and LEACH protocols. However, the time complexity of this protocol renders it impractical for sensor nodes.

Roy had proposed energy aware cluster based routing scheme (EACBRS) which selects energy efficient route to the sink while providing congestion control mechanism to minimize the re-transmissions [21]. In this protocol, after the CH election process, nodes send the sensed data to their cluster heads by following the congestion control mechanism where data are buffered in the memory for a period. Thus, no packets are dropped or number of re-transmissions are reduced considerably. In addition, CHs aggregate the data and send it to the sink to save energy. The proposed EACBRS was compared with distributed energy efficient clustering (DEEC) and energy efficient hybrid clustering (EEHC) protocols from the literature in terms of network lifetime and throughput. Average residual energy of nodes in EACBRS protocol is significantly higher than as compared to DEEC and EEHC protocols. However, the stability period of proposed protocol is very low as compared to other state of the art protocols such as SAERP and ERP in the literature. Besides, the protocol does not consider nodes mobility.

In table 3, comparison of hierarchical routing schemes using various metrics such as scalability, mobility, real-time support and robustness of the algorithms is given.

Table 3. Comparison of Hierarchical Routing Scheme

Scheme	Pros	Cons	Scalability	Mobility	Route Metric	Robust	Real-Time
SHEAR	Two level cluster formation with energy efficient path selection.	Waste of energy due to clustering and CH selection for every round.	Good	Not Considered	Residual energy with least cost path	Low	Not Considered
LLEAP	Three phase cluster and tree formation for balanced routing.	Insignificant improvement in network lifetime.	Good	Not considered	Multi-hop	Moderate	Moderate
ICP	Efficient clustering with fast setup time and lower overheads.	Delay and reliability are not investigated.	Good	Not considered	Multi-hop	Moderate	Not Considered
ERP	Efficient cluster formation using cohesion and cluster separation metrics.	Stability period is not optimal.	Good	Not Considered	Multi-hop	Low	Not Considered
SAERP	Clustering which increases	Highly complex	Good	Not	Multi-hop	Low	Not

	stability period while decreasing the instability period.	algorithm which is not feasible for sensor nodes.		Considered			Considered
EACBRS	Energy efficient route selection with congestion control.	Very short stability period.	Good	Not Considered	Weight (residual energy and distance)	Low	Not Considered

4. Future research directions

In the recent years a large number of energy efficient routing protocols for the WSNs have been developed. However, there is still a lot of work that has to be done, not only in the area of energy efficiency but also, in other areas. In this subsection, some ideas and interesting avenues for routing protocols are elaborated.

- Mobility support

Majority of the reviewed routing protocols in this paper assume that nodes in the network are static. However, recently, there is an increased interest in applications that support the mobility of the users. The topological changes due to mobility and expected minimum reliability must be considered when designing mobility aware routing protocols.

- Real life performance evaluation using test beds or deployment

Most of the reviewed protocols were evaluated through simulations only. However, it is important to evaluate these protocols in real environments.

- Load balancing

Most of the reviewed protocol in this paper lacks load balancing capability. This leads to premature dead of a portion of the network.

- Cross-layer cognitive radio based routing

Even though plenty of researches had been done on medium access control (MAC) for cognitive radio (CR) [22], [23], cognitive aware routing is still a fresh topic.

5. Conclusion

Wireless sensor networks had matured to a point where it has found its usage in civilian applications from pure military based applications in the past. However, energy efficiency is still one of the major concerns with the current generation sensor nodes as it had been in the past. Hence, various efforts had been taken to tackle this issue either through hardware or software optimizations. This paper focuses on reviewing efforts on the software side, specifically, in the network layer of the sensor nodes.

In this paper, we focus on the network structure based energy efficient routing protocols that have been developed for WSNs. Also, we presented pros and cons each protocols. In addition, possible future research directions were pinpointed as a guidance for researchers.

Conflict of Interest

The authors declare that there is no conflict of interests regarding the publication of this paper.

Acknowledgment

This work is not funded by any government or private organizations.

References

[1] N. A. Pantazis, S. A. Nikolidakis, and D. D. Vergados, “Energy-Efficient Routing Protocols in Wireless Sensor Networks: A Survey,” *IEEE Commun. Surv. Tutorials*, vol. 15, no. 2, pp. 551–591, 2013.

[2] I. F. Akyildiz, W. Su, Y. Sankarasubramaniam, and E. Cayirci, “Wireless sensor networks: a survey,” *Comput. Networks*, vol. 38, no. 4, pp. 393–422, 2002.

[3] R. Mahapatra, Y. Nijsure, G. Kaddoum, N. U. Hassan and C. Yuen, "Energy Efficiency Trade-off Mechanism Towards Wireless Green Communication: A Survey", IEEE Communications Surveys and Tutorials, Jan 2016, pp. 686 - 705.

[4] I. F. Akyildiz and M. C. Vuran, *Wireless Sensor Networks*, vol. 1. Wiley, 2010.

[5] Al-Karaki, J. N. and Kamal, A. E. (2004) ‘Routing Techniques in Wireless Sensor Networks: A Survey’, *IEEE Wireless Communications*, 11(6), pp. 6–28. doi: 10.5120/16346-5685.

[6] Jayashree, A., Biradar, G. S. and Mytri, V. D. (2012) ‘Review of Multipath Routing Protocols in Wireless Multimedia Sensor Network -A Survey’, *International Journal of Scientific & Engineering Research*, 3(9), pp. 1–9.

[7] Ehsan, S. and Hamdaoui, B. (2012) ‘A survey on energy-efficient routing techniques with QoS assurances for wireless multimedia sensor networks’, *IEEE Communications Surveys and Tutorials*, 14(2), pp. 265–278.

[8] Movassaghi, S., Abolhasan, M., Lipman, J., Smith, D. and Jamalipour, A. (2014) ‘Wireless

Body Area Networks: A Survey', *Ieee Communications Surveys and Tutorials*, 16(3), pp. 1658–1686.

[9] Alanazi, A. and Elleithy, K. (2015) 'Real-Time QoS Routing Protocols in Wireless Multimedia Sensor Networks: Study and Analysis', *Sensors*, 15(9), pp. 22209–22233.

[10] E. Amiri, H. Keshavarz, M. Alizadeh, M. Zamani, and T. Khodadadi, "Energy efficient routing in wireless sensor networks based on fuzzy ant colony optimization," *Int. J. Distrib. Sens. Networks*, p. 17, 2014.

[11] E. Ghadimi, O. Landsiedel, P. Soldati, S. Duquennoy, and M. Johansson, "Opportunistic Routing in Low Duty-Cycled Wireless Sensor Networks," *ACM Trans. Embed. Comput. Syst.*, vol. 0, no. 0, pp. 1–35, 2014.

[12] K. Machado, D. Rosário, E. Cerqueira, A. Loureiro, A. Neto, and J. de Souza, "A Routing Protocol Based on Energy and Link Quality for Internet of Things Applications," *Sensors*, vol. 13, no. 2, pp. 1942–1964, 2013.

[13] A. Jain and B. V. R. Reddy, "Ant Colony Optimization Based Orthogonal Directional Proactive–Reactive Routing Protocol for Wireless Sensor Networks," *Wirel. Pers. Commun.*, vol. 85, no. 1, pp. 179–205, 2015.

[14] M. A. Shah, G. Abbas, A. B. Dogar, and Z. Halim, "Scaling hierarchical clustering and energy aware routing for sensor networks," *Complex Adapt. Syst. Model.*, vol. 3, no. 1, p. 5, 2015.

[15] D. Kandris, P. Tsioumas, A. Tzes, G. Nikolakopoulos, and D. D. Vergados, "Power conservation through energy efficient routing in wireless sensor networks," *Sensors*, vol. 9, no. 9, pp. 7320–7342, 2009.

[16] B. M. Mohammad El-Basioni, S. M. Abd El-Kader, H. S. Eissa, and M. M. Zahra, "An optimized energy-aware routing protocol for wireless sensor network," *Egypt. Informatics J.*, vol. 12, no. 2, pp. 61–72, 2011.

[17] L. Kong *et al.*, "ICP: Instantaneous clustering protocol for wireless sensor networks," *Comput. Networks*, vol. 101, pp. 144–157, 2016.

[18] B. Attea and E. Khalil, "A new evolutionary based routing protocol for clustered heterogeneous wireless sensor networks," *Appl. Soft Comput.*, vol. 12, pp. 1950–1957, 2012.

[19] K. Lehre, "Time-complexity Analysis of Evolutionary Algorithms." pp. 1–60, 2011.

[20] E. A. Khalil and B. A. Attea, "Stable-Aware Evolutionary Routing Protocol for Wireless Sensor Networks," *Wirel. Pers. Commun.*, vol. 69, pp. 1799–1817, 2013.

[21] S. Roy, "Energy Aware Cluster based Routing Scheme for Wireless Sensor Network," *Found. Comput. Decis. Sci.*, vol. 40, no. 3, pp. 203–222, 2015.

[22] A. De Domenico, E. Calvanese Strinati, and M. G. Di Benedetto, "A survey on MAC strategies for cognitive radio networks," *IEEE Commun. Surv. Tutorials*, vol. 14, no. 1, pp. 21–44, 2012.

[23] C. Cormio and K. R. Chowdhury, "A survey on MAC protocols for cognitive radio networks," *Ad Hoc Networks*, vol. 7, no. 7, pp. 1315–1329, 2009.

DOI: 10.1051/matecconf/201714001006

60 GHz Milimeter-Wave Antennas for Point-to-Point 5G Communication System

[1]A.S.*Aishah*, [1]Beson Mohd Rashidi Che, [1]S.N.*Azemi*, [1]S.A.Al Junid

[1]Advance Communication Engineering,
Center of Excellent School of Computer and Communication Engineering.
Universiti Malaysia Perlis,
1st Floor, Pauh Putra Main Campus,
02600 Arau, Perlis,
Malaysia.
aishahshukre@gmail.com

Abstract. In this paper microstrip patch antenna for millimeter-wave is proposed. Evolution of shape microstrip antenna are designed which is from rectangular antenna to triangle antenna and changed to triangle with slot. The proposed antenna configuration achieved for covering 5G wireless system. The lowest return loss of the antenna is -29.23dB which is triangle with slot and the maximum gain obtained is 8 db at the 61.93 GHz for the triangle antenna. This antenna are suitable for the 5G wireless application for short range and high rate communication system.

1. Introduction

Nowadays, the 5G technology for wireless systems becomes attractive and huge demand from the end users. The new 5G wireless system starts to formulate in a more concrete method, they will be incorporated into the new 5G wireless system. The 4G wireless system have already been initiated in some of the countries and is going to be in other soon. However, major problem is power consumption still persist even with the presence of 4G system[1]. Therefore The need for 5G wireless system came in order to solve the issues and the requirement of high data rate on 5th generation wireless system is going on and expected to be initiated by 2020, and the significant growth in the customers demand for wireless communication using handsets has created the need for important development of antenna design as a basic part of any wireless system [2].

The user and service provides demand wireless unit with antenna that small and compact, cost effective and low profile. This paper proposed, a microstrip antenna capable to fulfil most of the wireless system requirement using milimeter-wave frequency. The frequency spectrum band for milimeter-wave lies between 30GHz-300GHz in electromagnetic spectrum[5]. The free space wavelength in this band is order of 1 milimeter to 10 milimeter thus permit the use of small size of antenna. The path loss proportional to frequency square so at this high frequency band path loss is high. In 2001, Federal Communication Commission (FCC) released the unlicensed milimeter-wave band around the 60GHz frequency[1],[4] This band is of immense interest due to the availability of universal unlicensed spectrum around 60GHz for short-range communication system such as indoor and underground communication and high speed wireless application [2],[3] Antennas for communication system at 60GHz shuld have a high bandwidth to achieve high data rate and should have a high gain to compensate the high path loss ath these frequencies [4]. The advantages of microstrip antenna for point-to-point operate at 60GHz is proposed which has a fast data rate and reliable gain. The gain is enhanced by inserting the slot into the patch and the most optimized results are discussed in the following sections . Microstrip antenna for wireless system application in operate 60GHz are proposed which has a faster data rate and high gain. The gain is enhanced by inserting the slot into the patch and the most optimized results are discussed in the following sections.

2. Essential of 5g Milimeter-Wave Antenna

Three antenna shapes are proposed in this project, which is rectangular antenna, triangle antenna and hexagon antenna, label with dimension in figure 1. The antenna is designed on a RogerRT5880 substrate of thickness (0.127mm) and relative permittivity of 2.2, loss tangent of 0.0009 and CST Microwave Studio is used for simulation proposed. The design of basic a basic patch antenna begins with the calculation of the dimension or antenna design specifications. The operating frequency of the antenna must be determined before calculation can begin. For this design we need to choose an operating frequency of 60 GHz.

The width of the antenna *(W)* and the length (L) of both feed from the centre of the path were two tuneable parameters. The line feed is attaching a (SMA) connector at the bottom. The quarter-wave transformer is used as matching network and allocated in the middle of 50 ohm feed-line and patch design. Basically the antenna is designed using a transmission line model. For patch antenna design, a rectangular patch antenna will be designed as a limitation of time to study and optimization.

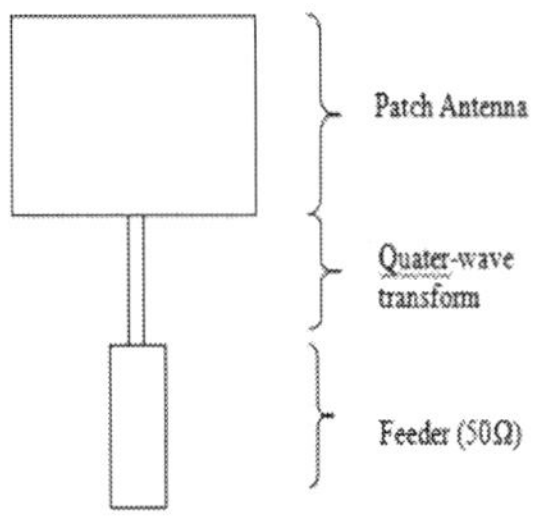

Figure 1: Single Patch antenna using Quarter-wave

The design procedure was to first estimate the width (W) and length *(L)* of the patch antenna. Below show the calculation by using the equation from (1) to (4) a parameter of the antenna, it will carry out εreff = effective dielectric constant, W = patch width, *L* = patch length, *ΔL* = patch length extension.

W = Width patch

$$\frac{Co}{2fo}\sqrt{\frac{2}{\varepsilon r+1}} \quad (1)$$

ε_{reef} = effective dielectric constant

$$\frac{\varepsilon r+1}{2}+\frac{\varepsilon r+1}{2}\left[1+12\frac{h}{w}\right]^{\frac{1}{2}} \quad (2)$$

ΔL = Patch length extension

(3)

$$\left[\frac{(\varepsilon reff+0.3)\left(\frac{w}{h}+0.264\right)}{(\varepsilon reff-0.258)\left(\frac{w}{h}+0.8\right)}\right]$$

L = Length

$$\frac{\lambda}{2\sqrt{\varepsilon reff}}-2\Delta L \quad (4)$$

From the formula, theoretically calculated width and length of the top radiating patch for 60 GHz resonance are found to be 2mm and 0.5mm, respectively. However, the dimension have been adjust and optimized to meet requirement of the resonant frequency and other characteristics. The design parameter are obtained from several parametric studies and suitable patch and slot size are selected for high gain. The primary design parameter are listed in Table 1.

Table 1 : Initial Design Paremeter

Parameter	Value
Thickness (h)	0.127 mm
Dielectric constant (ε_r)	2.20
Loss tangent	0.0009
Patch Size (W x L)	1.84 x 1.50 mm^2
Ground Plane (W_g x L_g)	3.6 x 4.3 mm^2
Total Antenna Profile	3.6 x 4.3 x 0.127 mm^2
Substrate	Roger RT5880

The geometry of the three shape microstrip antenna designed for 60 Ghz is shown in fig. 3. The antenna has overall dimension of only 1.8 x 1.50mm^2 . A copper plate as a ground with dimension 3.6 x 4.3 mm^2 and thickness is 0.035mm.

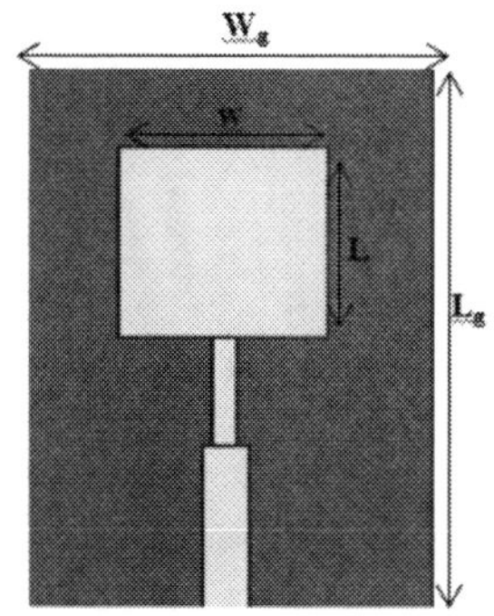

Figure 2: Geometry of the antenna

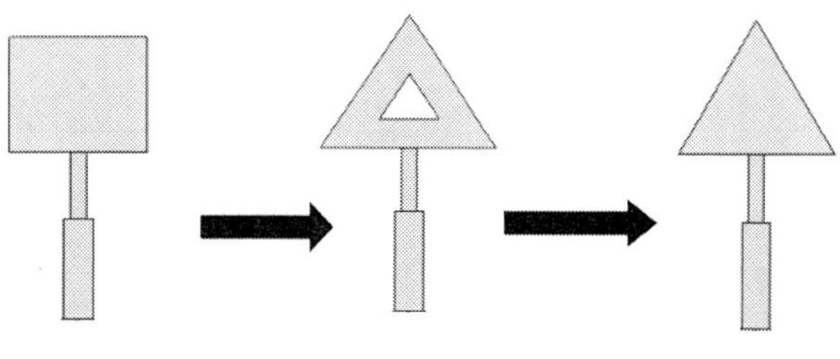

Figure 3 : Evolution shape of the antenna

3. Simulation And Result

The simulation of the proposed antenna is performed using Computer Simulation Technology (CST) Microwave Studio commercial software program. The simulated results of the reflection coefficients $|S_{11}|$ for the proposed milimeter-wave antenna are illustrated in Fig. 4. It is apparent that the proposed antenna can cover millimeter-wave bands of 60 GHz for $|S_{11}|$ less than -10. The simulated maximum realized gain of the proposed antenna . A stable gain for rectangular antenna with a value 7.32 dB in the frequency at 60 GHz and with return loss -20.89 dB is observed and gain at 8 db for triangle antenna with frequency at 61.93 GHz with -10.155 db return loss, meanwhile for trianglewith slot antenna the gain is 7.75 db at frequency at 56 GHz and return loss is -29.23 db. Simulated results demonstrate that the antenna is characterized by omnidirectional patterns.

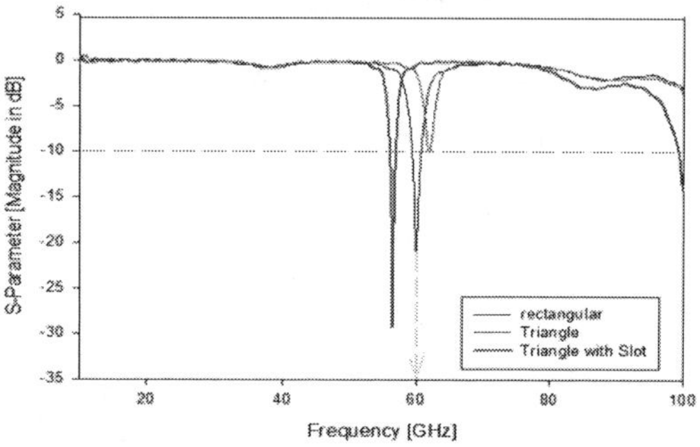

Figure 4: Simulated S_{11} characteristics of the three different antennas [magnitude in dB]

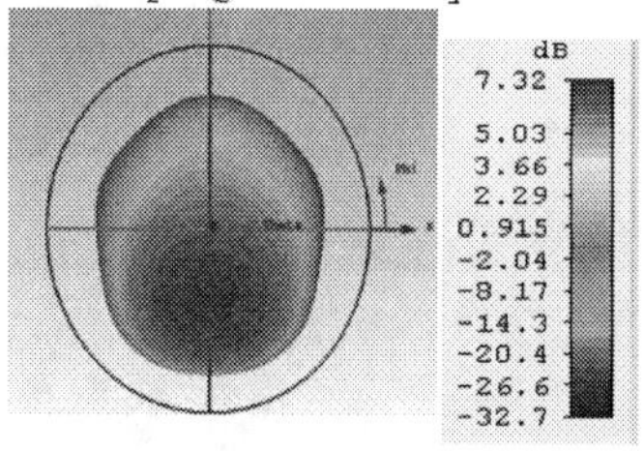

Figure 5a : 3D radiation pattern for rectangular antenna

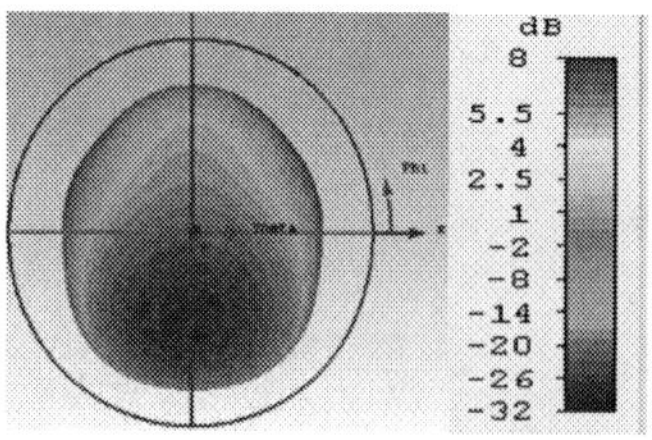

Figure 5b : 3D radiation pattern for triangle antenna

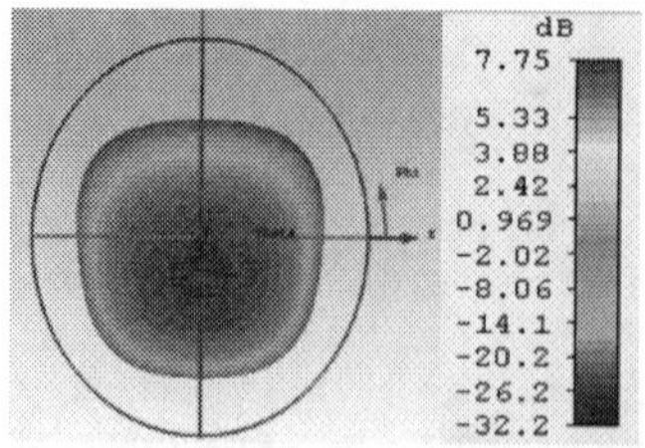

Figure 5c: 3D radiation pattern for triangle with slot

Table 2 : The Final Performance Antenna Configuration

Antenna Shape	Return Loss in db	Gain in db	Frequency range (GHz)
Rectangular	-20.89	7.3	60 GHz
Triangle	-10.155	8	61.93 GHz
Triangle with slot	-29.23	7.55	56.35

Table 2 summarizes the final antenna performances. The perforances of the proposed antenna such as return loss, gain and frequency rang.Gain is highly sensitive to change of the slot, truncated, and shape of the antenna. Thus, the slot of the antenna are adjusted to get the optimized result as high gain and return loss of the antenna.

4. Conclusion

In this paper, the design of a milimeter-wave antenna has been proposed for a short range and high speed rate wireless application 60 GHz. The antenna configuration is designed and analysed by using CST software based on the finite element method. From the anaylsis, its can be concluded that the different shape of the antenna can influence the return loss, gain and frequency. The triangle with slot antenna has a good result of return loss ,S_{11} compare to rectangular and triangle antenna, meanwhile for the best gain is rectangular antenna. As we can see, the result of simulation antenna are achieved for the 5G wireless communication.

Acknowleddgements

This work was supported by High Ministry of Eduacation Malaysia under the Fundamental Research Grant Scheme (FRGS) #9003-00556

References

1. Alam, M. S., Islam, M. T., Misran, N., & Mandeep, J. S. (2013). A Wideband Microstrip Patch Antenna for 60 GHz Wireless Applications. *Elektronika ir Elektrotechnika*, *19*(9), 65-70.

2. Kathuria, N., & Vashisht, S. (2016, March). Dual-band printed slot antenna for the 5G wireless communication network. In *Wireless Communications, Signal Processing and Networking (WiSPNET), International Conference on* (pp. 1815-1817). IEEE.

3. Smulders, P. (2002). Exploiting the 60 GHz band for local wireless multimedia access: Prospects and future directions. *IEEE communications magazine*, *40*(1), 140-147

4. Biglarbegian, B., Fakharzadeh, M., Nezhad-Ahmadi, M. R., & Safavi-Naeini, S. (2010, July). Optimized patch array antenna for 60 GHz wireless applications. In *Antennas and Propagation Society International Symposium (APSURSI), 2010 IEEE* (pp. 1-4). IEEE.

5. Goyal, R. K., & Sharma, K. K. (2016, September). T-slotted microstrip patch antenna for 5G Wi-Fi network. In *Advances in Computing, Communications and Informatics (ICACCI), 2016 International Conference on* (pp. 2684-2687). IEEE

6. Hannachi, C., & Tatu, S. O. (2016). Performance comparison of 60 GHz printed patch antennas with different geometrical shapes using miniature hybrid microwave integrated circuits technology. *IET Microwaves, Antennas & Propagation*.

7. Garg, R., & Long, S. A. (1988). An improved formula for the resonant frequencies of the triangular microstrip patch antenna. *IEEE transactions on antennas and propagation*, *36*(4), 570.

8. Karnfelt, C., Hallbjorner, P., Zirath, H., & Alping, A. (2006). High gain active microstrip antenna for 60-GHz WLAN/WPAN applications. *IEEE Transactions on Microwave Theory and Techniques*, *54*(6), 2593-2603.

9. Li, W. (2007, February). Performance of Ultra-Wideband Transmission over 60GHz WPAN Channel. In *Wireless Pervasive Computing, 2007. ISWPC'07. 2nd International Symposium on*. IEEE.

10. Md.Mamunur Rasid, Saddam Hossain, "Antenna Solution for Milimeter wave Mobile Communication (MWMC): 5G, " *International Journal of Scientific Research Engineering & Technology (IJSRET)*, ISSN 2278-0882, vol. 3, issue 8, 2014

MATEC Web of Conferences **140**, 01007 (2017)
ICEESI 2017
DOI: 10.1051/matecconf/201714001007

Simulation Realization of 2-D Wavelength/Time System Utilizing MDW Code for OCDMA System

M. S. A. Azura, Beson Mohd Rashidi Che, S. A. Al Junid, *R.* Endut, and *N.* Ali

Advanced Communication Engineering,
Centre of Excellence,
School of Computer and Communication Engineering,
Universiti Malaysia Perlis,
Kangar, 01000, Perlis
Malaysia.
E-mail: amyazura92@gmail.com

Abstract. This paper presented a realization of Wavelength/Time (W/T) Two-Dimensional Modified Double Weight (2-D MDW) code for Optical Code Division Multiple Access (OCDMA) system based on Spectral Amplitude Coding (SAC) approach. The MDW code has capability to suppress Phase-Induce Intensity Noise (PIIN) and minimizing the Multiple Access Interference (MAI) noises. At the permissible BER 10^{-9}, the 2-D MDW (APD) had shown minimum effective received power (*Psr*) = -71 dBm that can be obtained at the receiver side as compared to 2-D MDW (PIN) only received -61 dBm. The results show that 2-D MDW(APD) has better performance in achieving same BER with longer optical fiber length and with less effective received power (*Psr*). Also, the BER from the result shows that MDW code has capability to suppress PIIN ad MAI.

1 Introduction

Optical Code Division Multiple Access (OCDMA) technology is one of the talented technologies to execute all-optical networks. It is group of multiplexing and internetworking technologies that encodes/decodes signal through employing straightforward and cost-effective inactive optical component such as the signal multiplexing, routing and switching can be implement easily [1]. OCDMA can be realized in one-dimension (1-D) and two-dimension (2-D) encoding through spatial/spectral domain. The execution assessment of OCDMA framework in view of wavelength/time (W/T) code has been investigated by measuring the estimations of bit error rates for various number of active users [2, 3]. The OCDMA systems undergo from assured noise such as phase induce-intensity noise (PIIN) shot noise and thermal noise [4, 5]. Besides, multiple access interference (MAI) is the major performance degradation mainly when a large number of users are occupying in the OCDMA systems. Then, the most significant consideration is the code design for decreasing contribution at the MAI to the optical power receives [6]. In OCDMA systems has several techniques can be implement to eliminate MAI. Spectral Amplitude Coding (SAC) is the one can be used in this system. SAC schemes activate at bit rate, as a result the condition for receiver bandwidth is calm [7, 8]. Here, the performance of 2-D MDW codes for OCDMA systems to archive higher performance possible through suppressing PIIN and decreasing the MAI was analyses [9, 10].

2 Development of 2-D Modified Double Weight (MDW) Code

2-D MDW codes also signify by using the $M \times N$ matrix form. The 2-D MDW codes with M as a number of wavelengths, N as a temporary codes length, W as a weight, λ_a and λ_c are auto-correlation and cross-correlation values. There are denoting by $(M \, x \, N, W, \lambda_a, \lambda_c)$. The j_{th} user's 2-D codes $C^j_{M,N}$, in equation (1), is a matrix of M row vectors $d^j_{k,N}$ related to the temporary spreading; $d^j_{1,N} = c^j_{k,1}, c^j_{k,2}, \dots, c^j_{k,N-1}, c^j_{k,N}$ where $c^j_{k,i} \in \{0,1\}$ and k is the emitted wavelength of $k \in \{1,\dots,M\}$ [5, 7, 9].

$$C^j_{M,N} = \begin{bmatrix} d^j_{1,N} \\ d^j_{2,N} \\ \cdot \\ \cdot \\ \cdot \\ d^j_{M-1,N} \\ d^j_{M,N} \end{bmatrix} \quad (1)$$

The signals $r_{k,N}(t)$ are sum of the temporary spreading data of Fu user carried on the wavelength λ_k and expressed as $r_{k,N}(t) = \sum_{j=1}^{Fu} b^j_i(t) d^j_{k,N}$, where $b^j_i(t)$ is

i_{th} user data bit of j_{th} and Fu number of users. The M signals $r_{k,N}^{j}(t)$ are multiplexed and total signal $R_{M,N}(t)$ transmitted on optical fiber is expressed as $M \times N$ [5, 7, 9].

$$R_{M,N}(t) = \begin{bmatrix} r_{1,N(t)} \\ r_{2,N(t)} \\ \cdot \\ \cdot \\ \cdot \\ r_{M-1,N(t)} \\ r_{M,N(t)} \end{bmatrix} \quad (2)$$

2-D MDW optical CDMA network are made of M (transmitter) and N (receiver). $A_{g,h}$ correspond to the code where $g \in (1,2,3,\ldots,M-1)$ and $h \in (1,2,3,\ldots,N-1)$. X_g and Y_k are spectral and spatial encoding respectively. Table 1 shows the 2-D MDW code projections [5, 7, 9].

Table 1. 2-D MDW Code Projections $k_1 = 4$ and $k_2 = 2$.

$X_{g,h}/Y_g$	[000011011]	[011000110]	[110110000]
$\begin{bmatrix}0\\1\\1\end{bmatrix}$	$\begin{bmatrix}000000000\\000011011\\000011011\end{bmatrix}$	$\begin{bmatrix}000000000\\011000110\\011000110\end{bmatrix}$	$\begin{bmatrix}000000000\\110110000\\110110000\end{bmatrix}$
$\begin{bmatrix}1\\1\\0\end{bmatrix}$	$\begin{bmatrix}000011011\\000011011\\000000000\end{bmatrix}$	$\begin{bmatrix}011000110\\011000110\\000000000\end{bmatrix}$	$\begin{bmatrix}110110000\\110110000\\000000000\end{bmatrix}$

$X = [x_0, x_1, x_2, \ldots, x_{M-1}]$ and $Y = [y_0, y_1, y_2, \ldots, y_{N-1}]$ are the code sequence of 1-D MDW codes. The 2-D MDW cross-correlation can be derived through four characteristic matrices of $A^{(d)}$, where $d \in (0,1,2,3)$ are defined as:

$$A^0 = Y^T X \quad (3)$$

$$A^1 = Y^T \bar{X} \quad (4)$$

$$A^2 = \bar{Y}^T X \quad (5)$$

$$A^3 = \bar{Y}^T \bar{X} \quad (6)$$

The cross-correlation between $A^{(d)}$ and $A_{g,h}$ can be defined as:

$$R^{(d)}(g,h) = \sum_{i=0}^{M-1}\sum_{j=0}^{N-1} a_{ij}^{(d)} \, a_{(i+g)(j+h)} \quad (7)$$

Where $a_{ij}^{(d)}$ is $(i,j)_{th}$ of $A^{(d)}$ and $a_{(i+g)(j+h)}$ is $(i,j)_{th}$ of $A_{g,h}$; $g \in (1,2,3,\ldots,M-1)$ and $h \in (1,2,3,\ldots,N-1)$.

Table 2 shows the cross-correlation of the 2-D MDW code generated from $R^{(d)}(g,h)$. From the table, $R^{(3)}(g,h)$ has nonzero value when $g \neq 0 \cap h \neq 0$. Furthermore, when $g \neq 0$ and $h \neq 0$, the values of $R^{(0)}(g,h)$, $R^{(1)}(g,h)$, $R^{(2)}(g,h)$ and $R^{(3)}(g,h)$ indicates the specific relationships. Cross-correlation function is defined by using $R^{(3)}(g,h)$ to eliminate influence due to $A_{g,h}$ from $R^{(0)}(g,h)$, $R^{(1)}(g,h)$, and $R^{(2)}(g,h)$ when $\neq 0 \cap h \neq 0$. New derived 2-D MDW cross-correlation function is expressed as;

$X_{g,h}$	$R^{(0)}(g,h)$	$R^{(1)}(g,h)$	$R^{(2)}(g,h)$	$R^{(3)}(g,h)$
$g = 0,$ $h = 0$	$k_1 k_2$	0	0	0
$g = 0,$ $h \neq 0$	k_1	k_1	0	0
$g \neq 0,$ $h = 0$	k_2	0	$k_2(k_1\text{-}1)$	0
$g \neq 0,$ $h \neq 0$	1	1	$k_1\text{-}1$	$k_1\text{-}1$

Table 2. 2-D MDW Cross-Correlation.

$$R^{(0)}(g,h) - R^{(1)}(g,h) - \frac{R^{(2)}(g,h)}{(k_1-1)} + \frac{R^{(3)}(g,h)}{(k_1-1)} = \begin{cases} k_1 k_2, & \text{for } g = 0 \text{ and } h = 1 \\ 0, & \text{otherwise} \end{cases} \quad (8)$$

3 2-D Wavelength/Time (W/T) Signal-to-Noise Analysis

Phase-induced intensity noise (PIIN), shot noise and thermal noise is three types of noise are considerate in performance analysis. To identify thermal lights produces the photodiode can be expressed as follows:

$$< i^2 > 2eIB + I^2 B\tau_c + \frac{4K_b T_n B}{R_L} \quad (9)$$

The noise power of I_{PIIN}^2 can be expressed as below.

$$\langle I_{PIIN}^2 \rangle = \frac{B_r \Re^2 P_{sr}^2}{M\Delta f k_2^2 (MN-1)^2} \begin{bmatrix} k_1 \, k_2 \, (MN-1)^2 + \\ [k_2 \, (W-1)(M-1)]^2 \end{bmatrix} \quad (10)$$

The short noise current can be written as follow:

$$\langle I_{short}^2 \rangle = 2eB_r \left\{ \frac{\Re^2 P_{sr}^2}{M k_2 (MN-1)} [k_1 \, k_2 (MN-1) + 2k_1 (W-1)(N-1) + 2k_2 (W-1)(M-1) + 4(W-1)(M-1)(N-1)] \right\} \quad (11)$$

The thermal noise power is:

$$\langle I_{thermal}^2 \rangle = \frac{4K_b \, T_n \, B_r}{R_L} \quad (12)$$

Total photocurrent output from receiver I_r:

$$\langle I_r^2 \rangle = \left[\frac{\Re \, P_{sr} \, k_1}{M} \right]^2 \quad (13)$$

The signal-to-noise ratio (SNR) equation at receiver is developed

$$SNR = \frac{I_r^2}{\langle I_{PIIN}^2 \rangle + \langle I_{short}^2 \rangle + \langle I_{thermal}^2 \rangle} \quad (14)$$

Therefore, bit error rate (BER) can be expressed in terms of the signal-to-noise ratio (SNR) [9]:

DOI: 10.1051/matecconf/201714001007

$$BER(M) = \frac{1}{2}erfc\left(\sqrt{\frac{SNR}{8}}\right) \quad (15)$$

where $erfc$ is:

$$erfc(x) = \frac{2}{\sqrt{\pi}}\int_x^\infty e^{(-z^2)}dz \quad (16)$$

4 Simulation Results and Discussions

Table 3. Table of link parameters.

Parameters Used in Numerical Calculation	
PD quantum efficiency	η=0.6
Spectral width of broadband light source	$\Delta\lambda$=30 nm(Δ=3.75 THz)
Operating wavelength	λ_0=1.55μm
Electrical bandwidth	B=311 MHz
Data transmission rate	R_b=622 Mbps
Receiver noise temperature	T_n=300 K
Receiver load resistor	R_L=1030 Ω
Boltzmann's constant	K_b=1.38x10^{-23} W/K/Hz
Electron charge	e=1.6021764x10^{-19} coulombs
Light velocity	C=3x10^8 m/s

Table 3 is the system parameters used to find the numerical results.

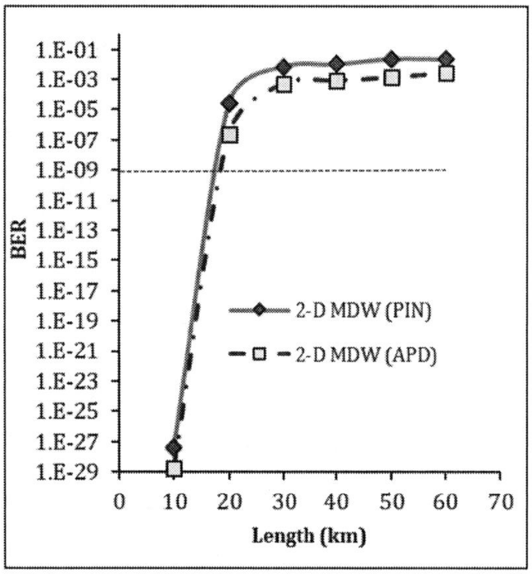

Fig. 1. BER versus optical fiber length for 2-D MDW codes.

Figure 1 demonstrates the graph of BER versus optical fiber length in kilometer (km) for the 2-D MDW codes. In this simulation used 622 Mbps of bit rate. The number of length use is from 10 km until 60 km. According to the simulation and the result get, the graph increase for both bit rate due to the increasing of the optical fiber length. The system performances BER 10^{-9} 2-D MDW (APD) at 19 km of fiber length have better performance compared to 2-D MDW (PIN) at fiber length 18 km. This is because of the distance described by the number of dispersion and the signal loss.

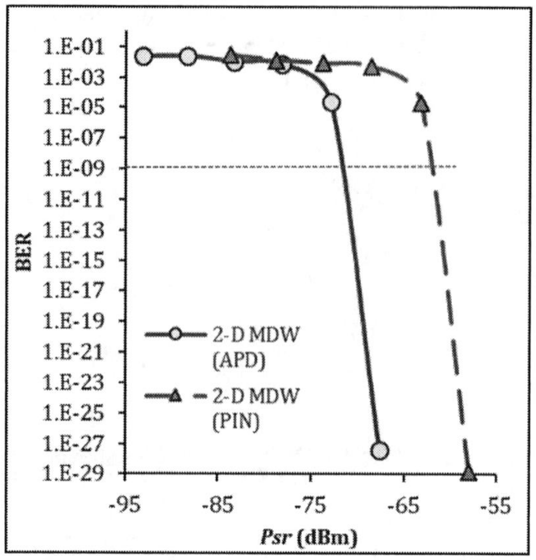

Fig. 2. BER versus effective received power (*Psr*) for 2-D MDW codes.

Figure 2 illustrates the curve of BER versus effective received power (*Psr*) for 2-D MDW codes. The values of effective received power (*Psr*) are varies from -95 dBm until -40 dBm. The number of bit rate used is 622Mbps. Then the length of optical fiber used is 10 km to 60 km. It can be seen that, 2-D MDW (APD) code requires -71 dBm more effective source power to achieve normal optical transmission requirements compared to 2-D MDW (PIN) requires only -61 dBm. This situation occurs when the losses of fiber optic cable make the *Psr* decrease while the BER is increase.

5 Conclusion

Resolution can be made from the numerical results, the *Psr* value for 2-D MDW (APD) codes is – 71 dBm while 2-D MDW (PIN) codes is -61 dBm. Therefore, photodiode APD are more effective compared to photodiode PIN because of photodiode APD are increase according their gain. In addition, the theoretical calculation demonstrates the number of BER 10^{-9} so, it's easily to suppress PIIN and eliminate MAI.

Acknowledgement

The author would like to acknowledge this work was supported by the High Ministry of Education Malaysia under the Fundamental Research Grant Scheme (FRGS) under grant #9003-00475.

References

1. C. B. M. Rashidi, S. A. Aljunid, F. Ghani, H. A. Fadhil, M. S. Anuar, A. R. Arief, "Cardinality Enrichment of Flexible Cross Correlation (FCC) Code for SAC-OCDMA System by Alleviation Interference Scheme (AIS)," *Optik-International*

DOI: 10.1051/matecconf/201714001007

Journal for Light and Electron Optics, **125,** 17, pp. 4889-4894 (2014)

2. V. Jyoti, R. S. Kaler, "Design and Implement of 2-Dimensional Wavelength/Time Codes for OCDMA", *Optik,* pp. 851-857 (2011)
3. A. R. Arief Jamil Abdullah, S. A. Aljunid, A. M. Safar, J. M. Nordin, R. B. Ahmad, "Mitigation of multiple access interference using two-dimensional modified double weight codes for optical code division multiple access systems*", Fiber Optics and Optical Communication*, **51**, 6 (2012)
4. C.C. Yang, "The application of spectral-amplitude-coding optical CDMA in passive optical networks", *Optical Fiber Technology*, **14**, pp. 134-142 (2007)
5. A. R. Arief, S. A. Aljunid, M. S. Anuar, M. N. Junita, R. B. Ahmad, F. Ghani, "Phase-Induce Intensity Noise Suppression Wavelength-Time Modified Double Weight Codes Optical Code Division Multiple Access Systems", *3rd International Conference on Photonics,* (2012)
6. C. B. M. Rashidi, S. A. Aljunid, M. S. Anuar, Hilal A. Fadhil, F. Ghani, "Performance Analysis of a New Class of Codes with Flexible Cross Correlation for SAC-OCDMA System", *Journal of Theoritical and Applied Information Technology*, **61** (2014)
7. A. R. Arief, S. A. Aljunid, M. S. Anuar, M. N. Junita, R. B. Ahmad, F. Ghani, "Enhanced performance of new family modified double weight codes spectral amplitude coding optical CDMA system network", *IEEE International Conference in Control System, Computing and Engineering (ICCSCE)*, pp. 488-494 (2011)
8. C. C. Yang, "Hybrid Wavelength-Division-Multiplexing/Spectral-Amplitude-Coding Optical CDMA System", *IEEE Photonics Technology Letters*, **17,** 6, pp. 1343-1345, (2005)
9. A. R. Arief, S. A. Aljunid, M. S. Anuar, M. N. Junita, R. B. Ahmad, "Cardinality Enhancement of Spectral/Spatial Modified Double Weight Code Optical Code Division Multi-Access System by PIIN Suppression", *Optik - International Journal for Light and Electron Optics*, pp. 3786-3793 (2013)
10. N. Din Keraf, S. A. Aljunid, A. R. Arief, M. S. Anuar, C. B. M. Rashidi, P. Ehkan, M. N. Nurol, "An Optimal Cardinality of Wavelength/Time Incoherent OCDMA System using 2-D Hybrid FCC-MDW code", *International Conference on Electronic Design (ICED),* (2014)

MATEC Web of Conferences **140**, 01008 (2017)
ICEESI 2017
DOI: 10.1051/matecconf/201714001008

Performance Comparison Of Triangle Antenna of 60 GHz for 5G Wireless Communication Network

[1]A.S.*Aishah*, [1]Che Beson Mohd Rashidi, [1]S.N.*Azemi*, [1]*Aljunid Syed A.*

[1]Advance Communication Engineering,
Center of Excellence School of Computer and Communication Engineering.
Universiti Malaysia Perlis,
1st Floor, Pauh Putra Main Campus,
02600 Arau, Perlis,
Malaysia.
aishahshukre@gmail.com

Abstract-In this paper microstrip triangle with slot antenna for 5G wireless communication network are proposed. The microstip triangle antenna is design and operating 60 GHz milimeter-wave frequency band and it's suitable for 5G wireless communication. The substrates are chosen in the design, which are RogerRT5880 with copper thickness 0.035 mm to analyze their effect toward milimeter-wave performance on the designed. The designed and analysis is performed by using CST Microwave Studio.The lowest return loss of the antenna is -24.75dB which is triangle with slot and the maximum gain obtained is 6.82 db at the 59.68GHz for this antenna. The antenna is considering the gain, return loss and size, the microstrip antenna can be a suitable candidate for the 5G wireless application for short range high speed communication.

1. Introduction

The current status of the 5G technology for wireless systems is very much in recent development stages.5G technology for wireless system will probably start to come to awareness around 2020 with deployment following on afterwards. The wireless communication is development of target microstrip antenna is a thrilling research interest nowadays and many techniques have been proposed to improve their performance. The 4G wireless communication systems have already been initiated in some of the countries and are going to be in other soon. However, the problem and challenges of spectrum scarcity and power consumption still persist even with the presence of 4G systems [1]. Therefore the need for 5G wireless system came in order to solve the issues and the requirement of high data rate and mobility to solve the challenges the research on 5th generation wireless system is going on the expected to be accomplished by 2020[1].

Most of the technologies to be used for 5G will commence to come in the systems used for 4G and then as the new 5G cellular system starts to formulate in a more concrete manner, they will incorporated into the new 5G cellular system[5]. The main issues with 5G technology is that there is such a extremely wide variation in the requirement, superfast downloads to small data requirements for IoT than any one system will not capable to meet these needs. Accordingly a layer approach is likely to be adopted. As well say, 5G is not just a mobile technology, it is ubiquitous access to high & low data rate services. Through the second half 2014, 5G technology include all sort of advance features which will make it most powerful and in huge demand in near future. Nevertheless, the several between 4G and 5G techniques from a user point of view is increased data rate and less power usage with better coverage. 5G system may take the wireless signals to a higher frequency range of 30 to 300 gigahertz (GHz) and will reduce the wavelength from centimeter to millimeter[2]. One of the challenges is the technology may face is attenuation of line of sight communication is not possible between transmitter and receiver. Proposed antenna could provide the communication for future 5G network application which has a faster data rate. Furthermore millimeter wave bands could relieve congestion and reduce demand for spectrum in frequency bands below 5 GHz. Microstrip antennas for wireless system application in operates 60GHz is proposed which has a faster data rate and high gain. The gain is enhanced by inserting the slot into the patch and the most optimized results are discussed in the following sections.

2. Antenna and Design Geometry

Based on the several parameter studies, the geometry of the triangle antenna for 60 GHz band are proposed is shown in the figure 2. The proposed antenna is compare between of a triangle patch antenna without slot and triangle antenna slot inside the patch.

DOI: 10.1051/matecconf/201714001008

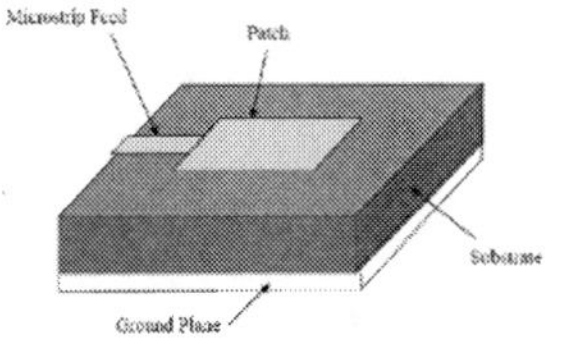

Figure 1:

Figure 1 : Basic microstip patch antenna

To compute the resonant frequency of an equilateral triangle patch, at the lowest-order resonant frequency f_r, the side length a, can be given by (1) and placed at the distance of h mm (substrate thickness) away from the ground plane. The chosen substrate material is 0.127mm thick RogerRT5880 dielectric board with a dielectric constant of 2.2 and loss tangent 0.0009. The printed circuit board (PCB) material has some advantage such as low dielectric torelance and loss, stable electric property against frequency and thus it is better choice for high frequency operation.

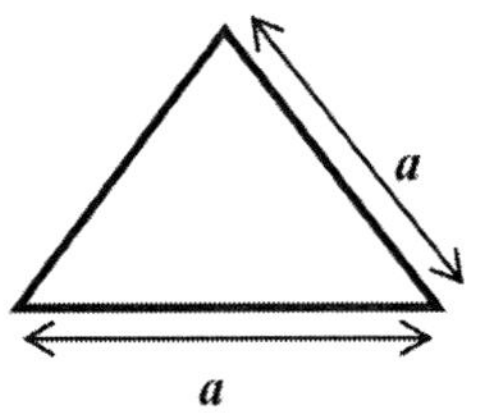

Figure 2 : Theoretical design parameter for triangle antenna.

$$a = \frac{2c}{3f_r\sqrt{\varepsilon_r}} \quad (1)$$

$$C = 3 \times 10^8 \quad (2)$$

Based on the formula above, theoritically calculated both side (a) of top radiating patch for 60 GHz resonance are found to be 2 mm and 0.5 mm, respectively. However, the dimension have been adjusted and optimized to meet requrement of the resonant frequency and other characteristics. The design paramaters are obtained from several parametric studies and suitable patch and slot size are selected for a high gain, wideband 60 GHz antenna.

Table 1 : Initial Design Paremeter

Parameter	Value
Thickness (h)	0.127 mm
Dielectric constant (ε_r)	2.20
Loss tangent	0.0009
Patch Size (a_1 x a_2)	2.02 x 2.2 mm^2
Ground Plane (W_g x L_g)	3.6 x 4.3 mm^2
Total Antenna Profile	3.6 x 4.3 x 0.127 mm^2
Substrate	Roger RT5880

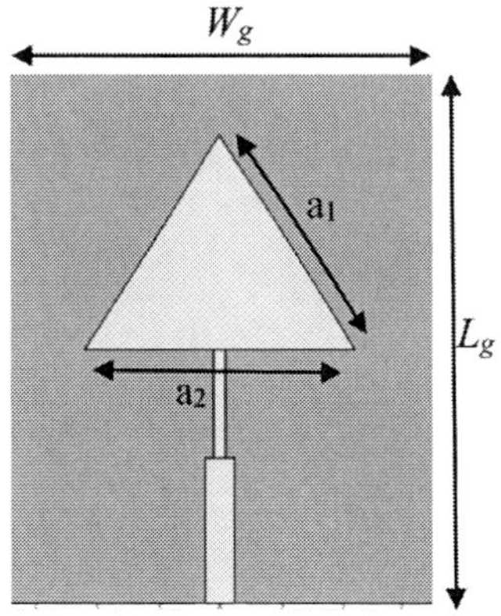

Figure 3 : Geometry and dimension of proposed antenna.

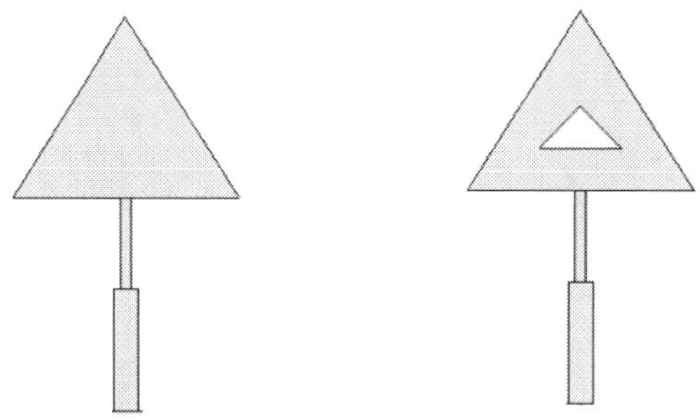

Figure 4 : Proposed design of triangle antenna without slot and with slot

3. Simulation and Result

The simulation of the proposed antenna is performed using Computer Simulation Technology (CST) Microwave Studio commercial software program. The simulated results of the reflection coefficients $|S_{11}|$ for the proposed milimeter-wave antenna are illustrated in Fig. 4. It is apparent that the proposed antenna can cover millimeter-wave bands of 60 GHz for $|S_{11}|$ less than -10. The simulated maximum realized gain of the proposed antenna . A stable gain for triangle without slot antenna with a value 4.55 dB in the frequency at 61.75 GHz and with return loss -14.76dB is observed and gain at 6.82 db for triangle antenna with slot frequency at 59.68 GHz with -24.75 db return loss, Simulated results demonstrate that the antenna is characterized by omnidirectional patterns.

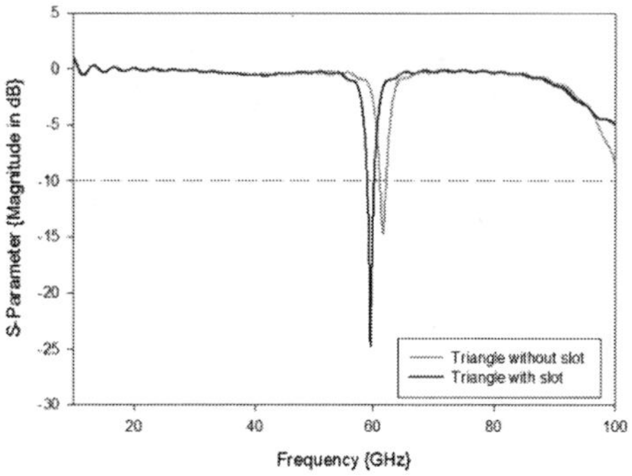

Figure 5 : Comparison of reflection coefficient S_{11} versus frequency between triangle without slot and with slot .

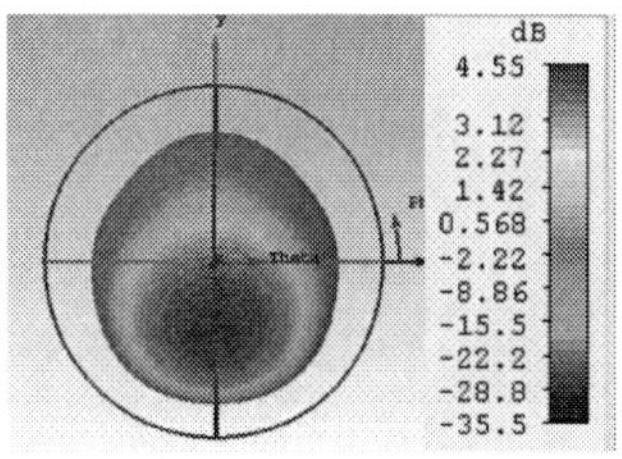

Figure 6(a) : Simulation radiation pattern plot for triangle without slot

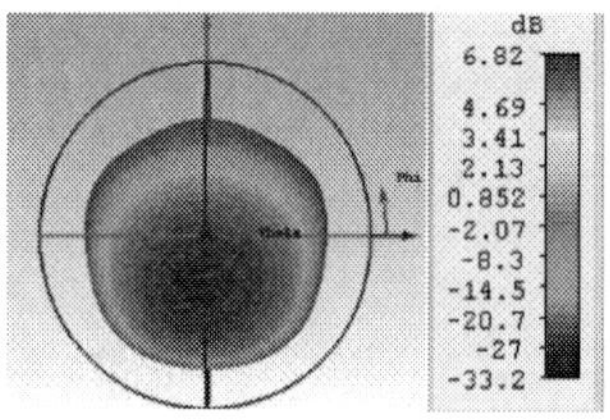

Figure 6(b) : Simulation radiation pattern plot for triangle with slot

4. Conclusion

In this paper, the triangle antenna for wireless application at 60 GHz is proposed. The antenna configuration is designed and analyzed by using the CST Studio Suite 2016 based on the finite element method. Several parametric studies have been performed to obtain a better combination of design parameters. From the anaylsis, its can be concluded that the insert the slot at the antenna can influence the return loss, gain and frequency. The triangle with slot antenna has a good result of return loss ,S_{11} compare to triangle without antenna. As we can see, the result of simulation antenna are achieved for the 5G wireless communication

Acknowleddgements

This work was supported by High Ministry of Eduacation Malaysia under the Fundamental Research Grant Scheme (FRGS) #9003-00556

References

1. Kathuria, N., & Vashisht, S. (2016, March). Dual-band printed slot antenna for the 5G wireless communication network. In *Wireless Communications, Signal Processing and Networking (WiSPNET), International Conference on* (pp. 1815-1817). IEEE.

2. Goyal, R. K., & Sharma, K. K. (2016, September). T-slotted microstrip patch antenna for 5G Wi-Fi network. In *Advances in Computing, Communications and Informatics (ICACCI), 2016 International Conference on* (pp. 2684-2687). IEEE

3. Smulders, P. (2002). Exploiting the 60 GHz band for local wireless multimedia access: Prospects and future directions. *IEEE communications magazine*, *40*(1), 140-147

4. Biglarbegian, B., Fakharzadeh, M., Nezhad-Ahmadi, M. R., & Safavi-Naeini, S. (2010, July). Optimized patch array antenna for 60 GHz wireless applications. In *Antennas and Propagation Society International Symposium (APSURSI), 2010 IEEE* (pp. 1-4). IEEE.

5. Alam, M. S., Islam, M. T., Misran, N., & Mandeep, J. S. (2013). A Wideband Microstrip Patch Antenna for 60 GHz Wireless Applications. *Elektronika ir Elektrotechnika*, *19*(9), 65-70.

DOI: 10.1051/matecconf/201714001008

6. Hannachi, C., & Tatu, S. O. (2016). Performance comparison of 60 GHz printed patch antennas with different geometrical shapes using miniature hybrid microwave integrated circuits technology. *IET Microwaves, Antennas & Propagation*.

7. Garg, R., & Long, S. A. (1988). An improved formula for the resonant frequencies of the triangular microstrip patch antenna. *IEEE transactions on antennas and propagation*, *36*(4), 570.

8. Karnfelt, C., Hallbjorner, P., Zirath, H., & Alping, A. (2006). High gain active microstrip antenna for 60-GHz WLAN/WPAN applications. *IEEE Transactions on Microwave Theory and Techniques*, *54*(6), 2593-2603.

9. Li, W. (2007, February). Performance of Ultra-Wideband Transmission over 60GHz WPAN Channel. In *Wireless Pervasive Computing, 2007. ISWPC'07. 2nd International Symposium on*. IEEE.

10. Md.Mamunur Rasid, Saddam Hossain, "Antenna Solution for Milimeter wave Mobile Communication (MWMC): 5G," *International Journal of Scientific Research Engineering & Technology (IJSRET)*, ISSN 2278-0882, vol. 3, issue 8, 2014

A COMPACT ANTENNA DESIGN FOR FIFTH GENERATION WIRELESS COMMUNICATION SYSTEM

[1]*N. N. Daud*, [1]*M. N. Osman*, [2]*M.R.Kamarudin*, [3]*A.R.Kram*, [4]*M.M.Azizan*

[1] Bioelectromagnetic Research Group (BioEM), School of Computer and Communication Engineering, Universiti Malaysia Perlis, 02600 Arau, Perlis, Malaysia.
[2] Centre for Electronic Warfare, Information and Cyber (EWIC) Cranfield Defence and Security, Cranfield University
[3] Faculty of Engineering, Department of Electrical Electronic, University Malaysia Sarawak
[4] School of Electrical System Engineering, Universiti Malaysia Perlis, 02600 Arau, Perlis, Malaysia

Abstract. This paper proposes a compact antenna design for wireless communication system that can operate at dual band frequencies which are 24.25 GHz and 38 GHz. The antenna with an overall size of 6.3 mm X 6.0 mm is printed on RogersRT5880 type of substrate with a thickness of 0.787 mm. The simulated reflection coefficient result is <-10 dB for both frequencies which makes it suitable for 5G applications. The proposed antenna has a stable gain value of 7.23 dBi at 24.25 GHz and 3.69 dBi at 38 GHz. All the simulation result was performed by using CST Microwave Studio Software. The result shows the feasibility of this antenna to be applied for future 5G application.

1 Introduction

Industry of mobile and wireless communication has widely grown over the last decade starting from analog to digital system like 2G (GSM) then towards high data rate wireless communication which is WCDMA or known as 3G. Evolution of the communication does not stop here, but it continues to grow in the presence of 3.5G (HSPA) and 4G or also known as LTE and LTE-A. Today, 5G pilot studies are going on and expected to be released in 2020 and beyond [1]. 5G or also familiarly known as millimeter wave is needed in order to overcome existing problems such as slow data rate and spectrum scarcity. Fifth generation also provides a better coverage compared to previous generation [2]. In order to fulfil all the requirements of the modern technologies, a new design of patch antenna is proposed in this paper. One of the major challenge to design a high frequency antenna is the size of the antenna become smaller thus the fabrication process become more challenging.

Nowadays, microstrip antenna plays a main role in the wireless communications industry. Microstrip antenna has several advantages such as low cost, light weight, ease of integration into arrays, and also produces the best result for mobile radio and wireless communication due to its planar configuration [3]. Moreover, microstrip antenna is perfectly suited for many applications that requires a low profile antenna because it can adapt to a given shape easily. The most important advantage of microstrip antenna is this type of antenna is capable of operating at dual and triple frequency operations [4]. Furthermore, microstrip antenna also easy to feed by using coaxial cable and mictrostrip lines.

In theory, an antenna that has dual band frequencies should operate with similar features at two separate frequencies in term of radiation and also its impedance matching [5]. Moreover, one of the advantages of implementing a dual frequency patch antenna is we can avoid the use of two different antennas when the system requires operation of two significantly dual different frequencies.

This proposed antenna could provide the communications for fifth generation wireless communication system applications which has a high data rate and better coverage compared to previous generations.

2 Antenna Design

The main objective of this research is to design a compact antenna that capable of operating at dual band frequencies of high frequency millimeter wave. Some basic calculation of the antenna will be considered in order to begin this research. The antenna was designed at RTRogers5880 as a substrate with a thickness of 0.787 mm, with relative permittivity $\varepsilon r = 2.2$ and loss tangent of 0.0009. In fact, this type of substrate shows a better performance compared with LTCC technology [6]. The patch of the antenna uses copper as a material with a dimension of 4.6 x 3.6 mm with the thickness of 0.035 mm. Moreover, copper is a material that high conductivity compared to others.

DOI: 10.1051/matecconf/201714001009

Figure 1 shows a front view and side view of a single antenna design with all the dimensions involved while the summary of antenna specification for fifth generation wireless communication is summarized in Table 1.

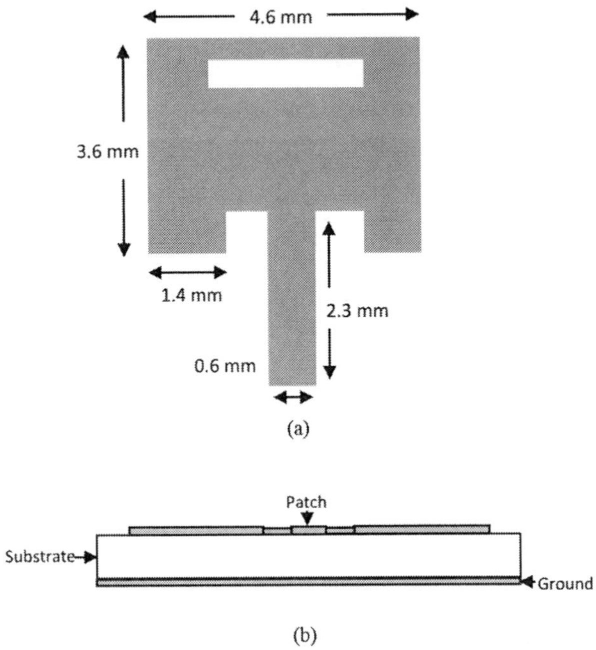

Fig 1. Front view of single antenna design (a) side view of single antenna (b).

Table 1. Antenna Specification

Shape	Rectangular
Frequency	24.25 GHz, 38 GHz
Dielectric constant, ε	2.2
Substrate	RogersRT5880
Thickness of substrate, h	0.787 mm

3 Result and Discussion

Design methodology for a rectangular patch microstrip antenna will be discussed in this section. Computer simulation technology (CST) microwave studio is the tools that have been used to analyze and optimize the performance of the antenna. Figure 2 depicts a parameter optimization of two important parameters which are XY and YZ.

Figure 3 indicates the simulated result of the reflection coefficient for the proposed millimeter wave antenna in five different values of XY which is 1.0 mm, 1.2 mm, 1.4 mm, 1.6 mm and lastly 1.8 mm. It is obvious that this compact antenna can cover both 24.25 GHz and 38 GHz frequencies with a result of S_{11} is -25.204 dB and -13.625 dB at the second band when the value of XY= 1.8 mm.

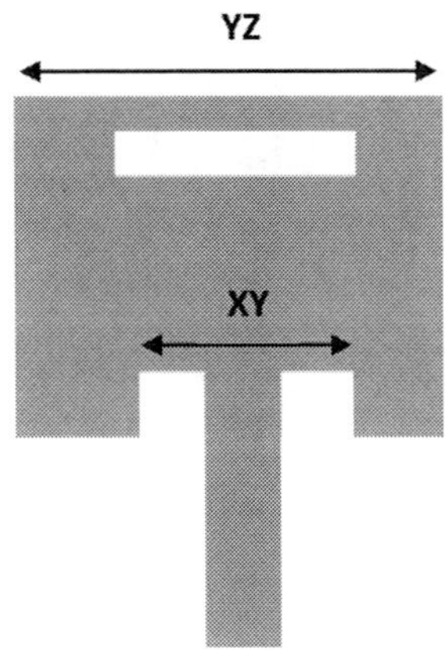

Fig 2. Parameter optimization

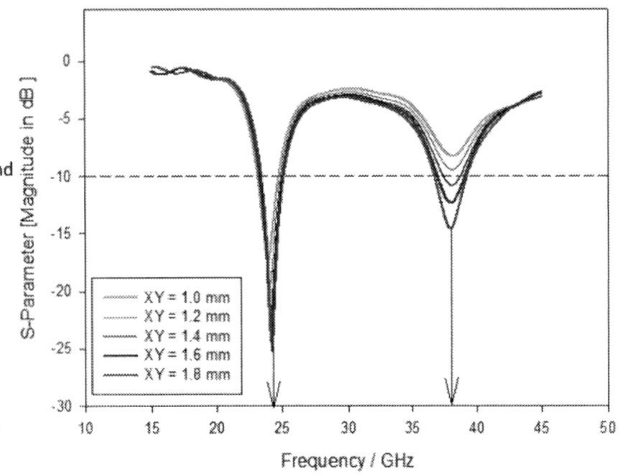

Fig 3. Reflection Coefficient for various XY

Principally, a good performance of antenna is when the frequency drop is less than -10 dB or < -10 dB. Reflection coefficient or normally known as S_{11} indicates a parameter that describes how much power is propagated from the antenna [7].

Moreover, in order to improve the frequency drop and the achieved gain of the desired antenna, the tuning of length of YZ is also required. There are also five different value needs to be adjusted which are when YZ is equal to 2.0 mm, 2.2 mm, 2.4 mm, 2.6 mm and 2.8 mm. The simulated of the reflection coefficient of the tuning process can be express in the next Figure 4.

 DOI: 10.1051/matecconf/201714001009

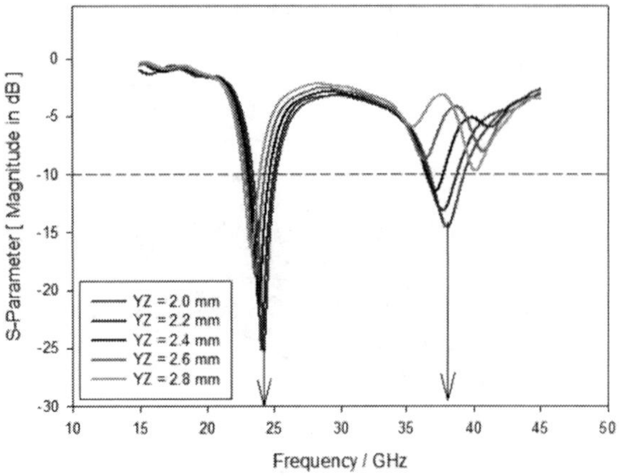

Fig. 4 Adjustable value of length YZ

Sequentially, Figure 5 illustrates the maximum gain achieved for (a) 24.25 GHz and (b) 38 GHz respectively. A stable gain of 7.23 dBi at 24.25 GHz and 3.69 dBi at 38 GHz was achieved. In fact, gain is a measurement of the ability of the antenna to direct the input power into radiation in a particular direction and is measured at the peak intensity.

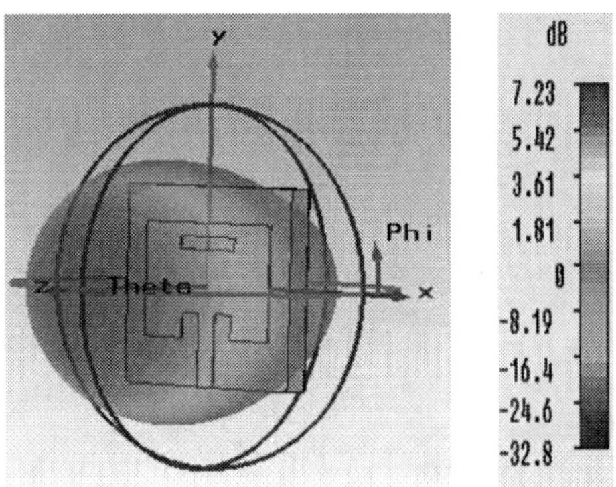

(a)

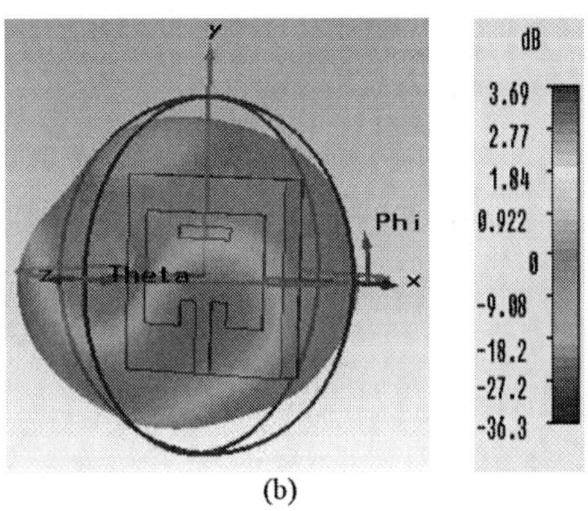

(b)

Fig 5. Maximum gain achieved for (a) 24.25 GHz and (b) 38 GHz

Next, Fig. 6 depicts the simulated radiation pattern the proposed antenna at dual band frequencies of 24.25 GHz and 38 GHz for future millimeter wave communication system. The total efficiency of this antenna is 75% which is good in order to operate in 5G applications.

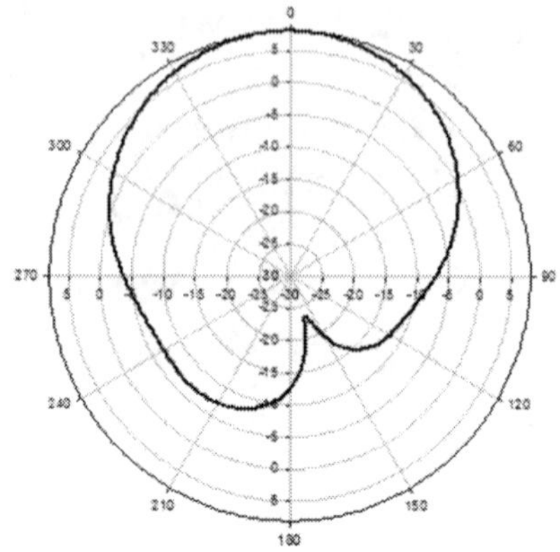

(a)

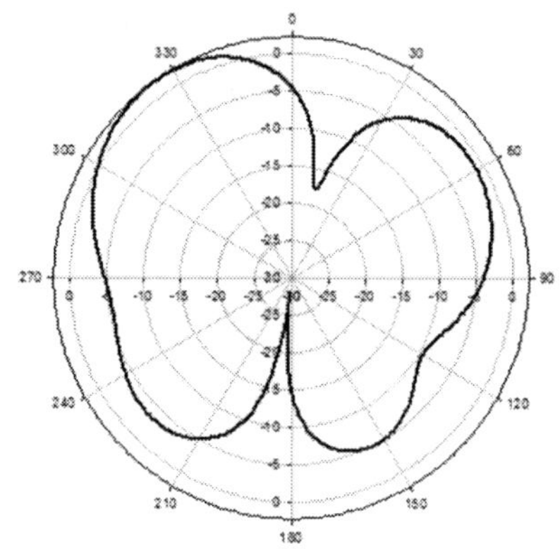

(b)

Fig. 6. Simulated Radiation pattern for (a) 24.25 GHz and (b) 38 GHz

4 Conclusion

A compact antenna design for fifth generation wireless communication system is proposed and discussed in this work. The proposed antenna operates successfully at two different frequencies which are 24.25 GHz and 38 GHz. The simulated antenna shows better performance in term of frequency drop, gain achieved and radiation pattern as well. This antenna has physical size with a compact size, simple structure and high efficiency to cover all the requirement of 5G. All these properties make it a good candidate for future millimeter wave application.

DOI: 10.1051/matecconf/201714001009

Acknowledgement

The authors would like to deeply thank University Malaysia Perlis (UniMAP) for financial support which to enable this research to be carried out. Once again, a big thanks to all members of School of Computer and Communication Engineering especially the members of ENAC Lab of UniMAP.

References

1. Gampala, G., & Reddy, C. J. (2016, March). Design of millimeter wave antenna arrays for 5G cellular applications using FEKO. In *2016 IEEE/ACES International Conference on Wireless Information Technology and Systems (ICWITS) and Applied Computational Electromagnetics (ACES)* (pp. 1-2). IEEE.
2. Kumar, A., & Kapoor, P. (2016, May). Design and performance evaluation of a dual-band antenna for the 5G mobile communication. In *Recent Trends in Electronics, Information & Communication Technology (RTEICT), IEEE International Conference on* (pp. 2034-2036). IEEE.
3. Yaseen, M. T. (2013). *Design of Dual-Band Antenna for Wireless Communication Applications* (Doctoral dissertation, University of Technology).
4. https://www.slideshare.net/totitarek/microstrip-antenna-15491076
5. Maci, S., & Gentili, G. B. (1997). Dual-frequency patch antennas. *IEEE Antennas and Propagation Magazine*, *39*(6), 13-20.
6. Outerelo, D. A., Alejos, A. V., Sanchez, M. G., & Isasa, M. V. (2015, July). Microstrip Antenna for 5G Broadband Communications: Overview of Design Issues. In *2015 IEEE International Symposium on Antennas and Propagation & USNC/URSI National Radio Science Meeting* (pp. 2443-2444). IEEE.
7. http://www.antenna-theory.com/basics/main.php

DOI: 10.1051/matecconf/201714001010

Basic structure of square and circle for defected waveguide structure

Shu Jia Chin[1] and *Mohamad Zoinol Abidin* Abd. Aziz[1,*]

[1]Centre for Telecommunication Research and Innovation (CeTRI), Faculty of Electronic and Computer Engineering, Universiti Teknikal Malaysia Melaka (UTeM), 76100 Durian Tunggal, Melaka, Malaysia

Abstract. Many existing waveguide designs focus on the application such as antenna and filter. Besides that, metamaterial used to design with waveguide is usually conducted by loading method. The configuration such as the walls or planes of waveguide is rarely proposed and designed. The motivation of this paper is to introduce the Defected Waveguide Structure (DWS) in waveguide to operate in Ultrawideband (UWB) frequency range. DWS is designed by using basic structure of square and circle. The square and circle are designed at all the walls of waveguide in patch. Hybrid design by mixing the square and circle is also proposed. The performance of DWS towards waveguide is analysed by using the transmission coefficient (S21) and reflection coefficient (S11). The filter circuit is constructed for modelling purpose to determine the value of inductance and capacitance. Copper waveguide acts as high pass filter at the frequency larger than cut off frequency 2.76GHz. The square DWS design works as band pass filter with narrowest bandwidth of 5.19GHz. The circle DWS design achieves the widest bandwidth of 6.55GHz for the pass band performance. Hybrid design improves the bandwidth slightly compare to square DWS design by 0.54GHz.

1 Introduction

Waveguide plays important role in signal transmission. It has low loss and high-power handling which is widely applied in communication system. Yet, it costs expensive and bulky in size. Substrate Integrated Waveguide (SIW) technology is then proposed to solve the problem of larger size of waveguide [1]. The waveguide concept is applied with the planar microstrip feed line to be implemented in miniaturized device.

Waveguide design with metamaterial also become popular topic for the researchers by using the loaded method [2-5]. For instance, metamaterial plates loaded with periodic arrays of complementary-split-ring-resonators is put inside waveguide for high power microwave generation [5]. However, it is found that the metamaterial design which contributes to the configuration of waveguide itself is seldom done. In [6], metamaterial is designed as anisotropic walls of waveguide where the configuration of waveguide itself is changed. Theoretical analytical approach is used to calculate the field distribution in waveguide [6].

Metamaterial receives great attention as it has been proposed to be useful in enhancing the performance of antenna [7-9]. Among all the types of metamaterial, defected structure is well known to design at ground plane and microstrip line [8-10]. It can be designed in slot or patch. The defected ground structure (DGS) at ground plane can improve return loss and achieve wider bandwidth when designs with microstrip antenna [8]. The defected microstrip structure (DMS) at microstrip line can perform as band stop filter [10]. However, a defected structure at the waveguide which can be called as defected waveguide structure (DWS) is not proposed yet which is the motivation of this paper. There is a room of opportunity and possible to introduce defect structure at waveguide instead of ground plane and microstrip feed line only.

As there is lack of information about shapes of DWS, this paper will propose the DWS design by using the basic structures which are square and circle. The square and circle are easy and simple to design. They are designed in patch form. The transmission coefficient (S21) and reflection coefficient (S11) are used to identify the performance of DWS of different structures towards waveguide. The operate frequency of waveguide is for Ultrawideband (UWB) applications.

2 Defected waveguide structure design

The rectangular waveguide is designed for the ease of fabrication. CST Microwave software is used to design and simulate waveguide. The waveguide dimension which is the width (a) and height (b) can be calculated from Equation (1) [4]. Cut off frequency (f_c) is used to identify the waveguide dimension. As the rectangular waveguide for this paper operates in UWB frequency range from 3.1GHz-10.6GHz, the lowest frequency is used to determine the waveguide dimension.

* Corresponding author: mohamadzoinol@utem.edu.my

DOI: 10.1051/matecconf/201714001010

$$f_c = \frac{1}{2\pi\sqrt{\mu\varepsilon}}\sqrt{\left(\frac{m\pi}{a}\right)^2 + \left(\frac{n\pi}{b}\right)^2} \qquad (1)$$

The rectangular waveguide is constructed to have copper layer at the inner wall of waveguide. The substrate used is FR4 with the specification listed in Table 1 below. The waveguide is designed with the width, a = 49mm and height, b = 52mm as shown in Figure 1. Two waveguide ports are located at both open-ended surface of waveguide.

Table 1. Specification of material used.

Specification of material	**Value**
Dielectric constant of substrate	4.4
Tangent loss of substrate	0.019
Thickness of substrate	1.6mm
Thickness of copper	0.035mm

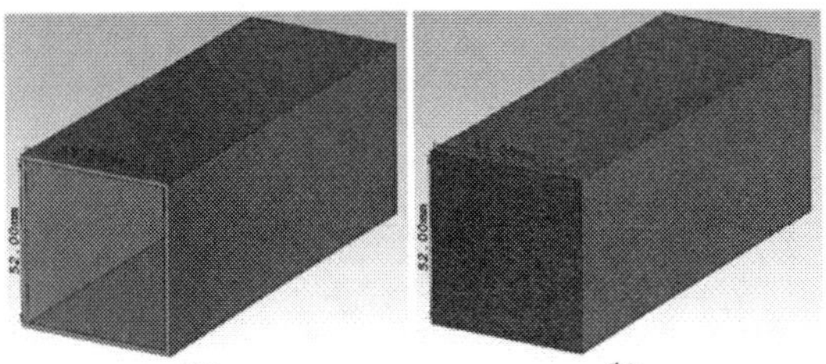

(a) (b)
Fig. 1. The view of (a) copper waveguide and (b) waveguide ports at both open-ended of waveguide.

The basic structure of square with dimension of 6mm x 6mm is then located at the inner wall of waveguide. Example of square DWS design in waveguide is showed in Figure 2. There are 150 of squares with three rows at both side walls and two rows at top and bottom walls in waveguide. Each row consists of 15 of squares. The gap separation between the structure is set to 3mm. The total length of waveguide is 142mm which is determined by the number of squares and gap separation used.

Fig. 2. The perspective view of square DWS design (Design B).

The squares can be then replaced by another shape which is circle structure. The circle has 3mm of radius. Other parameters such as gap separation for square DWS design are kept constant to observe the effects between these two different basic structures. Meanwhile, a hybrid design which combines the structures of square and circle together is also designed to investigate the performances for both structures can be worked after combining. They are mixed in horizontally and vertically. Figure 3 shows the configuration of square DWS design (Design B), circle DWS design (Design C) and mixed square and circle DWS design (Design D) in waveguide.

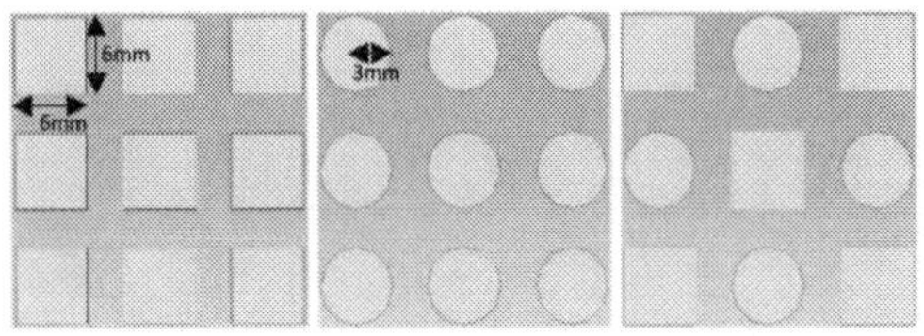

(a) (b) (c)
Fig. 3. The configuration of (a) Design B, (b) Design C and (c) Design D.

Filter circuit is also designed in Advanced Design System (ADS). The purpose of the filter circuit designed is used as circuit modelling to determine the components inductor (L) and capacitor (C) and their values. The S-parameters results of the filter circuit designed is optimized and matched with the S-parameters results that is provided by each design.

3 Result

The result of S-parameters (S11 and S21) is simulated for UWB frequency range from 2GHz-11GHz. The lower frequency range is simulated from 2GHz-6GHz while higher frequency range is from 6GHz-11GHz. For significant changes of S11 and S21 results, 2GHz-5GHz and 7GHz-11GHz are presented for lower and higher frequency range respectively.

Copper waveguide (Design A) performs as high pass filter at the frequency larger than cut off frequency. At the frequency that is larger than cut off frequency, the waveguide performs as pass band with S21 of 0dB and very low S11 of mostly -100dB. The cut off frequency for Design A occurs at 2.76GHz. The filter circuit is designed to demonstrate the behaviour of waveguide in UWB frequency range in Figure 5. High pass filter by using L = 475.4pH is performed at the lower frequency range. Low resistance only is used to show the transmission through waveguide at the higher frequency range with high S21 and low S11. This shows that transmission through copper waveguide can be operated with negligible loss at the frequency larger than the cut off frequency.

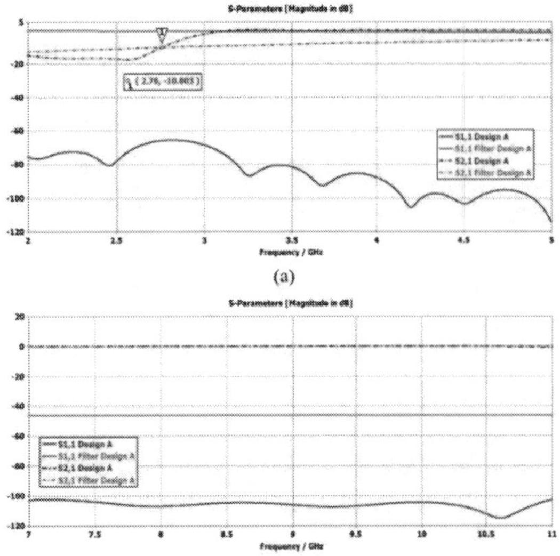

(a)

(b)

Fig. 4. The S-parameters result of copper waveguide (Design A) at (a) lower frequency range and (b) higher frequency range.

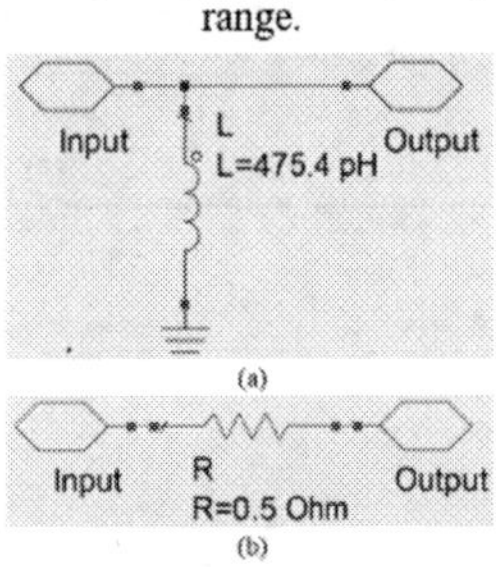

(a)

(b)

Fig. 5. The filter circuit of Design A at (a) lower frequency range and (b) higher frequency range.

Figure 6 shows the values of S11 is decreasing while S21 is increasing along the frequency for each design. Compare to Design A, the waveguide with square, circle and mixed square and circle DWS (Design B, Design C and Design D) also perform as high pass filter. They can operate at the frequency around 3GHz with low S11 and high S21. The operate frequency for all the DWS designs are near to cut off frequency of Design A. Design B starts to operate at the 2.81GHz which is the nearest to the cut off frequency. Design C starts at farthest frequency at 2.95GHz. Meanwhile, hybrid design of Design D operates at the middle frequency between Design B and Design C at 2.87GHz. It performs half of the characteristics from both Design B and Design C. All Design B, C and D also shows the lowest S21 and highest S21 at 2GHz and 5GHz respectively.

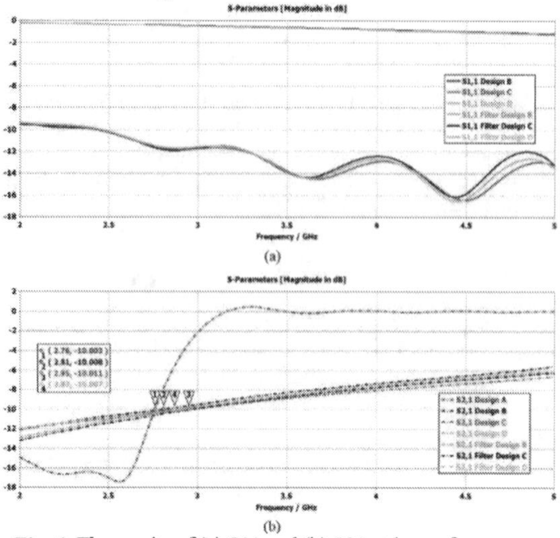

(a)

(b)

Fig. 6. The results of (a) S11 and (b) S21 at lower frequency range.

The high pass filter is presented by using the L components for Design B, Design C and Design D as well. As all designs show almost similar results for the S11 and S21, the filter circuits are designed to have slightly change in the value of L components. It is noticed that the higher the operate frequency, the smaller the value of the L component. Design C operates at the highest frequency at 2.95GHz with the smallest value of L components used, L = 446pH.

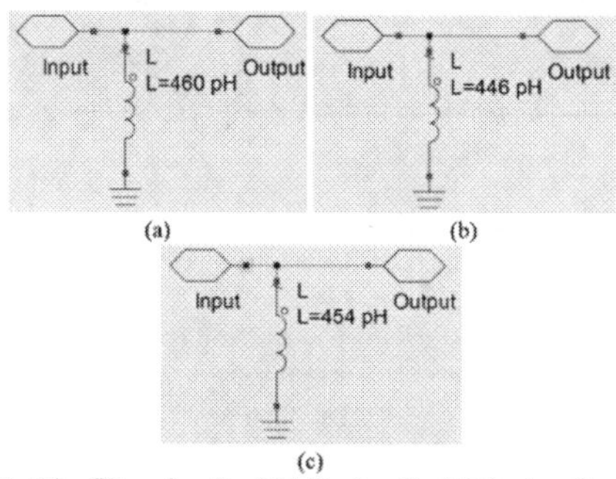

(a) (b)

(c)

Fig. 7. The filter circuit of (a) Design B, (b) Design C and (c) Design D.

Design A still operates at higher frequency from 7GHz-11GHz with high S21 in Figure 4. However, all the DWS designs shows stop band performance at higher frequency. High S11 and low S21 are showed at higher frequency for each DWS design. Design B has lowest S21 of -34.64dB occurs at 9.94GHz. Design C achieves lowest S21 of -23.87dB at higher frequency of 10.45GHz. The hybrid Design D has lowest S21 of -26.76dB at highest frequency of 10.6GHz. All the DWS designs have highest S21 at lower frequency from 7GHz-7.14GHz. The highest S21 is showed by Design C with -4.4dB at 7.14GHz.

DOI: 10.1051/matecconf/201714001010

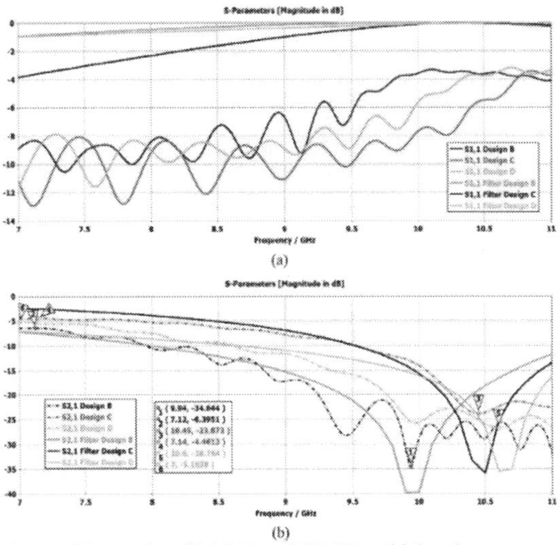

Fig. 8. The results of (a) S11 and (b) S21 at higher frequency range.

The filter circuit of DWS designs for the higher frequency range consists of one C and LC series tank which are connected in parallel form. The C contributes to stop frequency of each design which presents more than -10dB for S21 while the LC series tank shows the lowest S21. The LC series tank of Design B has L = 410pH and C = 624fF with C = 628fF. Design C has the smallest C = 210fF and LC series tank with L = 752pH and C = 308fF. Largest C = 1.25pF and LC series tank with L = 685pH and C = 326fF are used for Design D.

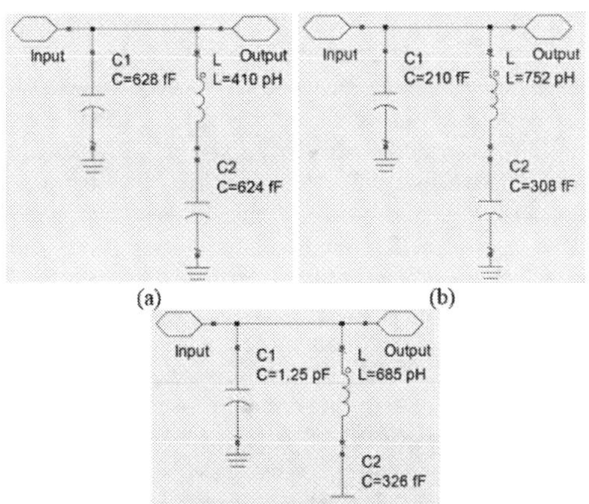

Fig. 9. The filter circuit of (a) Design B, (b) Design C and (c) Design D.

All the DWS designs operate at middle of UWB frequency range. Design B shows the narrowest bandwidth for the pass band performance. It achieves bandwidth of 5.19GHz from 2.81GHz-8GHz. Design C performs as band pass filter in the widest bandwidth of 6.55GHz from 2.95GHz-9.5GHz. It also shows the highest S21 of -4.46dB at 7.14GHz. The hybrid Design D combines both the characteristics from both Design B and Design C. At higher frequency, it shows moderate performance between both designs obviously. It achieves S21 of -10dB S21 at 8.6GHz in between 8GHz and 9.5GHz for Design B and Design C respectively. The bandwidth of Design D is improved from 5.19GHz to 5.73GHz compares to Design B. The lowest S21 of Design D occurs at highest frequency of 10.6GHz with -26.76dB.

Conclusion

The Defected Waveguide Structure (DWS) with basic structures of square and circle is presented. This proposed waveguide is designed to works as UWB application. Copper waveguide can operate at the frequency larger than cut off frequency and works as high pass filter. Square DWS design presents narrowest bandwidth of 5.19GHz for pass band performance. Circle DWS design performs as band pass filter in widest bandwidth of 6.55GHz from 2.95GHz-9.5GHz. Hybrid design by mixing both square and circle DWS design improves in bandwidth and achieves bandwidth of 5.73GHz compares to square DWS design. The monopole antenna will be designed to locate inside the waveguide with DWS in future.

References

1. Y. D. Wu, G. H. Li, Y. Wei and M. Tong, Progress In Electromagnetics Research Letters, **60**, 81–88, (2016)
2. K. Uyama, S. Nishimura, H. Deguchi and M. Tsuji, *International Conference on Electromagnetics in Advanced Applications (ICEAA)*, 728-731, (2016)
3. T. Rowe, P. Forbes, J. H. Booske and N. Behdad, IEEE Transactions on Plasma Science, **45**, 654-664, (2017)
4. M. A. Samad and A. K. Hamid, *International Conference on Electronic Devices, Systems and Applications (ICEDSA)*, 1-4, (2016)
5. J. Hummelt, Xueying Lu, Haoran Xu, M. Shapiro and R. Temkin, *IEEE International Vacuum Electronics Conference (IVEC)*, 1-3, (2016)
6. B. Byrne, N. Raveu, N. Capet, G. Le Fur and L. Duchesne, *IEEE International Symposium on Antennas and Propagation (APSURSI)*, 1091-1092, (2016)
7. A. Kandwal, *IEEE 5th Asia-Pacific Conference on Antennas and Propagation (APCAP)*, 277-278, (2016)
8. G. S. Kunturkar and P. L. Zade, *International Conference on Communications and Signal Processing (ICCSP)*, 0281-0285, (2015)
9. S. Gite and D. Niture, *International Conference on Information Processing (ICIP)*, 198-200, (2015)
10. J. Vijayakrishnan and R. P. Dwivedi, ARPN Journal of Engineering and Applied Sciences, **10**, (2015)

MATEC Web of Conferences **140**, 01011 (2017) DOI: 10.1051/matecconf/201714001011
ICEESI 2017

Traffic Adaptive Synchronized Cluster Based MAC Protocol for Cognitive Radio Ad Hoc Network

Sahelee Sultana[1], *Amran* Hossain[1], *Md. Obaidur* Rahman[1,*], and *Muzammil* Jusoh[2]

[1]Department of Computer Science and Engineering, Dhaka University of Engineering & Technology, Gazipur, Bangladesh
[2]School of Computer & Communication Engineering, Universiti Malaysia Perlis (UniMAP), Perlis, Malaysia

Abstract. In wireless communication, Cognitive Radio Network (CRN) is the contemporary research area to improve efficiency and spectrum utilization. It is structured with both licensed users and unlicensed users. In CRN, unlicensed users also called Cognitive Radio (CR) users are permitted to utilize the free/idle of licensed channels without harmful interference to licensed users. However, accessing idle channels is the big challenging issue due to licensed users' activities. A large number of cluster based MAC protocol have been proposed to solve this issue. In this paper, we have come up with a Traffic Adaptive Synchronized Cluster Based MAC Protocol for Cognitive Radio Ad Hoc Network, with the target of creating cluster structure more vigorous to the licensed users' channel re-occupancy actions, maximize throughput, and minimize switching delay, so that CR users be able to use the idle spectrum more efficiently. In our protocol, clusters are formed according to Cluster Identification Channel (CIC) and inter-communication is completed without gateway nodes. Finally, we have analysed and implemented our protocol through simulation and it provides better performance in terms of different performance metrics.

1 Introduction

In wireless communication, wireless users share a small range of radio spectrum in 2.4 GHz to 2.48 GHz and 5.0 GHz as an unlicensed band. Basically, ISM (Industrial, Scientific, Medical) band is globally fixed and this band is used by all wireless users such as Wireless Local Area Network (WLAN), Wireless Personal Area Network (WPAN), Wireless Sensor Network (WSN), Wireless Body Area Network (WBAN), and Wireless Fidelity (Wi-Fi) Networks comprises with Home networking. It is remarkable that most of the spectrums allocated to television and radio broadcast purposes share a huge amount of bandwidth and are kept idle most of the time; whereas mentioned wireless users share a small range of spectrum [1]. In reality the network gets overcrowded due to the limited number of channels and bandwidth. However, this limited bandwidth would not be adequate to provide better support for the wireless users in terms of Quality of Service (QoS), throughput, energy consumption, load balancing and delay. Thus, a solution is needed to overcome this problem. Cognitive Radio Network (CRN) is one of the probable solutions to overcome this type of problem.

Cognitive Radio (CR) is an auspicious technology for solving the problem of the coexisting of spectrum paucity for new applications [2]. Ultimately, CR is an intellectual transceiver, which can dynamically sense the network environment. It has ability to detect free/idle channels from spectrum and change its transmission parameters dynamically and access the temporally vacant spectrum opportunistically without unbearable interference to primary users (i.e., licensed users) [3].

Cognitive Radio Network (CRN) is an opportunistic network. It is organised with both primary user (PU) also called *licensed user* and secondary user (SU) also called *unlicensed user*, in which SUs consist CR-enabled radios and the PUs whose radios need not be CR-enabled [4]. In CRN, SUs are permitted to utilize free of licensed channels without any harmful interference to PUs [5]. In CRN, if two SUs want to communicate, both should active on at least one common channel. So, both the SUs can discover each other, exchange control message and then complete data communication. Since, SUs operates on dissimilar channels based on their channel availability, so channel rendezvous is the main challenging issue in CRN. It is observed that, there are different medium access control (MAC) protocol have been designed for channel rendezvous between two SUs in CRN. Cluster based MAC protocol is one of the popular protocol to rendezvous between to SUs.

Clustering is an operative management methodology in cognitive radio ad hoc networks (CRAHNs) [3]. This technique is used to improve scalability and the performance of CRAHNs. Clustering in CRAHNs involves grouping into clusters and the members of a cluster can communicate with other member of same cluster as well as another member of another cluster. However, in CRN, it is observed that, there are different cluster based approaches exist for channel rendezvous between two CRs. *Firstly*, in *CogMesh* approach [6], the authors proposed the neighbour discovery and cluster

*Corresponding author: orahman@duet.ac.bd

formation process. In this paper, when a new SU wants to join a cluster, the protocol just permits SU to join the cluster it meets first [6]. However, in this protocol, when SU is increased in a cluster, overall throughput is decreased and overhead ratio is increased accordingly. *Secondly*, in *Cluster-based MAC* approach [7], clustering is formed based on reserved values of neighbour nodes. Reserved values are also computed using three factors such as capacity, stability degree and quality of each link among the neighbours. It is the channel centric and common control channel (CCC) based protocol [7]. However, this approach is comparatively very complex to form cluster and inter cluster communication is impossible due to lack of gateway node(s). In this paper, we emphasis on three special issues, *firstly*, we propose a new MAC protocol named as Traffic Adaptive Synchronized Cluster Based MAC Protocol for CRAHNs. The core aim of this protocol is to create the cluster construction more robust to PU actions so as to use the free radio spectrum more competently, *secondly*, we also propose a new cluster formation algorithm based on CIC that deals with the network topology management and maintenance. Therefore, proposed algorithm can maximize the cluster throughput and maintain the cluster stability and *thirdly*, it significantly improves the performance under high traffic loads.

As a whole our proposed work makes the following contributions:

1. Design a new efficient MAC protocol to create the clusters that work powerfully and also more vigorous to PUs activities and design an efficient cluster formation algorithm based on cluster identification channel CIC (described later).
2. Minimize switching delay of SUs and minimize the contention of same channel and also maximize channel utilization for intra-communication using traffic adaptive technique.
3. Target to maximize throughput and easily build up inter-communication without gateway node(s).

The rest of the paper is organized as follows: in Section 2 the related works and problem statements are presented. Section 3 describes the proposed MAC protocol operation and performance evaluation is described in Section 4, and Section 5 concludes the paper.

2 Related works and problem statement

In this section we describe a number of cluster based MAC protocols those have been proposed for CRAHNs to solve the channel rendezvous problem such as- *CogMesh approach* [6], *cluster-based MAC approach* [7], *spectrum opportunity based clustering (SOC) approach* [8], *stability-capacity-adaptive routing approach* [9] and *distributed message passing approach* [10].

2.1 CogMesh MAC approach

CogMesh approach [6], is a decentralized and distributed cluster based MAC protocol.This protocol consists of two periods such as: (i) *guaranteed access* and *random access* periods, where the *guaranteed access* period is used for data communication in and between clusters, and the *random access* period is used for control message exchange. In CogMesh, clusters are interconnected in two ways such as (i) two Cluster Heads (CHs) are connected by one gateway node, and (ii) CHs are connected by two gateway nodes when no node is 1-hop neighbor of two CHs [6]. However, in this protocol, if a new SU wants to join a cluster, the protocol just agrees SU to join the cluster it meets first and gateway node must exist for communication. Though, in this protocol, when SU is increased in a cluster, overall throughput is decreased, overhead ratio is increased and switching delay is increased accordingly.

2.2 Cluster-based MAC approach

Cluster-based MAC approach [7], is an approach, which main aim is to make the cluster structure more vigorous to PU activities. In this approach, clustering is formed based on available channels, geographical position and experienced statistics and reserved values of neighbour nodes [7]. Reserved values are also computed using three factors such as capacity, stability degree and quality of each link among the neighbours. However, this approach is comparatively very complex to form cluster and inter cluster communication is impossible due to lack of contiguous cluster gateway node(s).

2.3 SOC approach

In *SOC* approach [8], neighbouring SUs are grouped with similar channel availability in the similar cluster. In this approach SUs individually compute their cluster memberships by solving the maximum edge biclique problem [11] and every SU broadcast the computed cluster membership statistics to their neighbors, and update cluster affiliations consequently and then new cluster information is rebroadcasted to SUs. Though, in this protocol generates large number of clusters with small cluster size. For that reason channels switching delay and overhead ratio is increased. In addition inter-cluster coordination problem may arise, when CH randomly selects a channel.

2.4 Stability capacity adaptive routing approach

Stability capacity adaptive routing approach [9], is a routing scheme for a multi hop high-mobility CRN. This scheme considers the path stability and node capacity. In this protocol, inter-cluster control channels and gateway nodes are selected from the CHs, considering the (i) average delay of control information transmission between two CHs, and (ii) the total throughput of control channels [9]. However, clustering is formed based on the radio link availability. In this approach, CH and control channel selection is major issue and switching overhead is increased due to PU activities.

DOI: 10.1051/matecconf/201714001011

2.5 Distributed message passing approach

Distributed message passing approach [10], is distributed clustering approach which is used affinity propagation technique [10]. In this approach, nodes interchange data containing local network information with their direct neighbors until a high excellence set of CHs and a well-organized cluster structure emerges. Though, in this approach clustering is formed based on affinity dissemination message passing method. Therefore, the result of this approach is, it produce smaller number of cluster with large cluster size. This approach only focus on the clustering structure design, but how to implement them in MAC protocol is not clear.

3 Proposed cluster-based MAC protocol

3.1 System model

We consider a cognitive radio ad hoc network with N number of channels and there are a total of M SUs in the network and each SU is equipped with a single radio. It is assumed that the radio can be used for listening to the control signals in a Common Control Channel of ISM band, and on the other hand while transmitting/receiving data the SU can have an access to the other available channels. At a given time, a SU can only sense the radio spectrum and generates the *Available Channel List (ACL).* A node can communicate with its neighbours when they are regulated to the same channel. For i^{th} *SU*, denoted by SU_i, the set of available channel list sensed by SU_i is C_i={ch_1,ch_2,ch_3...ch_n}, where n is the total number of available channels detected by SU_i. Here, we assume that the node mobility is slow and the channel accessibility at each node variations at a relatively low rate, such that the network topology does not alter during the cluster creation process [7].

3.2 Protocol operation

In this section we describe the proposed protocol operation. Our proposed protocol operation is divided into a sequence of periods those combinely defines a superframe. The periods of the proposed superframe are (1) Spectrum Sensing Period (2) Beacon & Cluster Formation Period (3) Data Exchange Period. The arrangement of the superframe is shown in Fig.1.

←— Fixed Part	—→	←— Dynamic part —→
Spectrum Sensing Period (1)	Beacon & Cluster Formation Period (2)	Data Exchange Period (3)

Fig. 1. Superframe Structure

The description of sequences are described in following sections. It is to note the superframe consists of a fixed part having the periods 1 and 2 and a dynamic part with period 3. The duration of dynamic periods depend on the number of available channels n amongst all the nodes.

3.2.1 Spectrum Sensing and Available Channel List Sorting

Initially, when a SU appears in the network, the node scan the channel spectrum and upon detecting available channels creates an *Available Channel List (ACL).* Thereafter, sort the *ACL* in ascending order, which is mandatory for cluster formation. Suppose after spectrum sensing, nodes 1, 2, 3, 4, 5 and 6 determine {ch_3,ch_1,ch_2},{ch_1,ch_5,ch_2},{ch_2,ch_3},{ch_5,ch_4},{ch_5,ch_3} and {ch_4, ch_3, ch_5}as available channels, respectively and after sorting, the sorted *ACL* of nodes 1, 2, 3, 4, 5 and 6 will be {ch_1, ch_2, ch_3},{ch_1, ch_2, ch_5},{ch_2, ch_3},{ch_4, ch_5}, {ch_3, ch_5} and {ch_3, ch_4, ch_5} respectively. We have defined the first channel of the sorted *ACL* as *Cluster Identification Channel (CIC).* So, ch_1 will be the *CIC* of nodes 1 and 2, ch_2 will be the *CIC* of node 3, ch_3 will be the *CIC* of nodes 5 and 6, and ch_4 will be the *CIC* of node 4 respectively.

3.2.2 Beacon Period in Common Control Channel (CCC)

After spectrum sensing, all the SUs tune their radio to a Common Control Channel (CCC) and we assume that the CCC would be a frequency channel from the unlicensed band. The motivation of using the CCC in unlicensed band is due to its always anytime availability, since there is no chance of occupying the CCC by the PU at any instance. Therefore, the CCC can be used for the beacon period of the superframe (as mentioned in Fig. 1). Furthermore, in the beacon period all the nodes of the network synchronize their clocks [12], exchange their control messages and share their respective *ACLs*. Hence, all the nodes of the network become aware about its corresponding cluster and cluster neighbours.

3.2.2.1 Cluster formation

According to our cluster formation algorithm, at most number of cluster would be N. Here, N is the total number of channels. Consider a network with 15 SUs, 6 PUs and total channel number N=6 as shown in Fig. 2. The nodes sorted *ACL* is shown in table 1. In the cluster formation algorithm, according to *CIC,* the network is divided into 4 clusters. So, nodes 8, 14, 4, 5, and 15 will form cluster-1(i.e., *CIC=1*), nodes 11, 1, 2, 3 and 10 will form cluster-2 (i.e., *CIC=2*), nodes 6, 7, and 9 will form cluster-3(i.e., *CIC=3*), and nodes 12 and 13 will form cluster-4 (i.e., *CIC=4*), which is shown in Fig. 3.

3.2.2.2 Node join to cluster or re-clustering

In this section we discuss (i) how a new node joins to the network as well as to cluster (ii) how handle the PU activities if a PU is occupied any channel (iii) how handle the PU activities if PU is occupied the *CIC.*

DOI: 10.1051/matecconf/201714001011

Table 1. Sorted ACL of all nodes.

No. of Nodes	Sorted Available Channel List (ACL)
1	$\{ch_2,ch_3,ch_5,ch_6\}$
2	$\{ch_2,ch_3,ch_5,ch_6\}$
3	$\{ch_2,ch_3,ch_5,ch_6\}$
4	$\{ch_1,ch_3,ch_5,ch_6\}$
5	$\{ch_1,ch_2,ch_3,ch_4,ch_6\}$
6	$\{ch_3,ch_5,ch_6\}$
7	$\{ch_3,ch_5,ch_6\}$
8	$\{ch_1,ch_3,ch_4,ch_5\}$
9	$\{ch_3,ch_4,ch_5,ch_6\}$
10	$\{ch_2,ch_3,ch_4,ch_5,ch_6\}$
11	$\{ch_2,ch_4,ch_5,ch_6\}$
12	$\{ch_4,ch_5,ch_6\}$
13	$\{ch_4,ch_5,ch_6\}$
14	$\{ch_1,ch_2,ch_3,ch_4,ch_5\}$
15	$\{ch_1,ch_3,ch_4,ch_6\}$

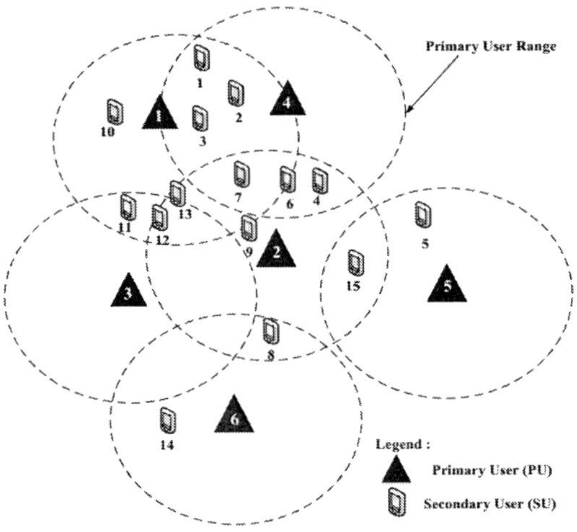

Fig. 2. Simple CRN architecture model

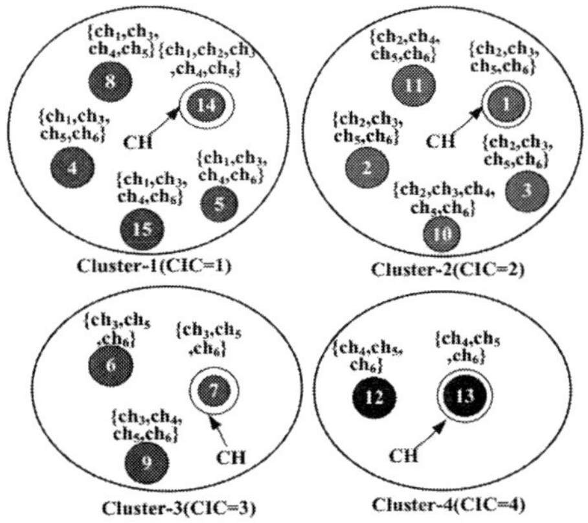

Fig. 3. Clusters after cluster formation

Firstly, when a new node wants to join to a cluster, it sends beacon on CCC and it will join a cluster according to *CIC*. *Secondly,* if a PU is occupied any channel instead of *CIC* or any occupied channel is free at any time. This means that when the sorted *ACL* at a node changes due to the PU, the node's neighbours have to be updated about its new channel list. *Thirdly,* if a PU is occupied *CIC* of a node(s) in same cluster. Then the node(s) resorted its *ACL*, and then it will join to that cluster according to new *CIC* and it is maintained by CH. Suppose in Fig.3, node 4 is the member of cluster-1 and its sorted *ACL* is $\{ch_1, ch_3, ch_5, ch_6\}$ so it's *CIC* is $\{ch_1\}$. Let a *PU* occupied the *CIC* of node 4 then the new sorted *ACL* of node 4 is $\{ch_3, ch_5, ch_6\}$, then the node 4 wants to join to the cluster-3, because its new *CIC* is now 3. Therefore, 4 sends the beacon on CCC and exchange its information with its neighbours. In this case, the *CH* of cluster-3 response the beacon of 4 and finally joint to the cluster-3. This procedure is called re-clustering. So the node 4 will be the new member of cluster-3.

3.2.3 Data Exchange Period

Data exchange period is dynamic part of our proposed protocol. It is called dynamic because the duration of dynamic periods depend on the number of available channels n amongst all the nodes. The node(s) divides time into n number of identical time slots of fixed duration 'T_C'. Here n is the number of available channels of a node. Each time slot is dedicated to one channel for data exchange. The time slots of a node may vary according to its channel availability. Hence, number of channel slot (N_{slot}) of a node is dynamic, which is depends on available channels n, so channel slots of a node at most N_{ch} slots. Therefore, number of channel slots of a node must be $N_{slot} \leq N_{ch}$. Data communication between nodes may be two types such as (i) Intra-communication and (ii) Inter-communication. The operations are described below.

3.2.3.1 Intra-communication

When nodes of same cluster are communicating with each other, this is called intra-communication. During

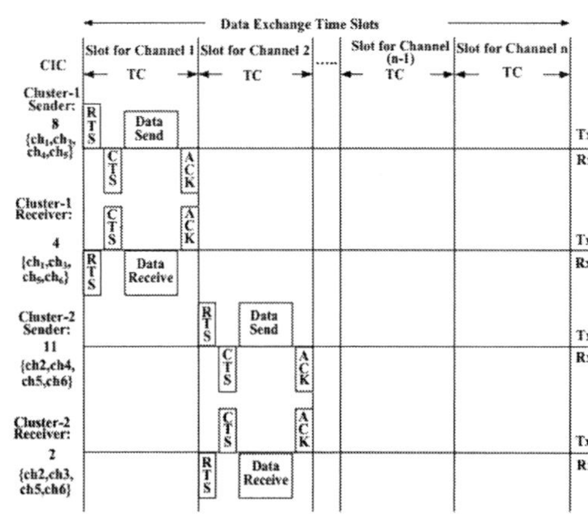

Fig. 4. Intra-communication scenario

 DOI: 10.1051/matecconf/201714001011

intra-communication, a node transmits and receives data packets on *CIC*, which is available on all nodes of the same cluster. Since each time slot is dedicated to one channel. However, in case of increasing the traffic load on *CIC*, then nodes exchange data packets on second available channel in sorted *ACL*, which is maintained by *CH*. The details intra-communication is illustrated in Fig. 4. Suppose according to Fig. 3 cluster-1 has five nodes *8, 4, 14, 15,* and *5* and *CIC* is $\{ch_1\}$. In Fig. 4, let, a sender node *8* wants to communicate with receiver node *4*, then *8* will start exchange control signals (i.e., *RTS* and *CTS)* similar to IEEE 802.11 DCF with node *4* on their *CIC* (i.e., ch_1) in time slot 1, then *8* sends data packets to node *4* and node *4* receives data packets from node *8* and finally node *4* sends acknowledgement (i.e., ACK) to node *8* for confirming successful communication. However, according to this procedure other cluster members are completed their communication on their *CIC* during intra-communication.

3.2.3.2 Inter-communication

When members of one cluster are communicating with members of other clusters, this is called inter-communication. It is also called cluster to cluster communication. Since, all nodes are synchronized and each node aware the information's about its neighbours and their sorted *ACL,* so during inter-communication, nodes transmits and receives data packets on particular time slot, which is assigned to a particular channel and available of both of them. Inter-communication is illustrated in Fig. 5. In Fig. 5, suppose according to Fig. 3 sender node *14* of cluster-1 wants to communicate with receiver node *11*of cluster-2. Then node *14* chooses one of the channels common between itself and its neighbour. Then node *14* waits for the time slot which represents the chosen channel. Since all nodes will be listening to that channel in that slot duration, then *14* will start exchange control signals (i.e., *RTS* and *CTS*) with its neighbour node *11* and data is exchanged after exchanging the control signals similar to IEEE 802.11 DCF and so on.

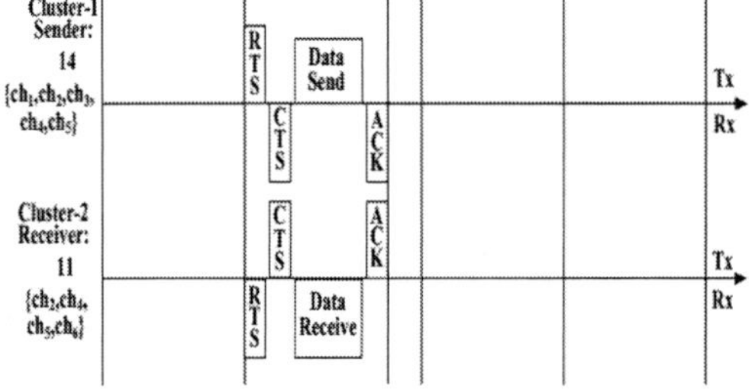

Fig. 5. Inter-communication scenario

4 Performance evaluation

We have demonstrated the performance evaluation and implemented of our proposed protocol and compare the results with that of a Custer-based MAC protocol. We have used our own simulation model developed in Java platform. In this simulation environment, 10 channels are used for modeling licensed channels for PUs and randomly some of the channels made free/unlicensed and set available to the SUs. Maximum 100 nodes are deployed in 600 x 600 m^2 area to make variation in number of PU and SU. The transmission range of each node is set to 250*m*. In our simulation, for evaluation the performance of proposed MAC protocol, and Cluster-based MAC, we have considered three performance metrics such as-, *Number of Cluster, Throughput, Packet Transmission Delay*.

4.1 Number of cluster formation

Fig. 6 shows the number of cluster formation of proposed MAC with the Cluster-based MAC protocols as the number of SUs are increased. From the Fig. 6 we can see that as the number of SUs nodes increases, the Cluster-based MAC protocol generates a large number of clusters. The main cause for this behaviour is that Cluster-based MAC is used link centric cluster formation algorithm, and on the other hand our proposed MAC protocol is used number of channel centric cluster formation algorithm. Cluster-based MAC uses multiple CCC, so increase number of cluster, increase the complexity as well as decrease channel rendezvous. In contrast, our proposed MAC does not use multiple CCC, it uses only CIC, so decrease number of cluster and it is steady which is shown in Fig. 6 as worst case and average case.

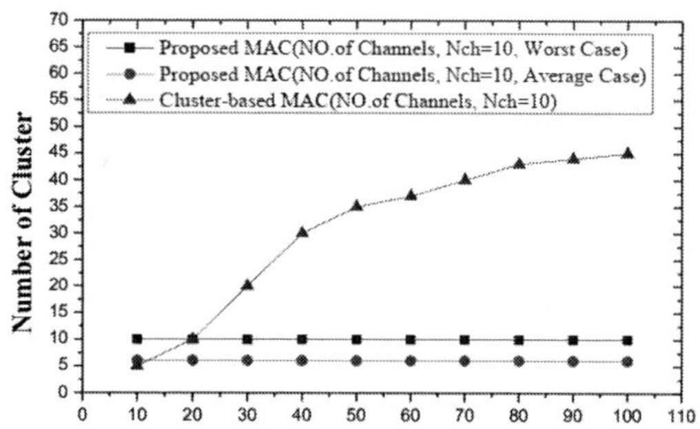

Fig. 6. Number of cluster Vs. Number of SUs nodes

4.2 Throughput performance

Fig. 7 demonstrate the aggregated throughput of proposed MAC with the Cluster-based MAC protocols as the network traffic is increased. The throughput of our

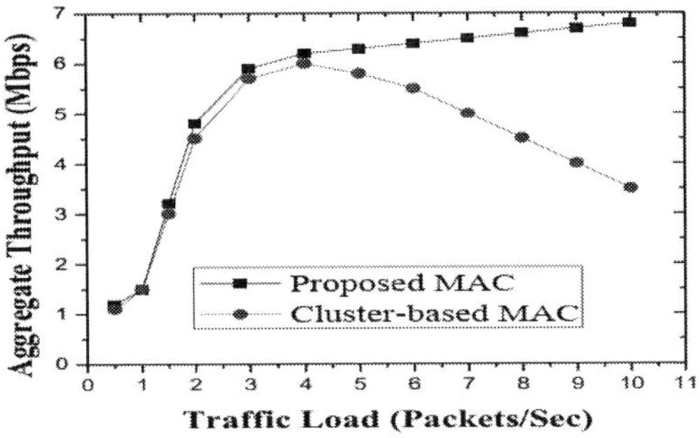

Fig. 7. Aggregated throughput Vs. traffic load

proposed MAC is significantly higher than Cluster-based MAC protocol. This is due to the fact that, Cluster-based MAC is CCC based solution, so bottleneck situation may occur in CCC. Therefore, throughput is decreased when network traffic is increased. In addition, inter cluster communication depends on the gateway nodes in Cluster-based MAC. Therefore, throughput may decrease lack of gateway node. On the other hand, our proposed protocol is channel id centric cluster based protocol, so communication takes place in the 1st channel of the sorted *ACL* list (i.e., *CIC*). Additionally, our proposed MAC does not depend on the CCC as well as gateway node during inter cluster communication. Therefore, throughput is increasing, when network traffic is increased.

4.3 Packet transmission delay

Fig. 8 demonstrate the packet transmission delay of proposed MAC with the Cluster-based MAC protocols as the network traffic is increased. The packet transmission deal of our proposed MAC is significantly lower than Cluster-based MAC protocol. This is due to the fact that, Cluster-based MAC is CCC based solution, so bottleneck situation may occur in CCC. Therefore, comparatively delay is increased when network traffic is increased. On the other hand, our proposed MAC is channel id centric cluster based protocol, so communication takes place in the 1st channel of the sorted *ACL* list (i.e., *CIC*). Additionally, our proposed

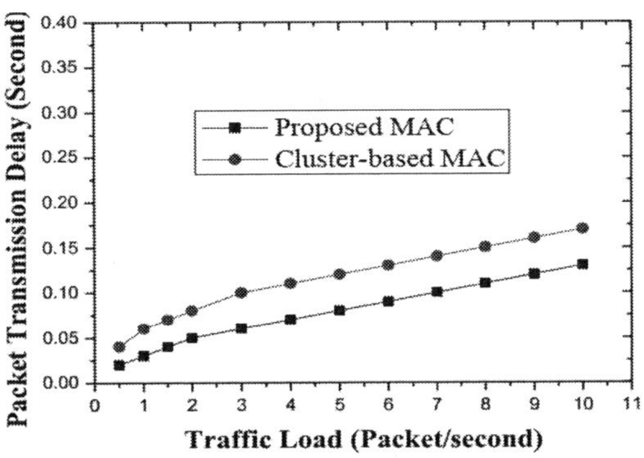

Fig. 8. Packet transmission delay Vs. traffic load

MAC does not depend on the CCC as well as gateway node during inter cluster communication. Therefore, comparatively delay is lower than Cluster-based. Moreover, due to more stable cluster structure proposed protocol has lower delay than Cluster-based MAC protocol.

5 Conclusion

In this paper we have presented a new cluster based MAC protocol for CRN using CIC based mechanism for SUs considering available channel list. The aim of proposed protocol is to create cluster structure more vigorous and use spectrum powerfully. Our proposed cluster formation algorithm is based on CIC and goals to maximize cluster throughput and maintain the cluster steadiness. Proposed MAC protocol algorithm ensures that maximum number of SUs can make channel rendezvous and successfully complete their data transmission with each other during intra and inter communication without gateway nodes(s). The simulation result show that proposed MAC can achieve lower number of clusters in the network, higher throughput, minimize packet transmission delay.

This work was collaboratively conducted by Universiti Malaysia Perlis (UniMAP) and by the Department of Computer Science and Engineering of Dhaka University of Engineering & Technology (DUET), Gazipur, Bangladesh and Dr. M. O. Rahman is the corresponding author.

References

1. A. Hossain, S. Sultana, M.O. Rahman, Receiver Initiated Multi-channel Medium Access Control Protocol for Cognitive Radio Network, 2nd International Conference on Electrical Information and Communication Technology (EICT), 350-355(2015)
2. C. Ghosh, S. Roy, D. Cavalcanti, Coexistence challenges for heterogeneous cognitive wireless networks in TV white spaces, IEEE Wireless Communications Journal, **18**, 22-31(2011)
3. J. Zhang, F. Yao, H. Zhao,Distributed Clustering in Cognitive Radio Ad Hoc Networks Using Soft-Constraint Affinity Propagation, Journal of Radio engineering, **21**, 785-794(2012)
4. N. Meghanathan, A survey on the communication protocols and security in cognitive radio networks, International Journal of Communication Networks and Information security (IJCNIS), **5**, 19-38(2013)
5. I. F. Akyildiz, W. Y. Lee, M. C. vuran, S. Mohanty, Next generation dynamic spectrum access/cognitive radio wireless networks: a survey, Journal of Computer Networks, **50**, 2127-2159(2006)
6. T. Chen, H. Zhang, G.M. Maggio, I. Chlamtac, CogMesh: A cluster-based cognitive radio network, In *Proceedings of IEEE DySPAN,* 168-178(2007)
7. X. Li, F. Hu, H. Zhang, X. Zhang, A Cluster-Based MAC Protocol for Cognitive Radio Ad Hoc

Networks, Journal of Wireless Personal Communications, **69**, 937-955(2013)
8. L. Lazos, S. Liu, M. Krunz, Spectrum Opportunity-Based Control Channel Assignment in Cognitive Radio Networks,In *Proceedings of IEEE SECON*(2009)
9. X. L. Huang, G. Wang, F. Hu, S. Kumar, Stability-capacity-adaptive routing for high mobility, multi-hop cognitive radio networks, IEEE Transactions on Vehicular Technology, **60**, 2714-2729(2011)
10. K. E. Baddour, O. Ureten, T.J. Willink, A distributed message-passing approach for clustering cognitive radio networks, Journal of Wireless Personal Communications, **57**, 119-133(2011)
11. R. Peeters, The maximum edge biclique problem is NP-complete, Journal of Discrete Applied Mathematics, **131**, 651-654(2003)
12. W. Ye, J. Heidemann, D. Estrin, An energy effiecient MAC protocol for wireless sensor networks, Journal of Wireless Sensor Networks, **1**, 59-69(2008)

MATEC Web of Conferences **140**, 01012 (2017)
ICEESI 2017 DOI: 10.1051/matecconf/201714001012

Lossy Compression of Biometric Images Implemented Using Floating Point DSP Processor

Azuwam Ali Alhadi Azuwam[1], *Muhammad Imran* Ahmad[1,*], *Mohd Nazrin* Md Isa[2], *Mohd Zaizu* Ilyas[1], *Raja Abdullah* Raja Ahmad[1]

[1]School of Computer and Communication Engineering, Universiti Malaysia Perlis, 02600 Arau, Perlis, Malaysia
[2]School of Microelectronic Engineerig, Universiti Malaysia Perlis, 02600 Arau, Perlis, Malaysia

Abstract. In this paper, several numbers of biometric images are compressed in order to reduce the number of bits needed in representing an image with conservation of image quality. Biometric images compression is important to solve the problem of efficiently transmitting data and storing large number of biometric images in low capacity of memory device. Biometric images are compressed using two techniques which are DCT and Quantization. The compression algorithm is implemented on general purpose computer and DSP processor in order to compare between both of them in terms processing time and evaluate the performance of this technique by measuring the difference between the original image and reconstructed image using PSNR, SSIM and MSE. Experimental results show DCT algorithm produces a high quality for reconstructed images with acceptable compression rate in terms of quality level is more than 50%. Furthermore, implementing the proposed algorithm using DSP board achieves better performance in terms of processing time compared with PC based.

1 Introduction

Image compression is the technique that is used to properly code digital images in order to minimize the number of bits needed in displaying an image. Advancement in digital technology has enhanced the communication media in which representation of visual information poses a vital role. High definition TV, videoconferencing, medical imaging, video wireless transmission, virtual reality, video telephony and video server are the emerging application of image compression [1]. If the number of users is large, the video signal and original biometric digital image often contains many information, hence, requires a large channel and storage capacity. The cost of implementation increases exponentially with the required capacity despite the improvement of the communication channels and storage capacity technology. However, compression techniques are important to overcome the shortcomings of the channel or storage capacity, which in turn reduce the data rate and retaining the quality of the image or video signal decoded.

Image compression techniques can reduce data to the extent where it is almost not obvious to the human eye and the utilization of statistical redundancies in the data [2]. Spatial and spectral redundancies are the two types of redundancies that ensue in images. The correlation that joins neighbouring pixels together causes spatial redundancy while spectral redundancy occur according to the correlation relating different color planes [3]. In the compression theory both spatial and spectral redundancies can be eliminated by utilizing transform coding or subband coding (Discrete Cosine Transform). Another type of redundancy called temporal redundancy which is due to the correlation of different frames in an array of images such as in video conferencing applications or broadcast images. Inter-frame coding also called motion compensated predictive coding MCPC can be used to remove temporal redundancies [4].

Several compression algorithms have been developed for image and video compression and some of these algorithms have become standards accepted for compressions such as MPEGs, JPEG, H.263 and H.261. These algorithms are computationally expensive to run in a standard processor because their execution is a block based method [5]. Discrete cosine transform DCT and Huffman coding are labelled the two most computational intensive functions or components and they are often employed using digital hardware to implement particular functions in the algorithm.

The implementation of video and image compression algorithms through the hardware posits a lower speed and density but performance is not flexible. However, recent technological improvements in software and semiconductor technology have enhanced the performance and lower the cost of floating point digital signal processor. Presently, digital signal processing DSP processor becomes an appealing resolution for executing video and image compression algorithm most particularly for low to medium video frame rate

* Corresponding author: m.imran@unimap.edu.my

applications. DSP processor has been largely used in graphic card and sound card most of the signal processing product. Speed rate, programmability and fast development cycle are the main benefit of using DSP processor. Compression can be implemented using high level programming language such as C and assembly language handles critical coding to further increase speed of execution.

Fixed point DSP and floating point DSP processor are the two types of programmable DSP available in the market. Fixed point DSP is of lower cost, high speed execution and large volume production. However, conversion of the algorithms into fixed point in order for adequate data scaling at every arithmetic step to avoid overflows and maintain accuracy can be very challenging. Floating point DSP is of higher cost in terms of price with the advantage in the algorithms mapping is reasonably simple and direct enabling a closer design cycle [6].

1.1 Biometric Image Compression

Image compression achieves reliable transmission and good image storage systems. The need to store, manage and transfer digitized biometric image, have increased tremendously especially for the 2D biometric data and video signal. The size of the image stored can be really high and take large part of the storage device. An ideal color image with 640 x 480 pixels resolution contains nearly a million elements meanwhile a 512x512 gray image may have 262144 components available for storage. It is also a time consuming process to transfer all the records from the biometric sensor to the running internet servers that classifies the process [7].

Biometrics image is the measurement and statistical analysis of people's physical attribute. It mainly identifies and get control to access any individuals or object of interest that are under surveillance. Biometric authentication usually authenticate with unique parameters of any individual's inherent physical traits. For example, if any government is launching program to gather the biometric grouping features particularly fingerprint, iris and facial patterns for its residents. Managing the large system storage can be challenging including database transmission over internet or designing a portable sensor device to carry. This forms an underlying reason to investigate the compression algorithm for several biometrics image based on JPEG compression techniques. It is also important to measure the compression performance of compressed biometric image using several measurement standards. Image generally dominates the major share of the communication bandwidth. Hence, efficient image compression technique is essential. The most focal objective of image compression is to eliminate redundancy and exclude irrelevancy. Redundancy helps to remove redundancy from the signal source and irrelevancy omits pixel values which are not noticeable by the human eye.

2 Propose Compression Method

There is a compression method based on a frequency transform that is able to compress biometric image which is Discrete Cosine Transform (DCT). This transform (DCT) contains unique features which allow for the creation of an efficient image compression. The main advantage of image transformation using DCT is the elimination of redundancy between neighbour pixels. Efficiency or performance of the transformation scheme can be directly measured by its ability to array input data into as few coefficients as possible [8]. Image compression can be implemented in both software and hardware. However, the use of hardware implementations will achieve faster processing than software implementation which is due to highly parallel algorithms that exist in specific hardware such as DSP processor.

In designing any compression algorithm, two most contrasting problems are reducing data rate and increasing processing time along with conservation of image quality. In this paper, DCT compression algorithm is implemented using floating point DSP processor in order to increase the processing time with preservation of image quality. In the implementation approach floating point DSP processor is used, because it has a special purpose internal module designed for highly extensive mathematical operation exist in the compression algorithm. It has a dedicated hardware to perform multiply and accumulate (MAC) operation and a floating point arithmetic unit which is able to speed up the processing time compare to general purpose computer.

2.1 Discrete Cosine Transform Algorithm

JPEG is a modern lossy/lossless compression technique for colour or grey scale static images. In a case where there is neighbouring similar coloured pixel, this compression works well on continuous tone images. JPEG utilizes many parameters to in order to enable the users to manipulate the amount of information lost (and in this way likewise the compression ratio) over a very wide range. In JPEG some visual information are permanently lost compared to other formats like GIF (Graphics Interchange Format), PPM (Predict, Probability Model) and PGM, It is however called a lossy compression technique. Knowing the data to dispose is the main requirement of JPEG. The JPEG encoder and decoder block diagram is given in figure 1.

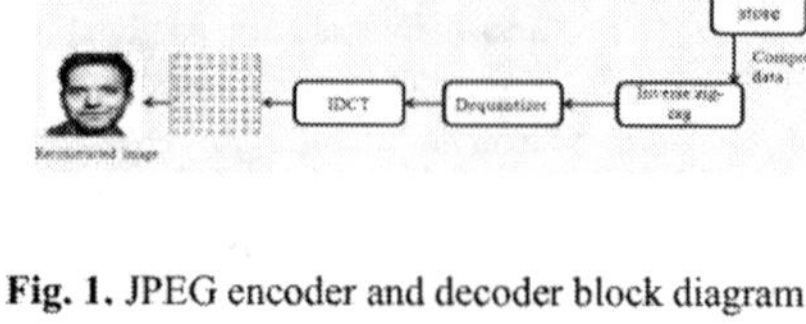

Fig. 1. JPEG encoder and decoder block diagram.

The original image as shown in figure 1 is usually divided into hundreds or even thousands of 8x8 blocks in JPEG compression technique. The pixels of each colour component are organized in groups of 8x8 pixels called data units. If the number of image rows or columns is not a multiple of 8, the bottom row and the rightmost column are duplicated as many time as necessary. The discrete cosine transform (DCT) is executed on every information unit to make 8x8 maps of frequency components. They are represented to the normal pixel value and progressive higher-frequency changes inside the gathering and each of the 64 frequency components in a data unit is reduced by a different number called its Quantization Coefficient (QC), and after that adjusted to a whole number. Quantization step size is calculated by the satisfactory visual quality of image. After quantization, the coefficients are organized as the frequency increases. For each non-zero DCT coefficients, JPEG reports all the zeros that precede the amount of bits desired to represent the largeness of the number. To consolidate the runs of zero, JPEG processes DCT coefficients in the zigzag pattern. The coefficient at the top left side are called DC coefficient while the other 63 coefficients are known as AC coefficients [9]. Entropy coding is used to code DC coefficients from previous block while AC coefficients are coded by entropy coding. JPEG committee delivers a standard table to be utilized for entropy coding.

The two dimensional DCT of image *f(x,y)* can be obtained from the to one dimensional DCT as expressed below.

$$F(u,v)=\frac{4}{N^2}\alpha(u)\alpha(v)\sum_{x=0}^{N-1}\sum_{y=0}^{N-1}f(x,y)\cdot\cos\left[\frac{(2x+1)u\pi}{2N}\right]\cdot\cos\left[\frac{(2y+1)v\pi}{2N}\right]$$

The two dimensional inverse DCT is given by

$$f(x,y)=\sum_{u=0}^{N-1}\sum_{v=0}^{N-1}\alpha(u)\alpha(v)F(u,v)\cdot\cos\left[\frac{(2x+1)u\pi}{2N}\right]\cdot\cos\left[\frac{(2y+1)v\pi}{2N}\right]$$

where

$$\alpha(u),\alpha(v)=\begin{cases}1/\sqrt{2} & \text{for } u,v=0\\ 1 & \text{for } u,v=1,2,\ldots,N-1\end{cases}$$

where u and v are coordinates in the transform domain. The sequential zeros that occurs within each and every block are exploited with zigzag pattern. It normally starts increasing from low-frequency to high frequency terms and the quantization will probably remove high frequency term in the quantization stage. This process will generally produce a more compressed output by positioning entropy for more order. The lower frequency components depict the steady luminance variations and are essential to the human visual system than the high frequency changes since human eye cannot note changes in higher frequency components. More zeros are expected to run after the quantization process where the important coefficients in the order of 8x8 block can be expected towards the end of the 8x8 block to enable more compression in the entropy [10]. The zigzag pattern is shown in figure 2. Data of 64 coefficients are rearranged in a zigzag pattern.

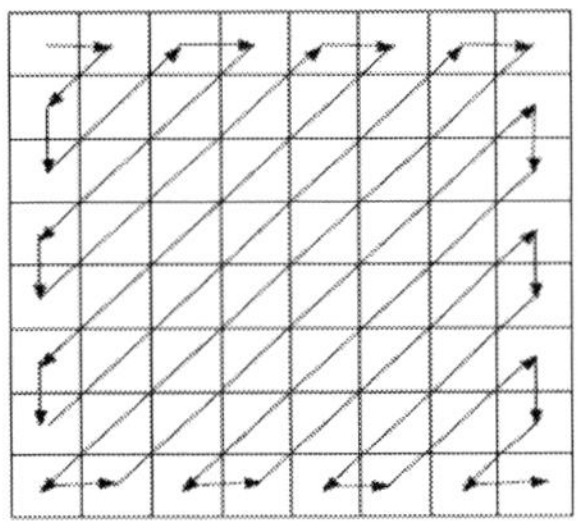

Fig. 2. Zigzag-sequence

2.2 Implementation of algorithm

After we have validated our design using simulation of C program, we will transfer our algorithm to floating point DSP processor using Code Composer Studio IDE development software. The algorithm is written in C, compile and load into DSP development board. Several biometric images for different traits and the DCT algorithm are implemented on this board. The coding part of the algorithm is done in the Code Composer Studio IDE which is done in the visual studio. Details of CCS development tool are given in figure 3 and figure 4.

The TMS320C6713 DSK board is connected to personal computer through a USB to debug and run the out file of the C code, as shown in figure 5. Communication with on-chip emulation support happens over the JTAG to monitor and control program execution. The C6713 DSK board also has JTAG interface through the USB port. Code composer studio can be used with a simulator (PC) or can be connected to a real DSP system and test the software on a real processor board (DSP).

3 Performance Analyses

The DCT image compression algorithm was developed using C program and implemented in DSP board running with floating point DSP processor. The performance is measured based on the several quality assessment of the reconstruct image.

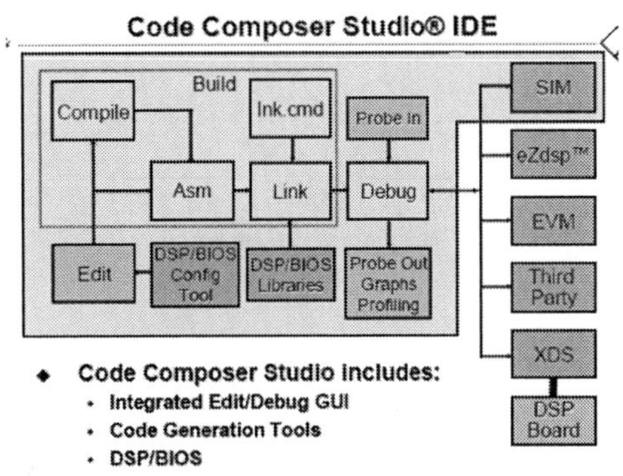

Fig. 3. Code Composer Studio Architecture

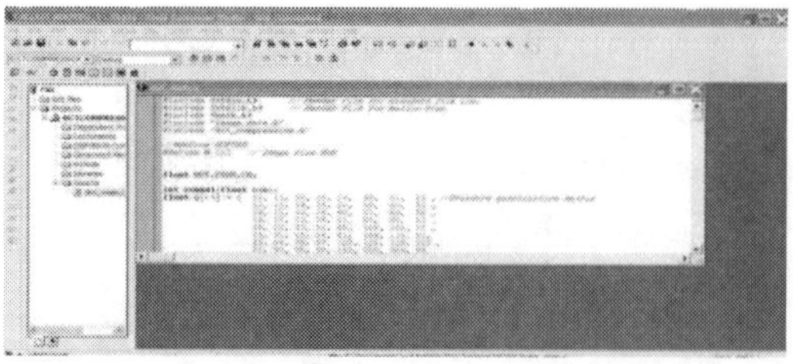

Fig. 4. Code Composer Studio design entry

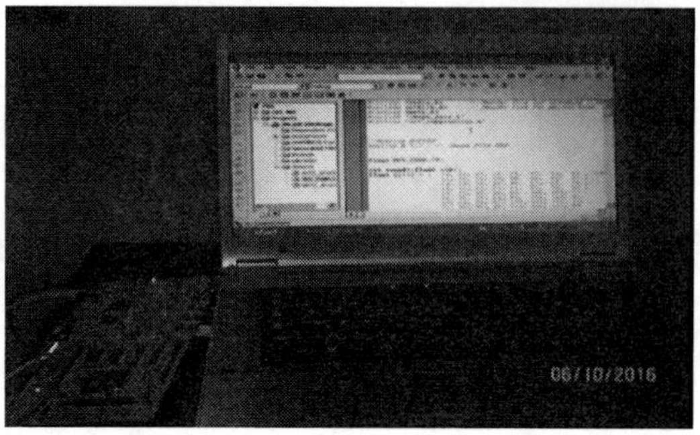

Fig. 5. Physical Connection of TMS320C6713 DSK board and PC through USB Cable

The performance in terms of speed also has been measure to show the advantage of DSP processor to perform compression algorithm. The DCT image compression algorithm was tested on several biometric images such as: facial, fingerprint, and palmprint to see how it operates on different types of biometric images. Biometric images were tested over five quality levels 90, 70, 50, 30, and 10. The results for the set of tested biometric images can be found in Table 1-3.The results are made in terms of Peak signal-to-noise ratio (PSNR), mean squared error (MSE), Bit Per Pixel (BBP), correlation (CORR) and Structural Similarity Index Metric (SSIM). Details of this analysis are shown as follows.

$$PSNR = 20log_{10}\frac{255}{RMSE}$$

$$MSE = \frac{1}{NM}\sum_{i=1}^{N}\sum_{j=1}^{M}\left(I_{ori}(i,j) - I_{rec}(i,j)\right)^2$$

$$CORR = \frac{\sum_{i=1}^{M}\sum_{j=1}^{N}(Ai,j - A^-)(Bi,j - B^-)}{\sqrt{\sum_{i=1}^{M}\sum_{j=1}^{N}(Ai,j - A^-)^2\sum_{i=1}^{M}\sum_{j=1}^{N}(Bi,j - B^-)^2}}$$

$$SSIM(x,y) = \frac{(2 \times \bar{x}\bar{y} + C1)(2 \times \sigma_{xy} + C2)}{(\sigma_x^2 + \sigma_y^2 + C2) \times ((\bar{x})^2 + (\bar{y})^2 + C1)}$$

$$BBP = \frac{total\ number\ of\ bit\ in\ the\ compressed\ image}{number\ of\ pixel}$$

In this experiment, table 1 shows the result of applying DCT compression algorithm on biometric face image with five quality levels which are 10, 30, 50, 70, and 90. The results are evaluated in terms of PSNR, SSIM, MSE, CORR and BBP. The best performance can be achieved with quality level 90. Fig. 6a illustrates original face image with reconstructed face images with qualities level 90 is shown in Fig. 6f.

In the second analysis, DCT compression algorithm is conducted on biometric fingerprint and palmprint image with different quality levels. These quality levels are from 10 to 90 scales. PSNR, SSIM, MSE, CORR and BBP are used to evaluate the difference between the original image and reconstructed image as shown in table 2 and table 3. Table 4 shows the performance analysis in term of processing speed. DSP processor offer better processing speeds compare to PC based method.

Table 1. Performance analysis of face image

Face					
Q	PSNR	MSE	SSIM	CORR	BPP
10	33.336	39.825	0.8748	0.99785	2.0335
30	38.805	9.7417	0.9649	0.99945	3.4255
50	41.183	5.5143	0.9795	0.99976	4.0198
70	43.859	2.9098	0.9887	0.99984	4.4965
90	48.956	0.8730	0.9962	0.99995	5.1158

Table 2. Performance analysis of fingerprint image

Fingerprint					
Q	PSNR	MSE	SSIM	CORR	BPP
10	27.879	121.45	0.9437	0.98122	5.9733
30	33.049	33.242	0.9833	0.99484	6.8471
50	35.533	18.618	0.9906	0.99711	7.021
70	38.087	10.582	0.9946	0.99836	7.1344
90	43.619	2.8928	0.998	0.99955	7.3322

Table 3. Performance analysis of palmprint image

Palmprint					
Q	PSNR	MSE	SSIM	CORR	BPP
10	35.983	16.396	0.9293	0.99713	0.6746
30	40.858	5.3368	0.9616	0.99905	1.5324
50	42.643	3.5403	0.9703	0.99937	2.1085
70	44.097	2.5313	0.976	0.99955	2.7718
90	47.082	1.2733	0.9857	0.99977	5.6057

Table 4. Comparison of processing time between DSP processor and PC based method.

	PC based	DSP processor
Face	35s	0.26s
Palmprint	40s	0.35s
Fingerprint	42s	0.46s

DOI: 10.1051/matecconf/201714001012

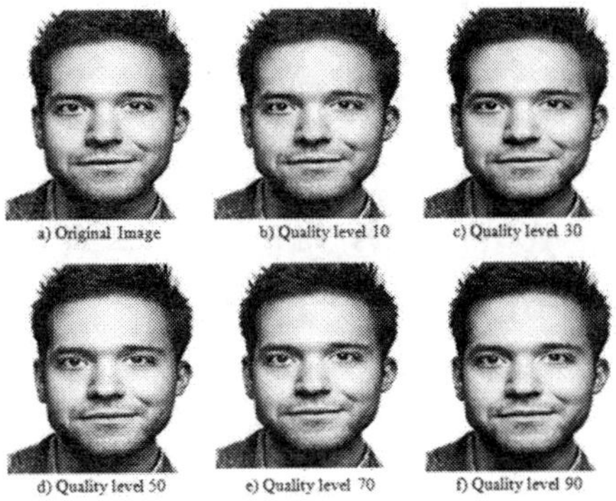

Fig. 6. Face image reconstruction with different quality level.

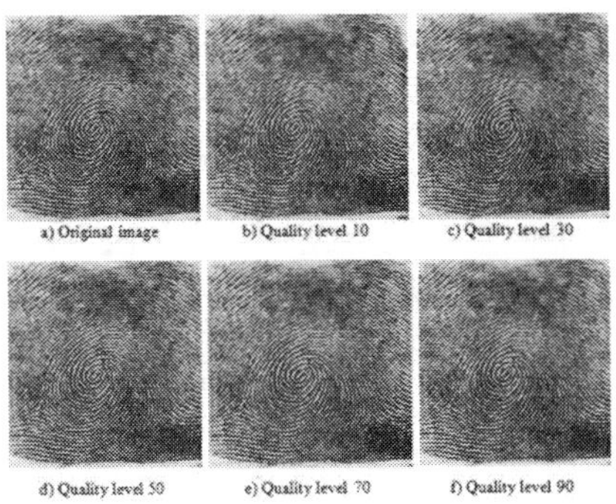

Fig. 7. Fingerprint image reconstruction with different quality level.

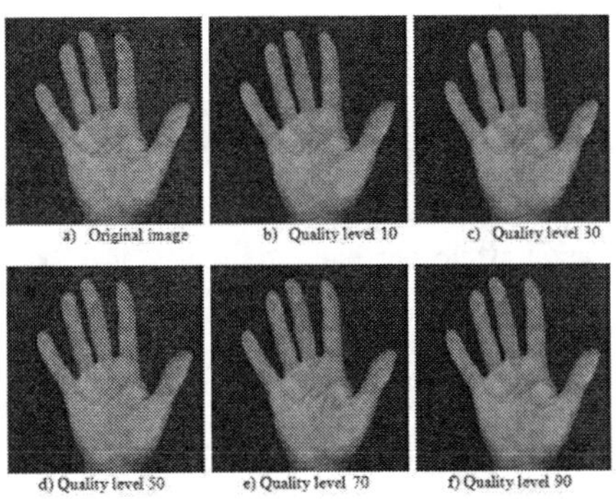

Fig. 8. Palmprint image reconstruction with different quality level.

3 Conclusions

In this paper, we have presented the implementation of DSP processor for 2D-DCT combined with quantization and zigzag process to compress biometric images such as facial, palm print, and fingerprint. In this paper, DCT is used to transform data from spatial domain to frequency domain because DCT is used for lossy compression to compact the most data into a few low frequency components. Several DCT components are discarded by using quantization process especially for high frequency components which do not cause loss in the visual impact and in zigzag process is used to consolidate relatively large runs of zeros. Biometric images compression was performed with five different quality levels namely mean MSE, BPP, PSNR, CORR and SSIM.

Acknowledgement

This research was supported by Ministry of Education Malaysia financial under the Fundamental Research Grant Scheme (FRGS) Grant No: 9003-00583.

References

1. Y. Chen, et al., Lossy image compression using PCA and contourlet transform, MATEC Web of Conferences, (2016).
2. S. Wang, et al., A surveillance video compression algorithm based on regional dictionary, MATEC Web of Conferences, (2016).
3. T. J. Chen, et al., JPEG2000 still image coding quality, Journal of Digital Imaging, vol. 26, pp. 866-874, (2013).
4. W. Zheng and Y. Liu, Research in a fast DCT algorithm based on JPEG, International Conference on Consumer Electronics, Communications and Networks, CECNet 2011, pp. 551-553, (2011).
5. N. I. Cho and S. U. Lee, DCT Algorithms for VLSI Parallel Implementations, IEEE Transactions on Acoustics, Speech, and Signal Processing, vol. 38, pp. 121-127, (1990).
6. Z. M. Yusof, et al., Real time implementation of baseline JPEG decoder using floating point DSP TMS320C31, Proceedings of IEEE Region 10 Annual International Conference, (2000).
7. A. Paul, et al., Iris image compression using wavelets transform coding, 2nd International Conference on Signal Processing and Integrated Networks, SPIN 2015, pp. 544-548, (2015).
8. Z. Li, et al., JPEG algorithm analysis and application in image compression encryption of digital chaos, Proceedings of International Conference on Instrumentation and Measurement, Computer, Communication and Control, IMCCC 2013, pp. 185-189, (2013).
9. R. C. Gonzalez, R. E. W., Digital Image Processing (2nd ed.). Prentice hall. (2002).
10. E. D. Kusuma and T. S. Widodo, FPGA implementation of pipelined 2D-DCT and quantization architecture for JPEG image compression, Proceedings of International Symposium on Information Technology - Visual Informatics, ITSim'10, (2010).

Analysis of Dielectric Properties On Agricultural Waste for Microwave Communication Application

Nurul Ain Zulkifli[1*], *Fwen Hoon* Wee[1], *Norfadila* Mahrom[1], *Been Seok* Yew[1], *Yeng Seng* Lee[1], *Siti Zuraidah* Ibrahim[1], *Anusha Leemsuthep* Am Phan[1]

[1] Bioelectromagnetic (BioEM), School of Computer and Communication Engineering, Universiti Malaysia Perlis, Arau, Perlis, Malaysia.

Abstract. This paper presents the analysis of dielectric properties of agricultural waste for microwave communication application such as microwave absorber and antenna. The residues products – rice straw, rice husk, banana leaves and sugar cane bagasse were studied in the range between 1-20GHz. Firstly, the 2 types of resins namely Epoxy der 331 and Polyamine clear hardener were mixed with the agricultural waste materials to produce the small size of agricultural waste sample. Then, the sample were measured using PNA network analyzer. The permittivity and tangent loss of different agricultural waste samples have been measured using dielectric probe technique. Besides, other objectives of this paper is to replace the conventional printed circuit board (PCB) using FR4, Taconic, and Roger material with the agricultural waste material. Besides that, the different percentage of filer for each agricultural waste materials were also investigated to specify the best material to be used as the substrate board and as the resonant material. the result shows the average of dielectric constants and the average of the tangent loss of agricultural waste materials.

1 Introduction

Dielectric properties measurement of agricultural waste materials is the main factor in defining the chemical-physical properties characterization associated to store and loss of energy [1]. Equation (1) shows the formula for permittivity (dielectric constant).

$$K = \frac{\varepsilon}{\varepsilon_o} = \varepsilon_r = \varepsilon'_r - j\varepsilon''_r \qquad (1)$$

Permittivity is known as dielectric constant of a material. the dielectric constant K is equal to the relative permittivity ε_r. Relative permittivity of a material refers to the ε'_r as the real part of dielectric constant and j ε''_r known as the imaginary part of permittivity. The energy storage capability in a material is known as ε'_r while ε''_r is familiar with the loss factor of a material [1–4].

Loss tangent measurement is another parameter used in determine the dielectric properties of a material. The formula of loss tangent is shown in equation (2) below. The relative "loss" of a material is the ratio of the energy lost ε''_r to the energy stored ε'_r [5–7].

$$\delta \tan = \frac{\varepsilon''_r}{\varepsilon'_r} \qquad (2)$$

Agricultural wastes are made up of organic compounds from living plants. Sugarcane bagasse [6, 8], rice husk [7], rice straw, and banana leaves [5], [9–11] are the residues products used in this work. These materials have potential to be used as an alternative material for microwave communication applications like antenna and microwave absorber. The high reading of dielectric constant of a material are suitable for antenna application while the high reading of loss tangent is suitable for microwave absorber used in anechoic chamber.

Agricultural waste material can be used as an alternative for microwave absorber application, antenna application, and for replacement the conventional printed circuit board (PCB).

Table 1 shows the percentage chemical elements of Agricultural waste materials. From the table, it shows that rice husk [12], banana leaves [13] and sugarcane bagasse [14] are suitable for microwave absorber application due to the high amount of carbon content. Oxygen content of sugarcane bagasse shows the highest reading among the other agricultural waste. This is the reason why the sugarcane bagasse is lighter in weight compared to others. Silica amount in agricultural waste material helps to reduce the loss tangent of a material. Low loss tangent of a material is suitable for antenna application.

* Corresponding author: nurulainzulkifli1993@gmail.com

 DOI: 10.1051/matecconf/201714001013

Table 1. Percentage chemical elements of agricultural waste material

Elements	Percentage Elements		
	Rice Husk	**Banana Leaves**	**Sugarcane Bagasse**
Carbon	35.77%	43.5%	17.89%
Hydrogen	5.06%	6.28%	-
Nitrogen	0.32%	1.31%	-
Sulphur	0.02%	0.36%	0.07%
Oxygen		41.3%	75.6%
Silicon dioxide	22.24%	-	-
Aluminium	-	-	1.47%
Silica	-	-	2.67%
Phosphorus	-	-	0.75%
Potassium	-	-	0.04%
Calcium	-	-	0.58%
Iron	-	-	0.4%

2 Method Used

This project is classified into 5 phases consisting collecting, grinding, mixing, sampling and analysing.

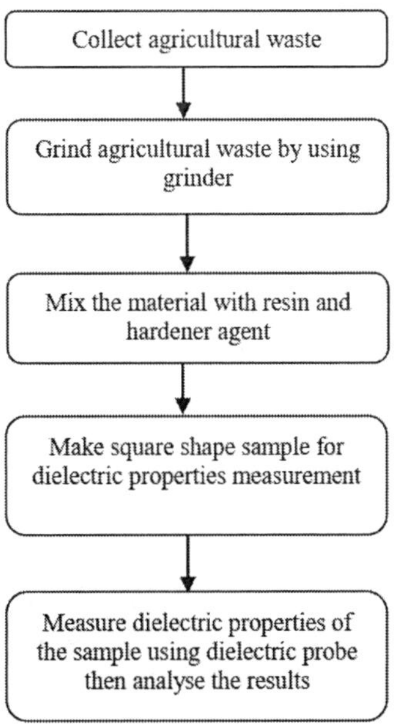

Fig. 1. steps of developing the sample of agricultural waste.

Figure 1 shows the steps for this projects. Firstly, the raw material for this project is collected. The agricultural waste that has been chosen are rice husk, rice straw, sugar cane bagasse and banana leaves. Rice husk and rice straw are collected from Kilang Beras BERNAS, Kuala Perlis, Perlis.

Fig. 2. Raw material of agricultural waste. a) rice husk. b) rice straw. c) sugarcane bagasse. d) dried banana leaves

Then, all the raw materials as shown in figure 2 have to grind by using grinder to get small particles. This is to make sure that all the materials can easily stick together with the chemical used. Figure 3 shows the image of grinder.

Fig. 3. Image of grinder

After that, all the materials were mixed together with the resin namely Epoxy Der 331. Resin is usually used as a glue for bonding the materials. Then to make the sample become solid, Polyamine Clear Hardener is added into the mixture. Five different sample of 10%, 20%, 30%, 40%, and 50% of filler for each materials is needed for this project. The mixture is required to be stirred evenly for more than 5 minutes to avoid air gap. Figure 4 shows the mixture is being stirred by using wood skewer stick. Next, figure 5 shows the mixture is then put in a rectangular mould to get the desired shape. Then, it will be left for two or three days in room temperature to ensure the mixture become hardened.

Fig. 4. Mixture is stirred using stick

Fig. 5. Rectangular mould

The analysis of dielectric properties will be take place after two days of moulding the mixture. This is to ensure that it is moulded firmly before proceeding to any processes. Figure 6 shows the sample of each agricultural waste material after taken out from the square mould.

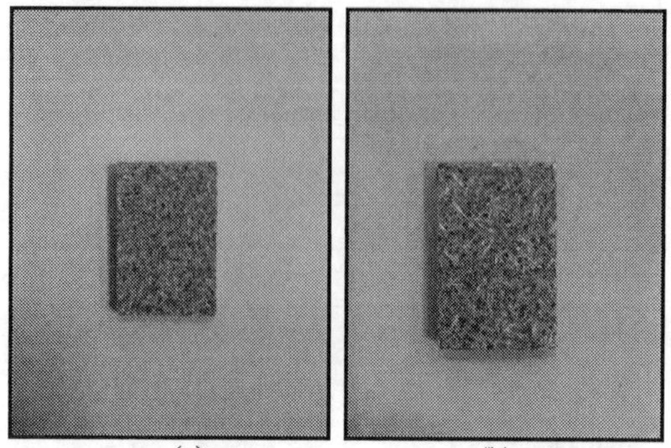

(a) (b)

(c) (d)

Fig. 6. Sample of (a) rice straw. (b) rice husk. (c) sugarcane bagasse. (d) banana leaves.

Figure 7 shows the Agilent dielectric probe technique that is used to analyse the dielectric properties of the material. This technique used Agilent Technologies 85070 measurement software. Before measuring the dielectric properties for each material, calibration is needed to avoid the error occurred during the measurement. Figure 8 shows the calibration tools for dielectric probe.

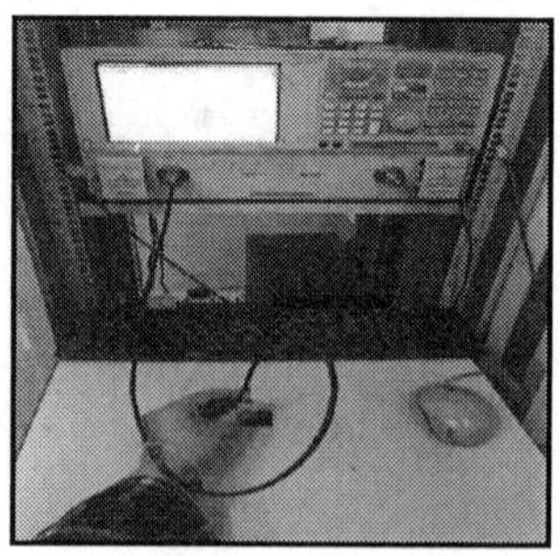

Fig. 7. Dielectric probe technique

DOI: 10.1051/matecconf/201714001013

Fig. 8. Calibration tools

3 Results and Discussion

Table 2 shows the comparison of dielectric constant between various percentage of rice husk, rice straw, sugarcane bagasse and banana leaves. The reading of dielectric properties is taken by using dielectric probe measurement technique[6]. In addition, table 3 shows the results of loss tangent with percentage 10%, 20%, 30%, 40% and 50% for four different agricultural waste. The overall results indicate that the higher the percentage of filler, the higher reading of dielectric constant and loss tangent. From the table 2 and 3, it shows that banana leaves are the best in term of the results of dielectric properties compared with other three agricultural waste sample. The results of dielectric constant and loss tangent show the unstable reading for rice husk, rice straw, and sugarcane bagasse.

Table 2. Average dielectric constant of different material with different percentage of filler.

Agricultural waste	Dielectric constant with different percentage of filler				
	10%	20%	30%	40%	50%
Rice husk	3.796	3.909	3.873	4.134	3.394
Rice straw	3.042	3.585	3.789	3.431	3.231
Sugarcane bagasse	3.447	3.736	3.422	3.078	2.791
Banana leaves	3.143	3.243	3.374	3.756	4.221

From the results in table 2, it shows that banana leaves with 50% of filler has the highest reading of dielectric constant. Hence, this result is suitable for antenna application. Furthermore, the overall analysis with 10% of filler for different agricultural waste material shows the reading of loss tangent is less than 0.1. This results are acceptable for antenna application due to the low loss tangent and high reading of dielectric constant. The general reading of 40% of filler for loss tangent for different agricultural waste is high compared with other percentage. Since the reading of loss tangent is high, so that the sample of different agricultural waste sample for 40% of filler has potential to be used as an alternative material for microwave absorber application. Microwave absorber application is used in Radio Frequency (RF) anechoic chamber.

Table 3. Average loss tangent of different material with different percentage of filler.

Agricultural waste	Loss tangent with different percentage of filler				
	10%	20%	30%	40%	50%
Rice husk	0.099	0.107	0.112	0.137	0.133
Rice straw	0.086	0.107	0.121	0.139	0.119
Sugarcane bagasse	0.094	0.105	0.125	0.135	0.129
Banana leaves	0.095	0.103	0.114	0.132	0.149

4 Conclusion

The investigation showed that different percentage of filler will give the different dielectric constant and different tangent loss performance. The variety type of agricultural waste material also brings it to different dielectric properties performance. Besides that, the high reading of dielectric constant is suitable for antenna application while the high reading of loss tangent is acceptable for microwave absorber application. The cost of the microwave absorber and other microwave communication application can be reduced by using agricultural waste as the main materials. This agricultural waste material is cheaper compared to the conventional materials. The existing materials for high frequency is expensive and not guarantee its performance and effectiveness. Lastly, the main material

 DOI: 10.1051/matecconf/201714001013

for the commercial antenna and absorbers are not environmental friendly due to the usage of 100% base chemical materials. Therefore, the usage of agricultural waste material is introduced to replace the conventional material to avoid the use of harmful chemicals and at the same time support the campaign of green technology.

5 Acknowledgement

The author would like to thank the Ministry of Higher Education of Malaysia (MOHE) for the financial support for this research work in terms of FRGS grant of 9003-00591.

References

[1] F. H. Wee, P. J. Soh, a. H. M. Suhaizal, H. Nornikman, and A. a. M. Ezanuddin, *Microw. Optoelectron. Conf. (IMOC), 2009 SBMO/IEEE MTT-S Int.*, no. I, pp. 183–187, 2009.

[2] A. Kundu and B. Gupta, *Int. Conf. Control. Instrumentation, Energy Commun. CIEC 2014*, no. 1980, pp. 480–484, 2014.

[3] O. V Tereshchenko, F. J. K. Buesink, and F. B. J. Leferink, pp. 1–4, 2011.

[4] W. F. Hoon, S. P. Jack, M. F. A. Malek, and N. Hassan, *Sci. Technol. Med. open access Publ.*, no. March 2017, pp. 91–112, 2012.

[5] R. Kaur, G. D. Aul, and V. Chawla, *Prog. Electromagn. Res. M*, vol. **43**, no. July, pp. 157–164, 2015.

[6] Z. Liyana *et al.*, *IEEE Symp. Wirel. Technol. Appl. ISWTA*, pp. 66–70, 2012.

[7] H. Nornikman, P. . Soh, a a H. Azremi, F. H. Wee, and M. F. Malek, *Piers Online*, vol. **5**, pp. 506–510, 2009.

[8] K. Ho, no. December, pp. 1–13, 2006.

[9] Z. S. Farhany *et al.*, *IEEE Symp. Wirel. Technol. Appl. ISWTA*, pp. 60–65, 2012.

[10] V. D. Kappel, L. H. Cazarolli, D. F. Pereira, F. H. Reginatto, and F. R. M. B. Silva, *Rev. Bras. Farmacogn. - BrazilianJournal Pharmacogn.*, vol. **23**, no. 4, pp. 706–715, 2013.

[11] E. R. K. Fernandes, C. Marangoni, S. H. W. Medeiros, O. Souza, N. Sellin, and Z. Industrial, *3rd Int. Conf. Ind. Hazard. waste Manag.*, pp. 1–8, 2012.

[12] H. Nornikman, F. Malek, P. J. Soh, A. A. H. Azremi, F. H. Wee, and A. Hasnain, *Prog. Electromagn. Res.*, vol. **104**, pp. 145–166, 2010.

[13] J. Y. Tock, C. L. Lai, K. T. Lee, K. T. Tan, and S. Bhatia, vol. **14**, pp. 798–805, 2010.

[14] P. B. Devnarain, D. R. Arnold, and S. B. Davis, *Congr. South African Sugar Technol. Assoc.*, pp. 477–489, 2002.

Controller Placement Algorithms in Software Defined Network - A Review of Trends and Challenges

Si-Kee Yoon[1], *ZIA* Khalib[1], *N* Yaakob[1, 2], and *A* Amir[1, 3]

[1]School of Computer and Communication Engineering, University Malaysia Perlis, Pauh Putra, 02600 Arau, Perlis
[2]School of Computer Science and IT, RMIT University, Melbourne, VIC 30000, Australia
[3]School of Information Technology, Monash University, Clayton, VIC 3800, Australia

Abstract. Traditional network architectures are complex to manage, comparatively static, rigid and difficult to make changes for new innovation. The proprietary devices in such architectures are based on manual configuration which are unwieldy and error-prone. Software Defined Network (SDN) which is described as a new network paradigm that decouple the control plane from data plane are capable to solve today's network issues and improve the network performance. Nevertheless, among so many challenges and research opportunity in SDN, Controller Placement Problem (CPP) is said to be the most important issues which can directly affect the overall network performance. Thus far, the issue regarding the CPP and its challenge has not been completely reviewed and discussed properly in any other papers. In this paper, we provide a comprehensive review on several optimized controller placement problem algorithms in SDN. This paper also highlights some limitations of the reviewed methods and also emphasizes on suitable approach to address the aforementioned problems.

1. Introduction

Software Defined Network (SDN) is an emerging network architecture which decoupled the control plane from the data plane, make forwarding decisions based on flow rather than destination based [1]. SDN help speed up the innovation process, lower the complexity and the cost on hardware implementation with its open sources features. Accordingly, network management become flexible and programmability compared to those conventional networking. SDN is currently highly implemented in many applications [1], such as data center (e.g. Google, Facebook etc.), cloud computing environment, Internet-of-thing (IoT) and wireless mesh networks (WSN). Fig. 1 shows the different of conventional and SDN architectures. Referring to the figure, SDN is made up of 3 layers which are data plane, control plane and application layer.

In recent years, Controller Placement Problem (CPP) is one of the hot topics in SDN research fields which can be denoted as NP-hard or known as facility location problem which normally regards the quantity and location of controllers in network [2]. According to the statements in [3-5], network performance may be affected due to inefficient controller placement strategy. Example consequences under poor controller selection such as long recovery time after failures.

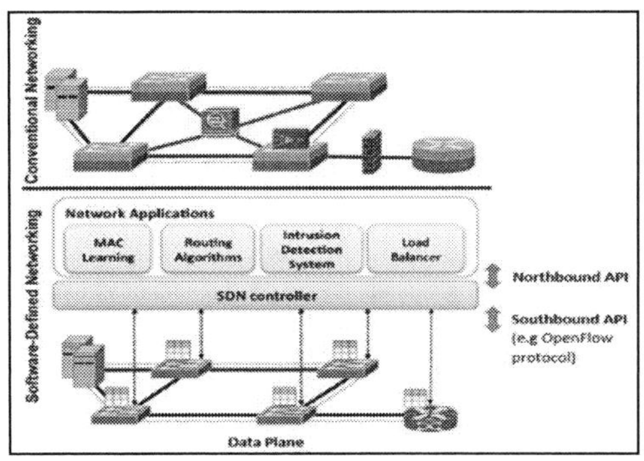

Fig. 1. Conventional networking versus Software Defined Network architecture.

The SDN control plane can be either in centralized or distributed [3]. Nevertheless, SDN deployment with a physically centralized controller has several limitations which may lead to a single point of failure for the entire network. To address these constraints, recent researches have supported implementation of multiple (distributed) SDN controllers. The remainder of this paper is structured as follows. In Section II, we discuss some of the important related works for controller placement in SDN and also provide performance comparisons between those works. Conclusion is pointed out in Section III.

* Corresponding author: zahereel@unimap.edu.my

2. Related Work

Several categories of controller placement approaches found in literature can be shown in Fig. 2. The following subsections describe each of the category in details.

2.1 Static Controller Placement

The mathematical formulation (1) is generally used to search for best placement of k controllers and n- node graphs, with the total number of placements:

$$\binom{n}{k} = \frac{n!}{(n-k)!k!} \qquad (1)$$

The first research on CPP was done by Heller et al. [2] which highlighted the importance to minimize average and worst case propagation latency in between the switch-to-controller as shows in formulation (2) and (3).

Average latency:

$$L_{avg}\ (S') = \frac{1}{n} \sum_{v \in V} \min_{(s \in S')} d\ (v, s) \qquad (2)$$

Worst case (maximum) latency:

$$Lwc\ (S') = \max_{(v \in V)} \min_{(s \in S')}\ d\ (v,\ s) \qquad (3)$$

However, this approach is a topology-based that is only limited to a certain topology and does not consider the capacity of controllers constraints, make it not really applicable in real network. The Capacitated Control Placement Problem (CCPP) in [4] is considering the additional load of controller. Cervello-Pastor et al. in [5] propose k-Criteria strategy is achieve best minimum number of controllers as compared to methods in [2]. The modified approach in [6] performs a single run time which is faster for clustering and more efficient in large-scale network compared to the iteration-based of capacitated k-Center algorithm in [5].

The works presented in [7-8] are natural-inspired metaheuristic optimization approaches. It is similar to human action in decisions making to solve CPP. Moreover, multiple controllers of hierarchical architecture [9-10] are designed to solve the single point of failure and computational complexity problem in large scale network. On the other hand, there are researches emphasis on the importance of reliability among switch and controller in CPP and several new metrics are proposed as stated in [11-12]. Others heuristic method are also proposed to solve CPP in [13-15]. Some methods in [16-17] use mathematical approaches to cater CPP. A new Integer Linear Programming (ILP) formulation is generated in [16] to find the minimum number of active controllers. In [17], a mathematical model named Failure Foresight Capacitated Controller Placement (FFCCP) presented in closest assignment when controller failures occurred.

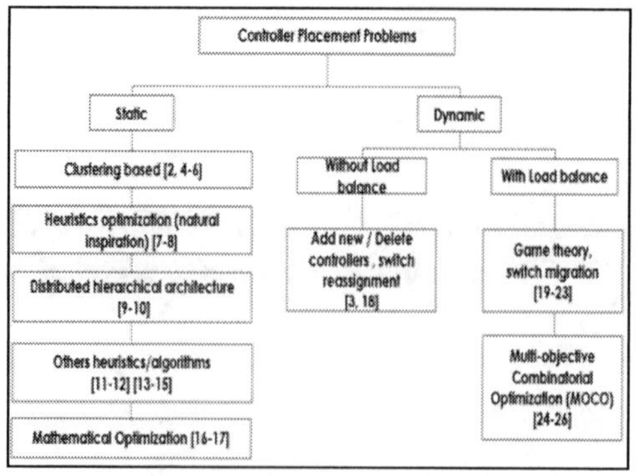

Fig. 2. Taxonomy of controller placement problem.

2.2 Dynamic Controller Placement

The previous section discussed all related works of static switch-to-controller mapping. This section will discuss, dynamic controller placement methods [3, 18-23] including those multi-objectives approaches that feasible in dynamic changes of large network as stated in [24-26].

Initially, dynamic controller provision problem (DCPP) is solved by simply dynamically changing the controller number and their locations as done in [3, 18] without load adaptation mechanism. This involves a dynamic adjustment of active controllers based on network conditions and instant switch–to–controller reassignment in case of controller failures. Dixi et al. [18] introduce "Elasticon" to dynamically alter controller resource pool and migrating switches to neighbour controller. Non-zero game theorem approach in [20] which performing load adaptation algorithms added on top of previous work ElastiCon for master controller reflection. The switch migration method as implemented in [19, 21-23] are some other enhanced DCPP approaches with load adaptation.

Furthermore, the Multi-objective Combinatorial Optimization (MOCO) approaches in [24-26] allow trade-off of multiple metrics to achieve overall optimization. Thus, users can decide a more suitable results based on their demands. Pareto-based Optimal Controller placement (POCO) is an exhaustive search actual optimal Pareto frontier which is generally used as reference value. POCO was then extended by using several heuristic algorithms to obtain approximated Pareto optimal placement to compute faster with acceptable accuracy in large network.

Table 1 shows the performance summary of the several methods discussed in this paper which is in static controller placement methods [2, 4-17]. Table 2 shows the performance summary of dynamic controller placement methods in [3, 18-26]. There are many heuristic algorithms proposed by considering various objective functions to solve controller placement problem.

DOI: 10.1051/matecconf/201714001014

Table 1. Performance comparison on static controller placement approaches.

Ref.	Techniques	Advantages	Disadvantages	Performance metrics / Objective Function							
				Flow setup time	Inter-controller latency	Switch-to-controller latency	Capacity controller	Inter-controller synchronization	Load balance	Reliability	Throughput
[2]	1. k-Center (max Lavg) 2. k-Median (min Lavg) (2012)	- Consider the distance between switch-to-controller - Trade-off number of controllers with latency minimization	- Number of controllers needed depends on the topology - High memory and time consumed for large networks			√					
[5]	3. Add k-Critical (2014)	- Consider network physical characteristic and load adaptation	- Shortest path distance for nodes-to-controller is not the best decision			√			√		
[4]	Capacitated k-Center strategy (2014)	- Capacity of controllers increased	- The average results achieved, but not the optimal one.			√	√		√		
[6]	Capacitated Density Cluster-Based CP (2017)	- Faster clustering - Can be extended for different objective solutions	- Partition cluster size can't as large as possible (capacity based)		√	√ (Propagation latency)	√			√ (Controller-loss switch)	
[7]	Network Clustering PSO (NCPSO) (2015)	- Overcome weakness of solely used of PSO in discrete problem. - Controller's load optimized	Not specified			√ (Propagation latency)	√		√		
[8]	PSO Capacitated Global Latency (2015)	- Better propagation latency, convergence time is achieved - Cognitive learning phases	- Controller numbers are given know in advance (not practical in real network)		√	√ (Propagation latency)	√				
[9]	Distributed hierarchical Architecture (2015)	- Computational complexity reduced (load distribution) - Flexibility enhanced (functionality distribution)	- Flow setup time and response time raised, higher complexity, more layering and clustering area needed as network size increased	√				√		√	√
[10]	Hierarchical controller load balancing (2016)	- To ensure load balancing below the given threshold	- 2 layer structure may not enough - Frequently (1s) controller's load checking less efficiency	√ (Flow request number)			√		√		
[11]	Expected (%) of control path loss (2014)	- Trade-off between reliability and latency for solving RCP.	- Given the topology network - Random generated failure probabilities		(Controller-controller adjacency)	√				√	

DOI: 10.1051/matecconf/201714001014

[12]	(Extended) Minimum five nine reliability for FTCP (2016)	- The operational probability of network components, controller's capacity, and different reliability levels are considered	- Minimum number of controllers achieved are limited to certain topology				√			√	
[13]	Partition and fairness minimization (2016)	- Network connectivity maintained even in failure cases - Fairly distribution switch-controller network	- Topology limited optimization result	√		√	√	√		√	
[14]	QoS guaranteed CPP (2015)	- QoS constraint (response time) never been concerned in previous work	- High controller overhead and response time increased - The result is topology based	√		√	√				
[15]	Multiple mapping with resilience constraints (2017)	- Better fairness and flexible load distribution - More stable towards dynamic traffic fluctuation	- Fixed and pre-defined controller placement even in dynamic traffic - Small network scale tested	√		√ (Shortest path select)	√				
[16]	New resilience ILP formulation (2016)	- Trade-off on latencies and maximum inter-controller load difference is considered	- Maximal latency increased if average latency minimized (nearest controller selection)		√	√			√		
[17]	Failure Foresight CCPP (2016)	- First work on worst-case latency minimization in between switches and their backup controllers	- Tested in small-scale network (at most 12 controllers)			√	√				

Table 2. Performance comparison on dynamic controller placement approaches.

Ref.	Techniques	Advantages	Disadvantages	Performance metrics / Objective Function							
				Flow setup time	Inter-controller latency	Switch-to-controller latency	Capacity controller	Inter-controller synchronization	Load balance	Reliability	Through put
[3]	Greedy Knapsack and Simulated Annealing (2013)	- Power saved as inactive controller in sleep stage - Flow setup time reduced - Low communication overhead	- Long convergence time of DCP-SA even better near optimal result - Network instability possibility	√	√			√			
[18]	ElastiCon (2013)	- Minimal influence on response time	- Possibility new load imbalance - Many RTT consumed before it start	√			√				
[19]	Crash-tolerant dynamic switch migration (2014)	- Resource utilization is enhanced - Response time reduced - Controller failover supported	- Scalability need improved - Response time increased as controller clustering needed	√					√		√
[20]	Non-Zero-Sum Game (2014)	- Simple, low complexity, minimum packet drops & delay - Topology independent, real time and utilization maximized	- Inconsistency possibility of each controller's routing table updating - Possible new load imbalance of local (closer) migration			√			√		

DOI: 10.1051/matecconf/201714001014

[21]	Flow-based dynamic switch migration (2015)	- New metric : weight of switch nodes (importance / throughput) Controller load release faster	- Small and medium network - QoS loss of the migrated switches not consider	√		√	√	√	√		√
[22]	Control mechanism distributed hopping algorithm (2017)	- No controllers added in migration - Lower average flow setup time - Better average network utilization	- Not enough just consider control load and the locality for choosing a target controller	√		√ (communication overhead)	√	√	√		
[23]	Switch migration-based decision-making (2017)	- Migration efficiency first considered	- Small scale network implementation	√					√		
[24]	Pareto capacitated k-Medoids (PCKM) (2015)	- Applicability in large network - Capacity controller included in the switch-controller assignment	- No single best optimal solution.			√	√		√		
[25]	Teaching-learning based optimizer (2016)	- Faster computation times - Less memory (search space)	- Only estimated optimal results are obtained not exact reading		√	√		√	√		
[26]	Multi-Criteria Decision Algorithm (2014)	- Trade-off multiple objectives before making decision - Failure cases are considered	- Number of controllers are given - No implemented in large and real network		√	√			√	√	

3. Discussion

SDN controller placement problem (CPP) can be categorized as static and dynamic methods. There are various performance metrics are discussed in most of the reviewed paper. In general, existing proposed methods are normally have a prior knowledge of the controller numbers to find the suitable placement in order to optimize the network performance. In another way, the optimal controller numbers are usually obtained on the condition as the specific network topology to be implemented are given in advance. Furthermore, the mapping decision based on dynamic traffic condition can be categorized into three types. First is single mapping, then second is multiple mapping where one switch assigns to multiple controllers in the case of any controller failure can distribute the loads to other controllers. Lastly, the third type of mapping decision is dynamic switch assignment to the controller that is periodically triggered by measuring value to avoid the congestion happen.

4. Conclusion

In conclusion, partition and fairness minimization heuristics give a fairer distribution of switch-to-controller allocation, but the results are limited to certain topologies. This paper discussed most performance metrics as compared to others static controller placement approaches. On the other hand, the flow-based dynamic switch migration algorithm is designed to migrate the boundary switches to its lightweight neighbour controller when there is flow burst. This paper performs most parameters among all others dynamic controller placement approaches. However, this topology only adopted for small and medium network size. Of all the reviewed methods, none have considered topology-aware controller placement based on network conditions that changes automatically. Majority research works just experimented their controller placement designs under particular kinds of network topologies environment. Most of the papers simply assume one controller failures at a time to make the simulation easier. For future works, variety network congestion scenarios should be included in controller placement designed such as multiple controllers, nodes or links are possible to fail at the same time. This is important to increase the accuracy and feasibility of the controller placement algorithm proposed. In addition, packet loss rate as one of the important QoS parameter to evaluate the network performance is also less being concerned in most of the current SDN controller placement.

The reported research in this paper is financially funded by the Ministry of Higher Education (MoHE) Malaysia through the Fundamental Research Grant Scheme (FRGS) and University Malaysia Perlis (UniMAP).

References

[1] T. Bakhshi, "State of the Art and Recent Research Advances in Software Defined Networking," Wirel. Commun. Mob. Comput., **2017**, 35 (2017)

[2] B. Heller, R. Sherwood, and N. Mckeown, "The Controller Placement Problem," SIGCOMM Comput. Commun. Rev., **42**, 6 (2012)

[3] M. F. Bari, A. R. Roy, S. R. Chowdhury, Q. Zhang, M. F. Zhani, R. Ahmed, and R. Boutaba, "Dynamic Controller Provisioning in Software Defined Networks," *Proceedings of the 9th International Conference on Network and Service Management (CNSM)*, 8 (2013)

[4] G. Yao, J. Bi, Y. Li, and L. Guo, "On the Capacitated Controller Placement Problem in Software Defined Networks," IEEE Commun. Lett., **18**, 4 (2014)

[5] Y. Jim and C. Cervell, "On the controller placement for designing a distributed SDN control layer," IFIP Netw. Conf., 9 (2014)

[6] J. Liao, H. Sun, J. Wang, Q. Qi, K. Li, and T. Li, "Density cluster based approach for controller placement problem in large-scale" Comput. Networks, **112**, 12 (2017)

[7] S. Liu, H. Wang, S. Yi, and F. Zhu, "NCPSO : A Solution of the Controller Placement Problem in Software Defined Networks," *Springer International Publishing*, 13 (2015)

[8] C. Gao, H. Wang, and F. Zhu, "A Particle Swarm Optimization Algorithm for Controller Placement Problem in Software Defined Network," *Springer International Publishing*, 11 (2015)

[9] P. D. Bhole and D. D. Puri, "Distributed Hierarchical Control Plane of Software Defined Networking," Int. Conf. CICN, 7 (2015)

[10] G. G. and F. A. H. Selvi, "Cooperative Load Balancing for Hierarchical SDN Controllers," IEEE 17th Int. Conf. HPSR, 6 (2016)

[11] H. U. Yannan, W. Wendong, G. Xiangyang, Q. U. E. Xirong, and C. Shiduan, "On Reliability-optimized Controller Placement for Software-Defined Networks," China Commun., **11**, 17 (2014)

[12] F. J. Ros and P. M. Ruiz, "On reliable controller placements in Software-Defined Networks," Comput. Commun., **77**, 11 (2016)

[13] D. M. F. Mattos, O. C. M. B. Duarte, and G. Pujolle, "A Resilient Distributed Controller for Software Defined Networking," IEEE ICC, 6 (2016)

[14] T. Y. Cheng, M. Wang, and X. Jia, "QoS-Guaranteed Controller Placement in SDN," IEEE GLOBECOM, 6 (2015)

[15] V. Sridharan, M. Gurusamy, and T. Truong-huu, "On Multiple Controller Mapping in Software Defined Networks with Resilience Constraints," IEEE Commun. Lett., 4 (2017)

[16] N. Perrot and T. Reynaudt, "Optimal Placement of Controllers in a Resilient SDN Architecture," 12th Int. Conf. DRCN, 7 (2016)

[17] B. Prakasa, R. Killi, and S. V. Rao, "Optimal Model for Failure Foresight Capacitated Controller Placement in Software-Defined Networks," IEEE Commun. Lett., **20**, 4 (2016)

[18] A. Dixit, F. Hao, S. Mukherjee, T. V Lakshman, and R. Kompella, "Towards an Elastic Distributed SDN Controller," SIGCOMM Comput. Commun. Rev., **43**, 6 (2013)

[19] C. Liang, R. Kawashima, and H. Matsuo, "Scalable and Crash-Tolerant Load Balancing Based on Switch Migration for Multiple OpenFlow Controllers," Second Int. Symp. Comput. Netw., 7 (2014)

[20] H. K. Rath, V. Revoori, S. M. Nadaf, and A. Simha, "Optimal Controller Placement in Software Defined Networks (SDN) using a Non-Zero-Sum Game," *Proceeding of IEEE International Symposium on a World of Wireless, Mobile and Multimedia Networks*, 6 (2014)

[21] L. Yao, P. Hong, W. Zhang, J. Li and D. Ni, "Controller Placement and Flow based Dynamic Management Problem towards SDN," IEEE ICCW, 6 (2015)

[22] G. Cheng, H. Chen, H. Hu, and Z. Wang, "Toward a Scalable SDN Control Mechanism via Switch Migration," China Commun., **14**, 14 (2017)

[23] C. A. N. Wang, B. O. Hu, S. Chen, D. Li, and B. I. N. Liu, "A Switch Migration-Based Decision-Making Scheme for Balancing Load in SDN," IEEE Access, **5**, 8 (2017)

[24] S. Lange, S. Gebert, J. Spoerhase, P. Rygielski, T. Zinner, S. Kounev, and P. Tran-Gia, "Specialized Heuristics for the Controller Placement Problem in Large Scale SDN Networks," 27th Int. Teletraffic Congr., 9 (2015)

[25] V. Ahmadi, A. Jalili, and M. Khorramizadeh, "Multi-Objective Controller Placement Problem : issues and solution by heuristics," IJCSIS, **14**, 7 (2016)

[26] E. Borcoci, R. Badea, S. G. Obreja, and M. Vochin, "On Multi-controller Placement Optimization in Software Defined Networking - based WANs," ICN 2015: Fourteenth Int. Conf. Networks, 6 (2015)

MATEC Web of Conferences **140**, 01015 (2017) DOI: 10.1051/matecconf/201714001015
ICEESI 2017

New catenated OFDM modulation scheme in zero cross correlation OCDMA at various number of user and effective power

N. M. Nawawi[1,*], *Anuar M. S.*[1], *M. N Junita*[1] and Che Beson Mohd Rashidi[1]

[1]Optical Research Group, Advanced Communication Engineering Centre of Excellence (ACE-COE), School of Computer & Communication Engineering, Universiti Malaysia Perlis (UNIMAP), 02600 Perlis, Malaysia.

Abstract. This paper proposes an integration of optical Code Division Multiple Access (OCDMA) with new catenated Orthogonal Frequency Division Multiplexing (OFDM) modulation scheme. This effective combination based on Zero Cross Correlation (ZCC) code can enhanced the system capacity and increased spectral efficiency by fully utilizing the available electrical bandwidth. We investigate the performance of the proposed system for various number of user, number of weight and effective power. The performance assessment is carried out by means of the signal to noise ratio (SNR) and bit error rate (BER) for up to five catenated OFDM bands transmitted simultaneously through optical link at 622 Mbps. More specifically, mathematical expressions for SNR and BER performance are derived. The corresponding numerical results are presented and compared with traditional OCDMA-ZCC system to verified the feasibility of the proposed system. The results show that with OCDMA/catenated-OFDM based on ZCC code provides 86% more number of permissible user for SNR of 15 dB. In addition, this integration provides higher receiver sensitivity; an approximately –22.5 dBm for 20 number of user with 8 number of weight. It is also found that, to accommodate more user, the system requires higher effective power at the receiver.

* Corresponding author: norizan1123@gmail.com

DOI: 10.1051/matecconf/201714001015

1. Introduction

OCDMA plays a vital role in the optical networks as this technology offers an attractive way of sharing massive bandwidth among users. It provides simplicity in all optical systems implementation [1], accurate time of arrival measurements, flexibility of user allocation, ability to support variable bit rate, busty traffic and security against unauthorized users [2]. A key limitation of this technology is the coding scheme. ZCC code is an invention code that can solve many issues in OCDMA including reduction of multiple access interference (MAI), higher permissible number of users, less system cost and reduce the complexity of overall system [3]

Quite recently, considerable attention has been paid to multiband or catenated OFDM technique [4][5] as a suitable solution for realizing next generation of optical access network. This technique transmits asynchronously multiple independent OFDM bands at different frequency. Besides, OFDM itself has several advantages such as high spectral efficiency due to subcarriers overlapping tolerate in the frequency domain [6]. Moreover, the effect of inter-symbol interference (ISI) and inter-carrier interference (ICI) are completely eliminated in OFDM by introducing the concepts of guard time and cyclic extension [7].

In this paper, new method of OFDM called catenated OFDM with ZCC is numerically demonstrated. The proposed system is compared with conventional OOK-ZCC to verify viability of the OCDMA system. In order to avoid interband interference issue, the guard band is introduced. The catenated OFDM signal bands are digitally generated using subband allocation technique in electrical domain. Adder has been used to concatenate more than one bands into electrical available bandwidth at different center frequency. Five bands OFDM signals and ZCC cordwords are used for mathematical analysis at different number of users, effective power and code weight. The results show the improvement in signal to noise ratio for higher number of user as compared to OOK-ZCC method.

The paper is organized as follows. The system description of the ZCC structure and, catenated OFDM and performance analysis are presented in section 2. Section 3 shows the analysis results and discussion and conclusions are presented in section 4.

2. System Description

The proposed of combine catenated OFDM-ZCC/OCDMA network is schematically shown in Fig. 1. It considers a hybrid technology with catenated data signal, ZCC encoder, external optical modulator and optical combiner at transmitter and optical passive splitter, ZCC decoder, photodetector and electrical filter at receiver. At transmitter, the modulated code sequences are multiplexed together by combiner before entering fiber optic channel. Meanwhile, at the receiver part, the code sequences are demultiplexed by splitter. An electrical filter is used to partition each of catenated OFDM band independently. The ZCC code structure and catenated OFDM signal are briefly presented in the next sub-section.

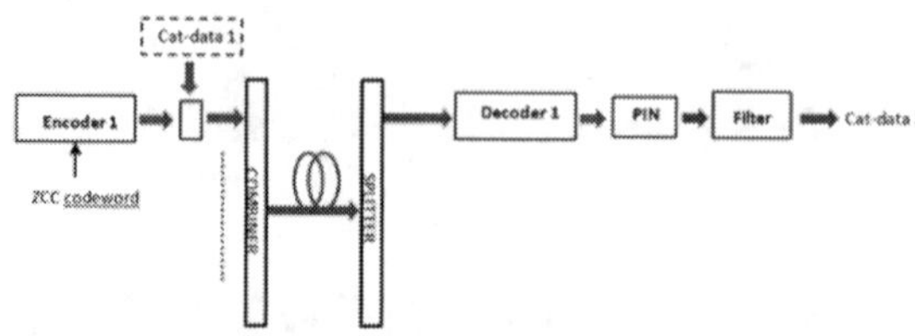

Fig. 1. Proposed network architecture

2.1 Zero Cross Correlation Code

In this section, the construction of ZCC code is explained. The matrix expression in equation 1 is the basic ZCC code

$$Z_w = \begin{bmatrix} A & B \\ C & D \end{bmatrix} \tag{1}$$

where matrix [A] is replication of matrix w-1, matrix [B] is matrix ($w\times 2w$), the diagonal matrix of ones with alternate of zero matrix ($w\times 1$) in between, matrix [C] consists of ($1\times w(w-1)$) matrix of zero and matrix [D] is ($1\times[0\ 1]$) which consist of matrix replication for w times [3]. The general equation for basic number of user K_B, weight w and code length L is given by:

$$K_B = w+1 \tag{2}$$

and

$$L = w(w+1) \tag{3}$$

From Equation 3, the relation between the code weight and the code length is proportional to each other. In order to adapt more users in the networks, the basic code, Z_w is transform into the mapping code, Z_m using mapping algorithm as shown in Equation 4.

$$Z_m = \begin{bmatrix} 0 & Z_{m-1} \\ Z_{m-1} & 0 \end{bmatrix} \quad (4)$$

Equation 5 and 6 are used to relate the number of user K_m, mapped code length L_m and mapping process m.

$$K_m = 2^m (K_B) \quad (5)$$

and

$$L_m = 2^m (L) \quad (6)$$

The structure of ZCC code explained in details in [8].

2.2 Catenated OFDM

The serial input data stream in QAM modulation format is shifted into parallel format. The parallel data is then mapped and transmitted by assigning one symbol for each carrier. An inverse Fourier transform is performed into mapped spectrum to find the corresponding time waveform.[9] To make sure any multipath interference not affected the orthogonality of the data channels, the cyclic prefix (guard period) can be added at the beginning of each symbol [10]. Finally, the parallel data with inphase and quadrature elements is shifted back into the serial symbols which is called OFDM signals.

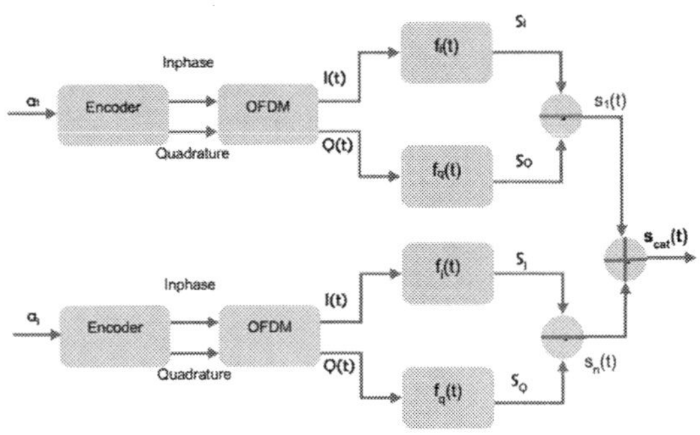

Fig. 2 Schematic design of catenated OFDM

The summed of IFFT modulated baseband OFDM signal is then called catenated OFDM signals [11]. This signal can be written as

$$s(t)_{cat} = \sum_{k=1}^{N-1} X_n[k].e^{(j2\pi f_k t)} \sum_{n=0}^{N_{sb}} m_n.e^{j2\pi n f_m t} \quad (7)$$

where m_n is the modulation index of the nth subband, while n is number of band at n=1, ..., N_{sb} and f_m is the sub-band frequency or also known as center frequency of first band. Here, we assume that the modulation index is identical for all the sub-bands. It calculated using $f_n = nf_m$. $X_n[k]$ denotes the OFDM information symbol and f_k is frequency of kth subcarrier. At the receiver, the detected photocurrent is express as

$$I_r = R.P_r.\sum_{k=1}^{N-1} X_n[k].e^{(j2\pi f_k t)}.\sum_{n=0}^{N_{sb}} m_n.e^{j2\pi n f_m t \, j2\pi n f_m t} \quad (8)$$

where R is the responsivity (R = 0.72) of the photodetector and P_r is the incident light power at the receiver.

2.3 Performance Analysis

The signal to noise ratio (SNR) is given by

$$SNR = \frac{I_r^2}{\sigma_{Total}^2} \quad (9)$$

For the noise variances, we only consider shot noise and thermal noise as written in equation (10).

$$\sigma_{Total}^2 = \sigma_{shot}^2 + \sigma_{thermal}^2 \quad (10)$$

$$\sigma_{Total}^2 = 2eI_r B + \frac{4kTB}{R_L} \quad (11)$$

where e is the electron charge which equal to $1.6x10^{-19} C$, I_r is the signal currents in the photodetector, R_L is load resistance $R_L = 1030\Omega$, T is an absolute temperature (T = *300 K)*, k is the Boltzmann's constant ($k = 1.38x10^{-23} J/K$), and B is the receiver bandwidth (311 MHz). The signal to noise ratio (SNR) is given by

$$SNR = \frac{(R.\frac{P_r w}{L}.\sum_{k=0}^{N_{sc}-1} x_k e^{j2\pi f_k t}.\sum_{n=1}^{N_{sb}} m_n e^{j2\pi n f_m t})^2}{2e(R.\frac{P_r w}{L})B + \frac{4kTB}{R_L}} \quad (12)$$

Thus, the bit error rate (BER) with 16-QAM can be obtained from SNR as below [12]:

$$BER = \frac{\sqrt{16}-1}{\sqrt{16}\log_2\sqrt{16}} erfc\sqrt{\frac{3\log_2 16}{2(15)} SNR} \qquad (13)$$

3. Result and Discussion

In this section, we compare the system performance obtained from theoretical results at 622 Mbps bit rate of the proposed technique with 5 bands to the conventional ZCC OCDMA system. Fig. 3 illustrates the variation of signal to noise ratio according to number of users. The code weight is 4 and effective power is fixed to −10dBm. It is clearly seen that the system with catenated OFDM performs better than the conventional OOK-ZCC system. At SNR of 15 dB, the proposed design has accommodated up to 210 number of user while for the case of OOK-ZCC only 30 number of user is permitted. This improves the SNR performance approximately to 86%. The performance enhancement for catenated OFDM-ZCC design is due to many of OFDM bands transmitted simultaneously for each optical channel, therefore improved the number of user. On the other hand, as many subcarriers used for each band, the signal power increased, thus signal to noise ratio increased. From this result, it is noting that the proposed method is an effective way to improve the capacity of optical channel. Moreover, it has higher spectral efficiency by fully utilizing the available electrical bandwidth for transmitting more bands at same channel.

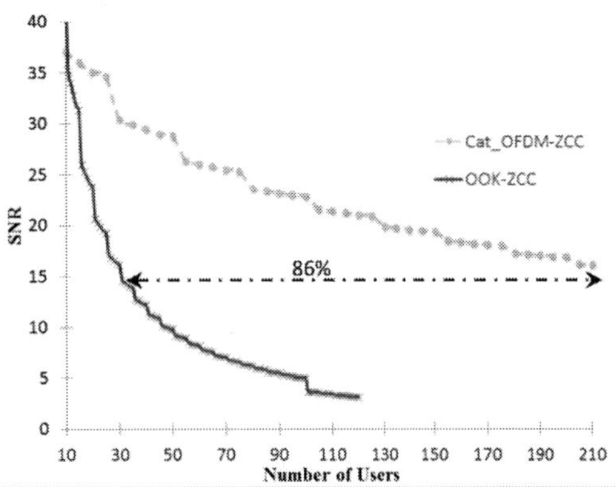

Fig. 3. SNR of catenated OFDM-ZCC and conventional OOK-ZCC according to number of user.

Fig. 4 shows the BER versus received optical power for the catenated OFDM with five bands at code weight, w of 2, 4, 6 and 8, and based on fixed number of user of 20. The acceptable BER limit ($10E^{-9}$) for w=2 reads −16.3 dBm whereas for w= 8 it is −22.5 dBm. This implies a gap of 6.2 dB for the 5 bands catenated OFDM-ZCC OCDMA. The higher the number of weight, the more effective power will be transmitted and less power required at photodetector. The BER declines as the effective power is increased due to the effect of an attenuation and lower magnitude of noise signals.

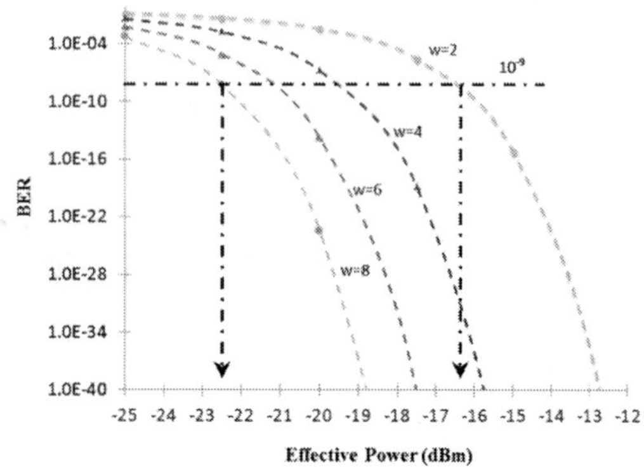

Fig. 4. BER performance with various number of weight according to effective power

In Fig. 5, the BER performance for different number of user is investigated. It is observed that at low effective power, the system with smaller number of user cause lower BER. Besides, as the number of user increases, the code length is increase correspondingly. From the graph, it shows that the effective power is -20 dBm for user = 15, -17 dBm for user = 30, -16 dBm for user = 45 and -14.7 dBm for user = 60. It has been found that, to accommodate more user, the system requires higher effective power at the receiver.

DOI: 10.1051/matecconf/201714001015

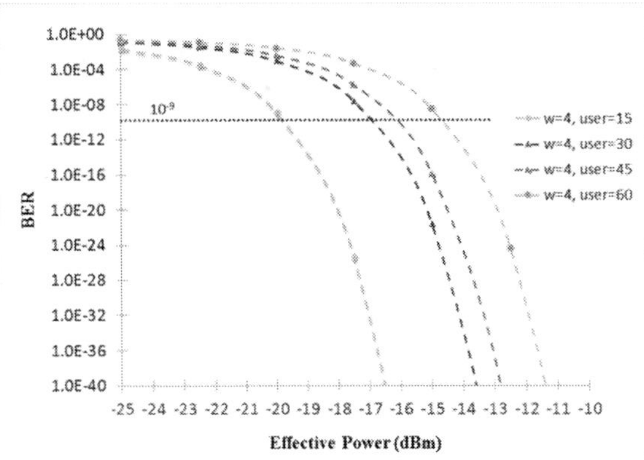

Fig. 5. BER performance according to effective power with different number of user

Conclusion

The effect of multiband OFDM modulation technique leading to a significant increase in total number of users in ZCC optical code division multiple access network. We achieve the high permissible number of user = 210 with five catenated OFDM bands. The improvement is approximately 86% compared to the traditional ZCC system. The receiver sensitivity improves by 6.2 dB at $10E^{-9}$ BER threshold for w = 8 as compared to w = 2 when the number of user is 20. Hence, when varies the number of users with constant code weight, the BER level also varies.

References

[1] C. Kandouci and A. Djebbari, "Optik Design of new hybrid wavelength hopping / time spreading codes for optical CDMA by combining OCC and BIBD ZCC codes," *Opt. - Int. J. Light Electron Opt.*, vol. 133, pp. 73–79, 2017.

[2] M. S. Ab-rahman, B. C. Ng, N. M. Ibrahim, and S. Shaari, "Development of Low-cost OCDMA Encoder based on Arrayed Waveguide Gratings (AWGs) and Optical Switches," pp. 657–660, 2008.

[3] M. S. Anuar, S. a. Aljunid, a. R. Arief, and N. M. Saad, "LED spectrum slicing for ZCC SAC-OCDMA coding system," *7th Int. Symp. High-Capacity Opt. Networks Enabling Technol. HONET 2010*, pp. 128–132, 2010.

[4] A. I. Technology, N. M. Nawawi, M. N. Junita, A. K. Rahman, S. A. Aljunid, and C. Engineering, "CATENATED OFDM MODULATION SCHEME," vol. 91, no. 1, 2016.

[5] C. Kim, S. Jung, S. Jung, S. Kang, and S. Han, "Optical Fiber Technology Inter-band interference suppression in multi-band OFDM-PON uplink transmission using window shaping," *Opt. Fiber Technol.*, vol. 27, pp. 35–40, 2016.

[6] N. Chide, S. Deshmukh, and P. P. B. Borole, "Implementation of OFDM System using IFFT and FFT," *J. Eng. Res. Appl.*, vol. 3, pp. 2009–2014, 2014.

[7] F. Almasoudi, K. Alatawi, and M. Matin, "1.05 Tb/s Optical-OFDM Using ROF over 3600 km," *Opt. Photonics J.*, vol. 3, no. September, pp. 318–323, 2013.

[8] M. S. Anuar, S. A. Aljunid, N. M. Saad, and S. M. Hamzah, "New design of spectral amplitude coding in OCDMA with zero cross-correlation," *Opt. Commun.*, vol. 282, no. 14, pp. 2659–2664, 2009.

[9] T. Wang, J. G. Proakis, and J. R. Zeidler, "Performance analysis of high QAM OFDM system over frequency selective time-varying fading channel," *IEEE Int. Symp. Pers. Indoor Mob. Radio Commun. PIMRC*, vol. 1, pp. 793–798, 2003.

[10] M. a Mohamed, a S. Samarah, and M. I. F. Allah, "Study Of Performance Parameters Effects On OFDM Systems," vol. 9, no. 3, pp. 410–417, 2012.

[11] V. Sharma, A. Shrivastav, A. Jain, and A. Panday, "BER performance of OFDM-BPSK , -QPSK , -QAM over AWGN Channel using Forward Error Correcting Code," *Int. J. Eng. Res. Appl.*, vol. 2, no. 3, pp. 1619–1624, 2012.

[12] J. Van Wyk, L. Linde, and S. Member, "Bit error probability for a M-ary QAM OFDM-based system," 2007.

MATEC Web of Conferences 140, 01016 (2017) DOI: 10.1051/matecconf/201714001016
ICEESI 2017

A Review on Recent T-way Combinatorial Testing Strategy

Nuraminah Ramli , *Rozmie Razif* Othman*, *Zahereel Ishwar* Abdul Khalib and *Muzammil* Jusoh

School of Computer and Communication, University of Malaysia Perlis

Abstract. T-way combinatorial testing aims to generate a smaller test suite size. The purpose of t-way combinatorial testing is to overcome exhaustive testing. Although many existing strategies have been developed for t-way combinatorial testing, study in this area is encouraging as it falls under NP-hard optimization problem. This paper focuses on the analysis of existing algorithms or tools for the past seven years. Taxonomy of combinatorial testing is proposed to ease the analysis. 20 algorithms or tools were analysed based on strategy approach, search technique, supported interaction and year published. 2015 was the most active year in which researchers developed t-way algorithms or tools. OTAT strategy and metaheuristic search technique are the most encouraging research areas for t-way combinatorial testing. There is a slight difference in the type of interaction support. However, uniform strength is the most utilized form of interaction from 2010 to the first quarter of 2017.

1 Introduction

Software application has become an integral part of our daily lives due to its numerous benefits. One's life can turn complicated if software applications fail to perform their tasks. Software failures can be prevented by implementing software testing activities. Exhaustive testing can be carried out to the software to ensure the software is bug-free. However software testers will be pushed to the limit if exhaustive testing is implemented. Enormous number of test cases needs to be performed. This situation is impossible to be exercised by software tester [1-2]. An increase in the number of test cases contributes to increased time of software development and cost. Combinatorial software testing is a type of testing technique to overcome the problem of exhaustive testing which involves interaction of parameters. This is because software failures are detected due to interaction of few input parameters, or known as t-way testing, where t is the interaction strength [1], [3-4].

To construct t-way combinatorial testing, Covering Array (CA) is used. It is the most active research in combinatorial testing [4]. Covering array can be defined as CA (N; t, k, v) where N represents the rows of the array, t is the strength, k is the degree and v is the order [3].

Many strategies and techniques have been developed to cater to t-way combinatorial testing. Optimization algorithm is supposed to assist in generating smaller test suite size or near optimum number of test cases. Generating the optimum number of test cases is NP-hard problem [6]. This situation has motivated many researchers in this area. Thus, it is the aim of this paper to investigate the existing algorithms or tools, current and potential knowledge related to t-way combinatorial testing. In addition, this paper also explores potential improvements in t-way testing. This paper focuses on studies published on combinatorial testing from 2010 to the first quarter of 2017. Published papers from various types of available databases were collected and analysed according to strategy approach, search technique, supported interaction and the year published.

This paper is organized as follows: In section 2, Combinatorial Software Testing Taxonomy is presented. Existing Algorithms and Tools are described in section 3 while analysis and discussion are highlighted in section 4. Lastly, section 5 delivers the conclusion.

2 Combinatorial Testing Taxonomy

Taxonomy of combinatorial software testing has been developed based on studies done between 2010 and 2017. Figure 1 presents taxonomy of combinatorial software testing. It can be divided into three main areas namely strategy approach, search technique and supported interaction.

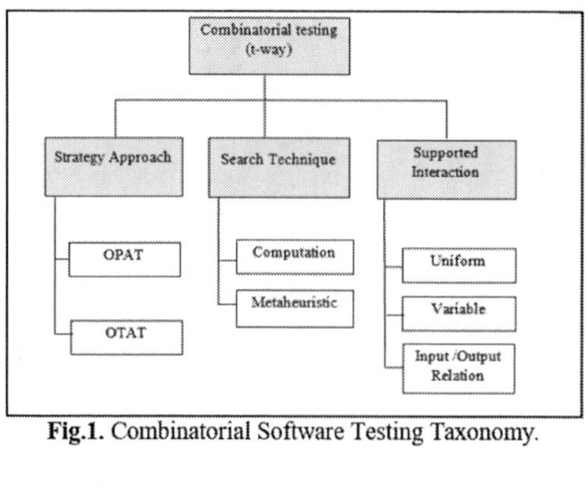

Fig.1. Combinatorial Software Testing Taxonomy.

* Corresponding author: rozmie@unimap.edu.my

2.1. Strategy Approach

Strategy approach can be categorized into two types of strategies. The first strategy is one-test-at-a-time (OTAT). OTAT strategies generate only one test case at one time. Each test case generated must cover at least one t-way combination or uncovered tuples. The best test case generated is the one that covers the most t-way combinations for all parameters. Once the test case is selected, it will be added into a final test suite. This is known as vertical extension. AETG [7] is the pioneer of OTAT strategy.

Another strategy is called one-parameter-at-a-time (OPAT). This strategy starts with t parameter. It produces test suites for t-wise combination. Next, another parameter and a new test are added (horizontal extension). During the horizontal extension, a new test data might be incorporated into test suite to cover all the uncovered t-wise combination. OPAT was popularized by IPOG [8] and its families.

2.2. Search Technique

The two types of search techniques are computational and metaheuristic. General computational approach is based on interaction possibilities. It is able to support very large configurations. The approach repeatedly searches combinations in the searching space until all combinations are covered. This is not practical and can be costly if there are too many combinations [9]. General computational approach is primarily uses greedy technique. Greedy technique constructs test cases that can cover as many uncovered combinations as possible. This technique has been quite effective in generating test cases. However, it is easily trapped in local optima [10].

Metaheuristic search technique generates a small number of rows in any type of covering arrays, support constraints and prioritization [7-8]. This approach starts with a random set of solutions. These solutions will undergo a series of processes to find the best test case using a fitness function. The process repeats until all combinations of parameter input are covered. Metaheuristic search technique has proven to produce an optimum test suite size compared to computational search technique. However, the limitation of metaheuristic is that it requires more time for test suite construction [9–11].

2.3. Supported Interactions

The type of interactions is also an important component in constructing combinatorial testing. There are three types of interactions to construct test cases namely uniform strength, variable strength and Input Output Based Relation (IOR).

Uniform strength or t-wise combines values of parameters with other parameters. The number of related parameters is based on the t (i.e. strength). For example, pair wise or 2-way testing involves combination of values for every two parameters. Pairwise is widely used by software testers. Research has found that faults are detected by smaller interactions (i.e. $t < 2$). Nonetheless, more research has been conducted and it is discovered that more faults can be detected by interactions greater than six [13]. The latest strategy has been developed to support the coverage strength of 14 [14].

Variable strength interaction testing is an extension of uniform strength. To ensure that all parameter combinations are covered at least once, software testers need to run testing at a higher strength or at the highest interaction level. Increasing the strength is unnecessary because that requires using more resources to do unimportant parameter combinations. In view of this problem, variable strength is introduced. Different level of strength can be set to different subset of parameters. This case is applicable for any application or system that needs to be run by multiple configurations [15].

Another form of interaction is IOR. It involves combinations of parameters that impact a particular output. It caters for any combination of related parameter inputs that affect certain output. IOR was initiated as not every application or system has the same characteristics and also to prevent redundancy of test cases or running unnecessary test cases [16]. Through implementation of IOR, both uniform and variable strength strategies are also applied [17]. Uniform strength, variable strength and IOR strategy are further explained in [18].

3 Existing Works

Many algorithms and tools have been explored to be used in t-way combinatorial testing. In this section, existing algorithms or tools developed from 2010 to 2017 are further examined. The existing algorithms and tools are identified according to the year published or their variant. Table 1 presents algorithms/tools published by year.

Table 1 Algorithms/tools published by year

Year	Algorithms / Tools
2010	Particle Swarm Optimization (PSO) • Pairwise PSO (2010) • Particle Swarm Test Generator (PSTG) (2010) • VS-PSTG (2011) • Discrete Particle Swarm Optimization (DPSO) (2015) • Swarm Intelligent Test Generator (SITG) (2015)
2011	• GTWay • Integrated T-Way Test Data Generation (ITTDG) • AURA
2012	Harmony Search • Harmony Search Strategy (HSS) • HS-PTSGT
2013	• DA-RO • DA-FO
2014	• General Variable Strength (GVS)
2015	• TCA • Cuckoo Search (CS) • Flower Strategy (FS)
2016	• High Level Hyperheuristic (HHH) • Artificial Bee Colony (ABC) • Ant Colony System (ACS)

2017	• Adaptive Teaching Learning Based Optimization (ATLBO)

Particle Swarm Optimization (PSO) is an algorithm for t-way combinatorial testing published in 2010. The algorithm is inspired by animal swarms hunting for food. Each individual in the swarm moves towards the best individual location and the best global location based on optimal solution that is calculated according to their position and velocity [19]. A few algorithms or tools have been developed based on PSO algorithm.

The generation of PSO algorithm for t-way testing starts with the development of pairwise PSO [19]. Researchers have developed two different algorithms based on OTAT and OPAT strategies respectively.

The utilization of PSO algorithm in previous research has encouraged the development of Particle Swarm Test Generator (PSTG) [20]. There are several reasons as to why PSO is chosen as the basic platform of PSTG including faster convergence rate attitude, only a few parameters are required to be controlled, can be easily applied to any optimization problem and lighter computational load. PSTG only supports uniform strength up to six levels. Next, VS-PSTG was developed to improve PSTG by enhancing variable strength interaction [21].

Discrete Particle Swarm Optimization (DPSO) [10] is another algorithm based on PSO and uses discrete particle swarm as its basis. The algorithm was designed by adopting S-PSO that uses set-based scheme as its discrete search space. The new DPSO improves performance by having two auxiliary strategies; particle re-initialization and additional evaluation of *gbest*. It also offers guidelines for parameter settings. DPSO is reported to be a promising improvement of PSO.

The last algorithm motivated by PSO between 2010 and 2017 is Swarm Intelligent Test Generator (SITG) [22]. SITG supports 2 to 6 levels of uniform strength and variable strength. SITG outperforms PSTG in some cases of uniform strength while exhibiting better results for variable strength.

As PSO is a metaheuristic algorithm, its entire variant follows a similar search technique. PSTG, V-PSTG, DPSO, SITG and Pairwise PSO (OTAT) use OTAT strategy approach while only Pairwise PSO (OPAT) applies OPAT strategy.

In 2011, three algorithms or tools that apply computational search technique were introduced. GTWay [23] uses backtracking concept. The algorithm needs to map actual data with symbolic representation. Then, it generates possible interactions of t-way pair. Once this is done, the strategy backtracks the uncovered t-way pairs. The t-way pairs are merged if they are able to be combined, complement the missing value of each other and if the new merge interaction can cover the most uncovered t-way pairs. If the pairs fail to merge, the strategy backtracks to the first defined values. GTWay is an OTAT strategy that supports uniform strength for t greater than six.

Another algorithm or tool is ITTDG [17]. It forms a list of candidate test data to be used in deciding a final test data. A visited tuples is put into the candidate list. Next, parameters that have a value that cover the most uncovered tuples is added to the candidate test data, one parameter at a time. In a situation where a tie occurs, the corresponding test data that contains the tied values are duplicated and put into the candidate list. Next, the weight for every test data is calculated. Test data with the highest weight holds the value of the most covered tuples. If a tie happens, the first round in the candidate list will be selected to be put in the final test suite and subsequently removed from the uncovered tuples list. This process is repeated until all uncovered tuples are eventually covered. ITTDG is able to support all three types of interaction; uniform, variable and IOR.

Similar to ITTDG, AURA [24] has the same ability in supporting all types of interactions. AURA starts by exploiting the combination of interactions created earlier to generate test suite. Then, actual data mapping algorithm is used to support symbolic values and actual data output.

Only one type of algorithm was published in 2012, which was Harmony Search (HS). It is the basis for two other algorithms, Harmony Search Strategy (HSS) and HS-PTSGT. HS was inspired by a musician tuning their music.

Another algorithm that uses metaheuristic search technique to generate t-way test suite is Harmony Search Strategy (HSS). It is based on Harmony Search (HS) algorithm. Similar to PSO, HS algorithm requires minimal computation with only a few parameter settings. HSS [14] can support uniform and variable interaction strength with high interaction strength, up to 14 and constraints. Besides HSS, HS-PTSGT [9] is a pairwise generator tool that was developed based on HS algorithm.

In 2013, DA-RO and DA-FO [25] emerged. Both algorithms are computational techniques and support all types of interactions. DA-FO starts with an empty test case which contains unfixed factors. Local density for each factor is determined to obtain the order of coverage requirements. A coverage requirement with the highest local density is selected to be fixed based on global density. For each combination in the test case where the values of factors are fixed, DA-RO will calculate the global density. The combination with the greatest global density is selected to fix the value of factors.

Unlike DA-RO, DA-FO algorithm produces a single test case by fixing values in order of factors. Priority numbers are defined to measure priority of different factors to determine the order of factors. Coverage requirements that hold high priorities and local densities will obtain a great factor density and will be chosen to fix the values. Similar to DA-RO, after fixing values, the local, global and factor densities may change due to re-modification of densities. DA-FO also suffers from the problem of tied values.

GVS [26] was the only algorithm published in 2014. This algorithm is inspired by GTWay. Similar to GTway, GVS is a computational technique and improves

DOI: 10.1051/matecconf/201714001016

the interaction by supporting all type of interactions. The algorithm chooses a don't care value sequentially. It generates one tuple at a time before the repetition starts. Each tuple is assigned with a don't care value to complete the test data. Next, GVS checks which test cases can cover the most uncovered tuples. The chosen test case is then placed in the final test suite and covered tuples are added in the covered tuples list.

In 2015, TCA, Cuckoo Search (CS) and Flower Strategy (FS) were published. They are metaheuristic algorithms.

TCA [27] is an algorithm that combines greedy Tabu search and random walk heuristic. Greedy Tabu search is used during initialization of test cases. TCA applies heuristic search technique to expand opportunities in covering the uncovered interactions. Through both techniques, the runtime for test suite generation is affected. Although TCA utilizes combination of search techniques, it still falls under the metaheuristic category because Tabu search is a metaheuristic algorithm. TCA supports 3-way constraint uniform strength.

Cuckoo Search [28] has been explored in t-way combinatorial testing as Cuckoo algorithm offers search capabilities using Levi flights [29] to update search space. It consists of a few parameters to be tuned [28]. Cuckoo Search supports uniform strength (up to three) and also variable strength.

Flower Strategy (FS) [30] evolved from the effectiveness of Flower Pollination Algorithm (FPA). FPA is a simple, flexible and requires light computation. It also strikes a balance between exploitation and exploration by employing levy flight. FS is a metaheuristic algorithm that supports uniform strength.

In 2016, hyper-heuristic strategy emerged. High Level Hyperheuristic (HHH) [31] is a pioneer in incorporating hyper-heuristic strategy in t-way combinatorial testing. It consists of high level metaheuristic Tabu Search and other four low level metaheuristic algorithms (i.e. Teaching Learning based Optimization, Global Neighborhood Algorithm, Particle Swarm Optimization, and Cuckoo Search Algorithm). The algorithm uses metaheuristic search technique and support uniform strength from level 2 to 6.

There are two proposed t-way testing strategies found in the literature. Artificial Bee Colony (ABC) [32] is proposed to generate an optimum test suite. This strategy was inspired by a group of bees searching for nectar for their hive. ABC was chosen by researchers as it has proven to be a great strategy in the combinatorial field.

Another proposed strategy found in 2016 is Ant Colony System (ACS) [33]. ACS is a variant of the Ant Colony Optimization (ACO) algorithm. It has successfully solved many combinatorial optimization problems. Ant Colony Algorithm (ACA)[34] and ACS [5] were developed to generate an optimum test suite size to cater uniform and variable strength respectively. The proposed ACS strategy is envisioned to serve all types of interactions especially IOR.

Adaptive TLBO (ATLBO) [35] is a metaheuristic t-way testing that emerged from Teaching Learning-based optimization (TLBO) algorithm in the first quarter of 2017. It is based on Mamdani fuzzy inference system. ATLBO is able to support uniform and variable strength.

4 Analysis and Discussion

This section will discuss the analysis from section 3, existing works and other related matters.

4.1. Analysis of Algorithms and Tools

20 algorithms or tools have been identified since 2010 until 2017. Figure 2 presents the number of uniform, variable and IOR supported interactions by year. The number of algorithms or tools exhibited excludes ABC algorithm as no type of suitable interaction for the algorithm was reported. From the graph, it can be inferred that uniform strength is the most favourable type of interaction within the seven years. IOR comes in second, while variable strength is last by only a slight difference. The number of algorithms and tools that support uniform and variable strength in 2015 is comparable. Six algorithms or tools were developed in 2015, the highest among all the years. The algorithm or tool development was reported to be the lowest in 2012, 2014 and 2017. However, as this report contains data as of the first quarter of 2017, it cannot be concluded that only one algorithm was produced in 2017.

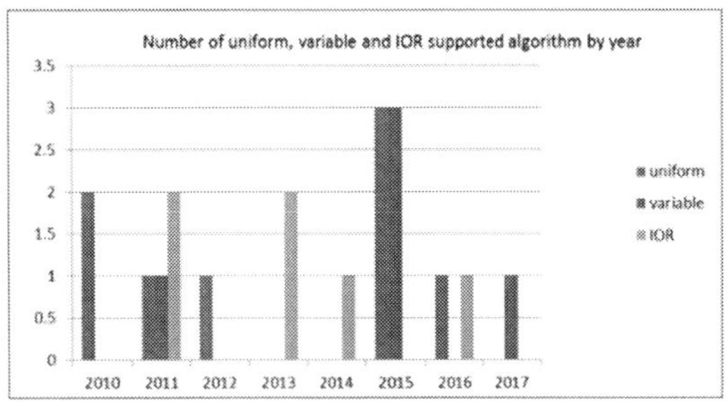

Fig. 2. Number of Uniform, Variable and IOR Supported Interaction by Year

Among these algorithms, PSO is the most widely used within seven years. Five algorithms were developed based on PSO including Pairwise PSO, PSTG, VS-PSTG, DPSO and SITG. PSO requires a smaller number of parameters to be tuned with a simple computational formula. PSO is still popular among researchers for this very reason. PSO has also proven to support uniform strength up to 6 levels and variable strength.

4.2. Analysis of Strategy Approach

There are two categories of strategy approach. From the literature, all 20 algorithms or tools developed employ OTAT strategy. Only PSO makes use of OPAT strategy. OTAT is more popular and favourable as a result of its ability to support higher configuration as opposed to OPAT. The latest trend in uniform strength is likely to offer higher strength such as HSS that supports 14 levels of strength. Most of the latest uniform strength based algorithm can support up to 6 levels of strength.

4.3. Analysis of Search Technique and Supported Interaction

Table 2 presents algorithms or tools based on their search technique and supported interactions. Six of them apply computational search technique, while the remaining 14 algorithms or tools use metaheuristic search technique. The capability of metaheuristic in solving optimization problem is well-known as it is used in Search-Based Software Engineering (SBSE) [36]. This could possibly be one of the reasons many researchers are interested in studying the metaheuristic search technique.

Among the three supported interactions, uniform strength is the most popular and the most widely used among researchers. In addition, algorithms and tools that support variable strength and IOR are also supporting uniform strength. This could be due to variable strength being an enhancement of uniform strength. While for the IOR, literature proves that IOR can support all types of interaction as mentioned by [18] and Table 2 shows that all six algorithms or tools supports both uniform and variable strength. The second most popular type of interaction is variable strength. 60% of the algorithms or tools support variable strength. Clearly, algorithms or tools that support IOR are the least developed type of interaction at only 30%. Algorithms or tools that support IOR fall under computation category. They were published between 2011 till 2014. It is apparent from the table that none of the metaheuristic algorithms support IOR. However, ACS is proposed to support all types of interactions especially IOR. ABC algorithm is also suggested. However, types of supported interaction are not reported.

Table 2. Summary of algorithms/tools

		Computational						Metaheuristic													
		GVS	GTWay	ITTDG	AURA	DA-RO	DA-FO	HHH	TCA	CS	PSO	PSTG	VS-PSTG	DPSO	HSS	HS-PTSGT	FS	SITG	ABC	ACS	ATLBO
Supported Interaction (strength)	Uniform	√	√	√	√	√	√	√	√	√	√	√	√	√	√	√	√	√	NR	√	√
	Variable	√	x	√	√	√	√	x	x	√	x	x	√	√	√	x	x	√	NR	√	√
	IOR	√	x	√	√	√	√	x	x	x	x	x	x	x	x	x	x	x	NR	√	x
		√ : Supported , x : Not Supported, NR : Not reported																			

5 Conclusion

T-way combinatorial testing can be explored through various ways or categories such as strategy approach, search technique or supported interactions. In strategy approach, OTAT is the dominant strategy chosen by the algorithms or tools. OPAT strategy is selected only by PSO algorithms (i.e. OPAT PSO). Metaheuristic search technique is the most widely used search technique as compared to computational technique. 70% of algorithms or tools use metaheuristic search technique. Out of three types of interactions, all 20 algorithms or tools is able to support uniform strength. HSS holds up highest strength at level 14. The number of algorithms or tools that support variable strength reached a peak in 2015 and this contributed to the highest number of algorithms or tools in that particular year. 58% of the algorithms or tools utilize metaheuristic search technique. In contrast to variable strength, only one proposed strategy that incorporates metaheuristic search technique is reported to support IOR (i.e. ACS).

References

[1] R. C. Bryce, Y. Lei, D. R. Kuhn, and R. N. Kacker, "Combinatorial Testing," *Handb. Res. Softw. Eng. Product. Technol. Implic. Glob.*,196–208 (2010).

[2] M. I. Younis, K. Z. Zamli, and R. R. Othman, "Effectiveness of the Cumulative vs . Normal Mode of Operation for Combinatorial Testing," in *IEEE Symposium on Industrial Electronics and Applications (ISIEA 2010)*, 350–354 (2010).

[3] R. Kuhn, R. Kacker, Y. Lei, and J. Hunter, "Combinatorial Software Testing," Computers, **42**, 8, 94–96, (2009).

[4] C. Nie and H. Leung, "A survey of combinatorial testing," ACM Comput. Surv., **43**, 2, 1–29 (2011).

[5] X. Chen, Q. Gu, A. Li, and D. Chen, "Variable strength interaction testing with an ant colony system approach," in *Asia-Pacific Softw. Eng. Conf. APSEC*, 160–167 (2009).

[6] M. Rahman, R. R. Othman, R. B. Ahmad, and M. Rahman, "Event Driven Input Sequence T-way Test Strategy Using Simulated Annealing," in *Fifth Int. Conf. on Intelligent Systems, Modelling and Simulation*, 663–667 (2014).

[7] D. M. Cohen, S. R. Dalal, M. L. Fredman, and G. C. Patton, "The AETG system: an approach to testing based on combinatorial design," IEEE Trans. Softw. Eng., **23**,7, 437–444 (1997).

[8] Y. Lei, R. Kacker, D. R. Kuhn, V. Okun, and J. Lawrence, "IPOG: A general strategy for T-way software testing," in *Proceedings of the Int. Symp and Workshop on Eng. of Comp. Based Systems*, 549–556 (2007).

[9] L. Y. Xiang, A. A. Alsewari, and K. Z. Zamli, "Pairwise Test Suite Generator Tool Based On Harmony Search Algorithm (HS-PTSGT)," NNGT Int. J. Artif. Intell.,**2**, 62–65 (2015).

[10] H. Wu, C. Nie, F. Kuo, H. Leung, and C. J. Colbourn, "A Discrete Particle Swarm Optimization for Covering Array Generation," IEEE Trans. Evol. Comput.,**19**,4,575–591, (2015).

[11] J. Torres-jimenez, C. V. Tamps, and C. V.

Tamps, "Survey of Covering Arrays," in *15th Int. Symp. on Symbolic and Numeric Algorithms for Scientific Computing*, 20–27 (2013).

[12] M. Rahman, R. R. Othman, R. B. Ahmad, and M. Rahman, "A Meta Heuristic Search based T-way Event Driven Input Sequence Test Case Generator," Int. J. Simul. Syst. Sci. Technol., **15**, 3, 65–71(2014).

[13] R. Kuhn, Y. Lei, and R. Kacker, "Practical Combinatorial Testing : Beyond Pairwise," IEEE IT Professional,**10**, 3,19–23 (2008).

[14] A. R. A. Alsewari and K. Z. Zamli, "Design and implementation of a harmony-search-based variable-strength t -way testing strategy with constraints support," Inf. Softw. Technol., **54**, 6, 553–568 (2012).

[15] M. B. Cohen, P. B. Gibbons, W. B. Mugridge, C. J. Colbourn, and J. S. Collofello, "A variable strength interaction testing of components," in *27th Annual Int. Comp. Software and Applications Conf.* (2003).

[16] P. J. Schroeder and B. Korel, "Black-box test reduction using input-output analysis," ACM SIGSOFT Softw. Eng. Notes, **25**, 5,173–177, (2000).

[17] R. R. Othman and K. Z. Zamli, "ITTDG : Integrated T-way test data generation strategy for interaction testing," Sci. Res. Essays, **6**,17, 3638–3648 (2011).

[18] R. Othman and K. Zamli, "T-Way Strategies and Its Applications for Combinatorial Testing," Int. J. New Comput. Archit. Their Appl., **1**,2,459–473 (2011).

[19] X. Chen, Q. Gu, J. Qi, and D. Chen, "Applying particle swarm optimization to pairwise testing," in *IEEE 34th Annual Computer Software and Applications Conference*, 107–116 (2010).

[20] B. S. Ahmed and K. Z. Zamli, "PSTG : A T-Way Strategy Adopting Particle Swarm Optimization," in *2010 Fourth Asia International Conference on Mathematical/Analytical Modelling and Computer Simulation*, 1–5 (2010).

[21] B. S. Ahmed and K. Z. Zamli, "A variable strength interaction test suites generation strategy using Particle Swarm Optimization," *J. Syst. Softw.*, vol. 84, pp. 2171–2185, 2011.

[22] K. Rabbi, Q. Mamun, and R. Islam, "An Efficient Particle Swarm Intelligence Based Strategy to Generate Optimum Test Data in T-way Testing," in *IEEE 10th Conf. on Industrial Electronics and App. (ICIEA)*, 123–128 (2015).

[23] K. Z. Zamli, M. F. J. Klaib, M. I. Younis, N. Ashidi, M. Isa, and R. Abdullah, "Design and implementation of a t-way test data generation strategy with automated execution tool support," Inf. Sci. (Ny)., **181**,9, 1741–1758 (2011).

[24] H. Y. Ong and K. Z. Zamli, "Development of interaction test suite generation strategy with input-output mapping supports," Sci. Res. Essays, **6**,16, 3418–3430 (2011).

[25] Z. Wang and H. He, "Generating Variable Strength Covering Array for Combinatorial Software Testing with Greedy Strategy," J. Softw., **8**,12, 3173–3181 (2013).

[26] R. R. Othman, N. Khamis, and K. Z. Zamli, "Variable Strength t-way Test Suite Generator with Constraints Support," Malaysian J. Comput. Sci.,**27**, 3, 204–217 (2014).

[27] J. Lin, C. Luo, S. Cai, K. Su, D. Hao, and L. Zhang, "TCA : An Efficient Two-Mode Meta-Heuristic Algorithm for Combinatorial Test Generation," in *30th IEEE/ACM International Conference on Automated Software Engineering* 494–505 (2015).

[28] B. S. Ahmed, T. S. Abdulsamad, and M. Y. Potrus, "Achievement of minimized combinatorial test suite for configuration-aware software functional testing using the Cuckoo Search algorithm," Inf. Softw. Technol., **66**,13–29 (2015).

[29] X. S. Yang and S. Deb, "Cuckoo search via Levy flights," in *World Congress on Nature and Biologically Inspired Computing,*210–214 (2009).

[30] A. B. Nasser, Y. A. Sariera, A. A. Alsewari, and K. Z. Zamli, "Assessing Optimization Based Strategies for t-way Test Suite Generation : The Case for Flower-based Strategy," in *IEEE Int. Conf. on Control System, Computing and Eng*, 150–155 (2015).

[31] K. Z. Zamli, B. Y. Alkazemi, and G. Kendall, "A Tabu Search hyper-heuristic strategy for t-way test suite generation," Appl. Soft Comput. J.,**44**, 57–74 (2016).

[32] M. Shaiful, A. Rashid, R. R. Othman, Z. R. Yahya, M. Zamri, and Z. Ahmad, "Implementation of Artificial Bee Colony Algorithm for T-way Testing," in *3rd Int. Conf. on Electronic Design (ICED)*, 591–594 (2016).

[33] N. Ramli, R. R. Othman, M. Shaiful, and A. Rashid, "Optimizing Combinatorial Input-Output Based Relations Testing using Ant Colony Algorithm," in *3rd Int. Conf. on Electronic Design (ICED)*, 586–590 (2016)

[34] T. Shiba, T. Tsuchiya, and T. Kikuno, "Using artificial life techniques to generate test cases for combinatorial testing," in *Proceedings of the 28th Annual Int. Comp. Soft. and App. Conf*, (2004).

[35] K. Z. Zamli, F. Din, S. Baharom, and B. S. Ahmed, "Engineering Applications of Arti fi cial Intelligence Fuzzy adaptive teaching learning-based optimization strategy for the problem of generating mixed strength t -way test suites," Eng. Appl. Artif. Intell., **59**, 35–50 (2017).

[36] M. Harman, Y. Jia, J. Krinke, W. B. Langdon, J. Petke, and Y. Zhang, "Search based software engineering for software product line engineering : a survey and directions for future work," in *15th Soft. Product Line Conference*, 5–18 (2014)

MATEC Web of Conferences **140**, 01017 (2017) DOI: 10.1051/matecconf/201714001017
ICEESI 2017

Helical Antenna sensor for Different Liquid Type Identification

Waleed Majdi Issa Ikhmis, *Saidatul Norlyana* Azemi, *Azremi Abdullah* Al-Hadi, *Ping Jack* Soh, *Mohd Faizal* Jamlos

Advanced Communication Engineering Centre (ACE) CoE, School of Computer and Communication Engineering, Universiti Malaysia Perlis (UniMAP), Pauh Putra, Arau 02600, Malaysia.

Abstract. There are variety of lliquid sensor measurement systems that available in the market nowadays. It is used to measure various parameter for wide range application or to address a specific application. However, most of the existing monitoring system still involved bulky equipment and required a quite amount or liquid to be tested. The results obtained time consuming and take a long time to obtain the desired results. To solve the problems, we develop and design a small sensor using antenna that is small, robust as well as reliable and efficient to work as a sensor for liquid sensor detection. It is efficient and reliable to obtain in-situ result. The helical antenna is design using CST Software and the fabricated antenna is measured. The helical antenna works on 2 GHz frequency is tested with three different type of liquids and it is shows different liquid has their own frequency shift.

1 Introduction

Sensor is a device that responds and detects to some type of input from the physical environment. The specific input could be heat, light, motion, moisture and pressure. The output is generally a signal that is converted to human-readable display at the sensor location or transmitted electronically over a network for further processing or reading [1-5]. Real time oil quality monitoring techniques help to protect important industry assets, minimize downtime and reduce maintenance costs. The measurement of a lubricant's complex permittivity is an effective indicator of the liquid degradation process and it can be useful in condition based maintenance (CBM) to select the most adequate liquid replacement maintenance schedules [4]. Following shows an example of the liquid sensor that available in the market.

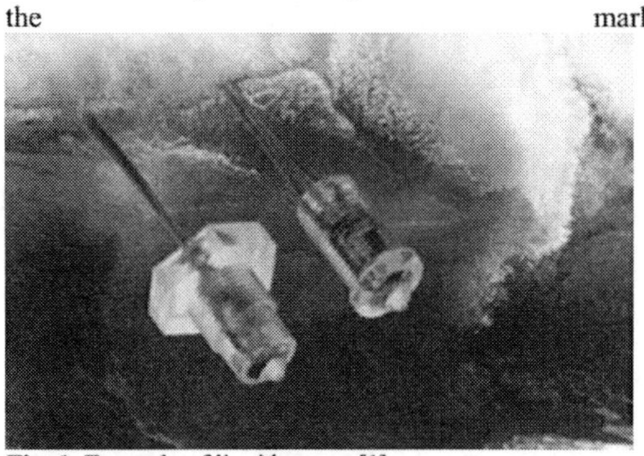

Fig. 1. Example of liquid sensor [1].

The family of liquid measurement systems is classified into different categories such as:

- Liquids or Solids Level Measurement
- Point or Continuous Level Measurement
- Electromagnetic or Electromechanical Level Measurement
- Contacting or Non-Contacting Level Measurement

Analytical, computational (finite element analysis) and experimental tests conducted has demonstrated that antennas can be used as a sensor. In addition, their straightforward integration with electronics has opened new fields of application such as smart sensing systems. Here, a helical antenna will be used as a sensor for this project because it is cheap, robust and chemical free. The example of helical antenna sensor shown in figure 2.

In this paper, a helical antenna is proposed to be used as sensor to detect a different type of liquid. It is cost effective wireless liquid type monitoring utilizing and helical antenna sensor. Different core materials with different effective permittivity will affect the radiation properties of the antenna. When the antenna core are loaded with liquids, the resonant frequency will sensitively change the antenna resonant frequency and return loss according to the permittivity value of the drug. This in turn increases the voltage standing wave ratio (VSWR) at the measuring frequency. The increase in VSWR affects the reflection coefficient as shown by (1), which in turn changes the return loss at the measuring frequency, as shown by (2) [6].

$$Reflection\ Coefficient\ |\tau| = \frac{|VSWR-1|}{|VSWR+1|} \quad (1)$$

$$Return\ Loss = -20\log|\tau| \quad (2)$$

2 Helical antenna sensor developments

Scope of this research was to develop a new method in antenna design in order to obtain the best performing of antenna with parametric studies. As for antenna specification, there are a few essential parameters in designing the antenna. As shown in figure 2 is the simple helical antenna design.

* Corresponding author: ikh.waleed@gmail.com

DOI: 10.1051/matecconf/201714001017

Following are the formulas of helical antenna [6]:

$$Characteristic\ Impedance\ (Z) = \frac{150}{\sqrt{C/\lambda}} \quad (3)$$

$$Antenna\ Diameter, (D) = \frac{\lambda}{\pi} \quad (4)$$

$$Spacing\ between\ coils\ (S) = \frac{C}{4} \quad (5)$$

$$Half\ power\ beamwidth\ (HPBW) = \frac{52}{\left(\frac{C}{\lambda}\right)\left(\sqrt{\frac{NS}{\lambda}}\right)} \quad (6)$$

$$Beamwidth\ for\ the\ 1st\ null: \frac{115}{\left(\frac{C}{\lambda}\right)\left(\sqrt{\frac{NS}{\lambda}}\right)} \quad (7)$$

$$Effective\ Aperture\ (Ae) = \frac{D\lambda^2}{4\pi} \quad (8)$$

Where:

D: Diameter of a turn on the helical antenna.
N: Number of turns on the helical antenna.
S: Vertical separation between turns for helical antenna.
C: Circumference of a turn on the helical antenna πD.

Table 1. Antenna Parameter.

Parameter	Value
λ (wavelength)	0.150
f (frequency)	2000 MHz
N (Number of Turns)	3
S (Spacing between coils)	0.25 cm
G (Antenna Gain)	6 dB
Z (Characteristic Impedance)	150Ω
D (Diameter)	4.77 cm
S (Spacing between coils)	3.75 cm
L (Length of wire)	8 cm
HPBW (Half Power BW)	60.0°
BWFN (BW first nulls)	133°
Ae (Effective Apertures)	0.0161 m

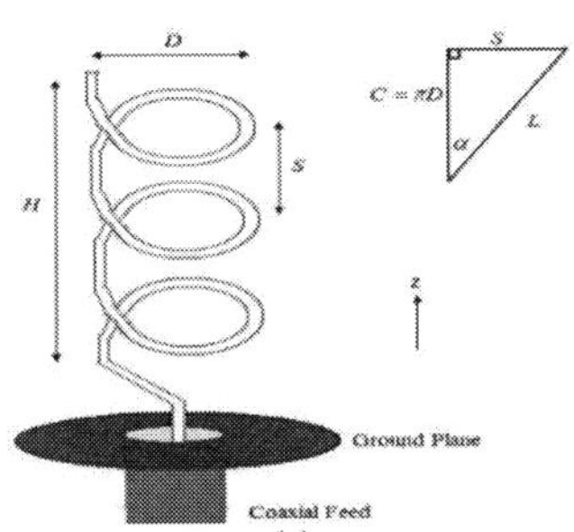

(a)

* Corresponding author: ikh.waleed@gmail.com

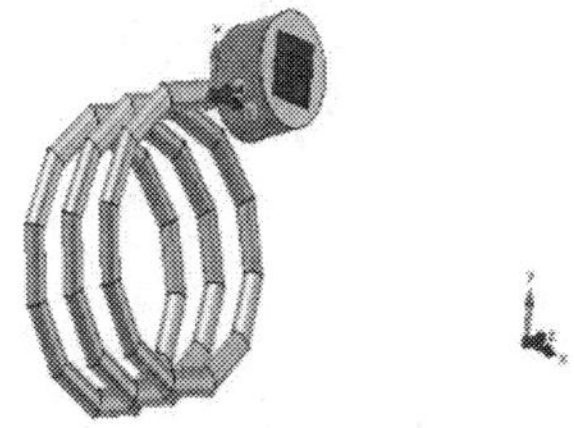
(b)

Fig. 2. Simple Helical Antenna. (a) Helical antenna (b) helical antenna design using CST (c) Helical antenna under test with an empty bottle.

3 Result and discussion

3.1 Helical antenna sensor results

The simulation and measured result for S_{11} are shown in figure 3 where both result simulation and measurement are approximately the same. The simulation and measured results are in good agreement with each other which is shows that the antenna operating at 2GHz. From the experiment, it is found out that the antenna copper wire angle of the design hardware will affect the result. These may happen because of manual fabrication and designing hardware where the angle is very difficult to make it 90° which is like a sharp angle as same in simulation design. The length of copper wire of antenna played a significant role because of the curved angle not sharply 90° and possibility length of copper wire is more than the simulation design. From the analysis, slightly different of the length and the curved angle antenna will affect the actual result. Nevertheless, the percentage different is only 1%. Return loss value of the measured result shows that it is in good value for the radiation signal where the insertion loss value is less than -10 dB.

DOI: 10.1051/matecconf/201714001017

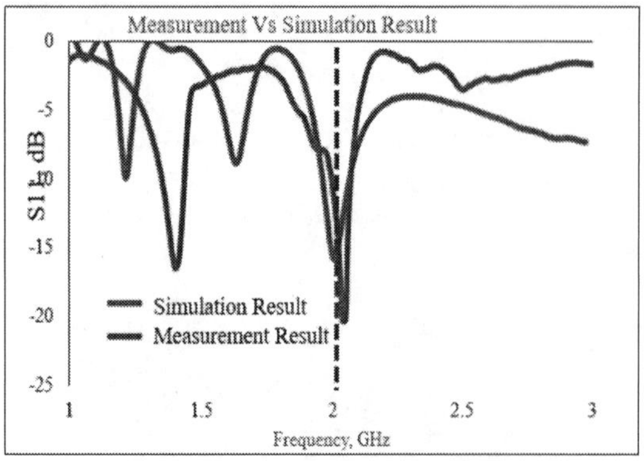

Fig. 3. S_{11} Measurement Vs Simulation Result

3.2 VSWR measurement Vs simulation result.

Shown in figure 4 is VSWR measurement versus simulation result. It is shows that, both of the results are approximately the same. VSWR is the ratio of the maximum voltage to the minimum voltage in standing wave pattern along the length of a transmission line structure. It is varies from 2GHz to (plus) infinity and is always positive. From the result, all the value is positive at the targeted frequency and it is approaching to 2 GHz.

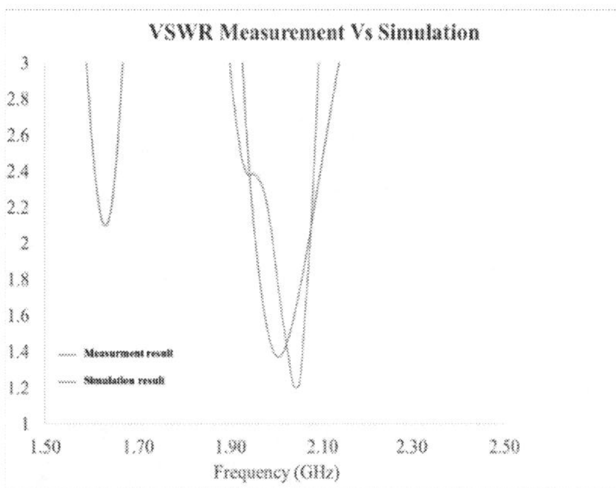

Fig. 4. VSWR Measurement Vs Simulation Result.

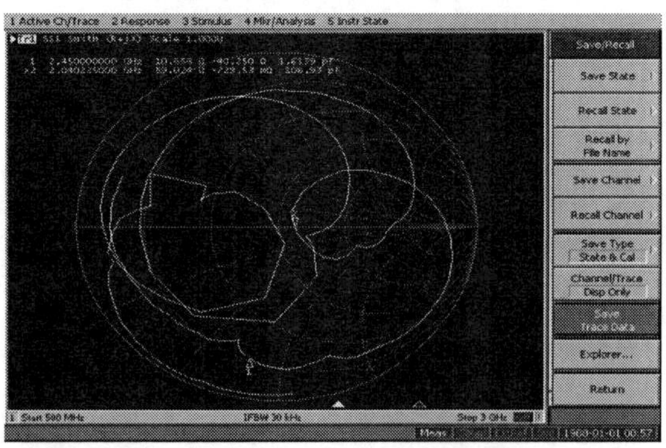

Fig.5. Smith chart at 2GHs frequency and 2.45GHz.

* Corresponding author: ikh.waleed@gmail.com

As shows in figure 5 is the result of impedance using network analyser. As can be seen when the frequency is at 2.45 GHz the value is 10.85Ω, but when the frequency 2GHZ the value is 59 Ω which is accurate because near to the 50Ω.

3.3 Gain

Antenna gain is the main performance which is combined the antennas directivity and electrical efficiency. It is shows the amount of energy radiated in a particular direction compared to the amount of energy directed in the same direction of an isotropic antenna [6]. For this helical antenna gain value is 6.46 dB, and pointing toward z plane, as shown in figure 6.

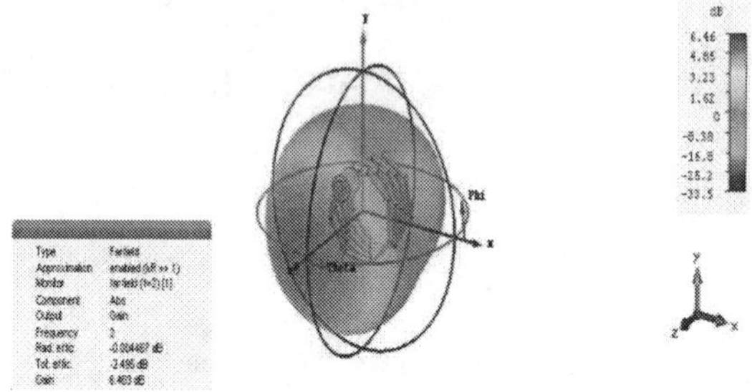

Fig. 6. Helical antenna radiation pattern, pointing toward z plane, the gain is 6.46 dB.

3.4 Summary of the measurements

The graph bellow as shows in figure 7 shows the measurements results for the chosen materials. The experiment results are compared with the approximation simulation performance whenever to verify that the design perform as presented. Once filling the bottle with liquid, the frequency will decreases depending on the type of the liquid.

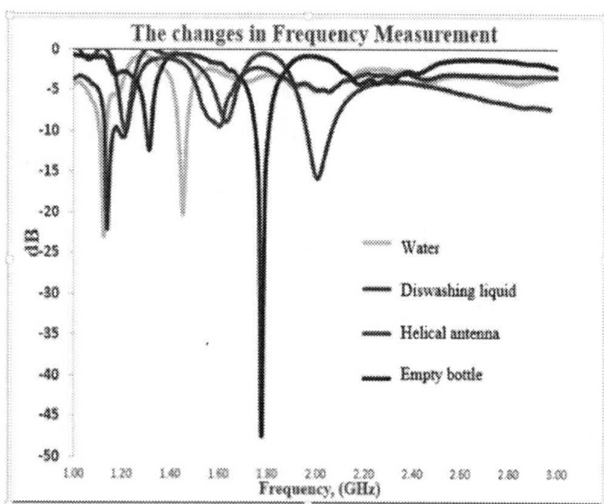

Fig. 7. The graph shows the measurement result.

Table 2. Summary of measurements results based on changes with resonant frequency.

Material	Frequency (GHz)	Bandwidth dB
Helical antenna	2GHz	-15dB
Helical antenna with empty bottle	1.8GHz	-47dB
Helical antenna with dishwashing liquid	1.2GHz	-22dB
Helical antenna with water	1.18GHz	-23dB

4 Conclusion

Based on the results that had been discussed in the paper, it had been proved that the objectives of this project have been achieved successfully and can be fabricated to measure the frequency for the liquids. Firstly, the design uses a small sensor using helical antenna that is small, suitable and robust. Secondly, the fabricated sensor antenna for liquid sensor detection that is efficient and reliable. Besides that, the helical antenna was evaluated and verified in order to overcome the disadvantages of current antenna which is bulky in size compared to their wavelength. This antenna proved that the size in term of wavelength is small enough compared than previous antenna design.

References

1. D. Fine et al., A robust nano_uidic membrane with tunable zero-order release for implantable dose speci_c drug delivery,*Lab Chip*, vol. 10, no. 22, pp. 3074-3083, Nov. 2010.
2. A. Oseev, R. Lucklum, M. Zubtsov, M.P. Schmidt, and S. Hirsch, SAW based phononic crystal liquid sensor-Periodic microfluidic channels approach. *Ultrasonics Symposium (IUS), 2016 IEEE International*, pp. 1-4. 2016
3. E. Bou-Maroun, J. Rossignol, B. De Fonseca, C. Lafarge, R.D. Gougeon, D. Stuerga, D. and P. Cayot, Feasibility of a microwave liquid sensor based on molecularly imprinted sol-gel polymer for the detection of iprodione fungicide. *Sensors and Actuators B: Chemica*l, 244, pp.24-30, 2017.
4. Ma, T.X., Wang, Y.S., Zhang, C. and Su, X.X., 2016. Theoretical research on a two-dimensional phoxonic crystal liquid sensor by utilizing surface optical and acoustic waves. *Sensors and Actuators A: Physical,* 242, pp.123-131, 2016.
5. M.A.M. Johari, A.M. Azize, R.M. Said, N.A. Ngatiman, and N. Zaine, Corn Oil Concentrations Detection for Food Industry Research Development by Using Application of Fiber Optic Liquid Sensor Concept. *In MATEC Web of Conferences* (Vol. 97, p. 01108). EDP Sciences, 2017.
6. C. A. Balanis, Antenna Theory: Analysis and Design, 3rd ed. New York,NY, USA: Wiley, 2005.

* Corresponding author: ikh.waleed@gmail.com

Performance Comparison of Energy Efficient Dynamic Transmission and Static Transmission Power in Static Mobility Node Wireless Ad-Hoc Network

Siti Asilah Yah[1,*], *Naimah* Yaakob[1], Ong Bi Lynn[1], *Mohammad Elshaikh* Elobaid[1], and *Wan Aida Nadia* Wan Abdullah[1]
[1] ENAC, School of Computer and Communication Engineering (SCCE), University Malaysia Perlis (UniMAP) 02600, Arau, Perlis, Malaysia

Abstract. Transmission power optimization in Wireless Ad-Hoc Network is an important thing in order to minimize the energy consumption for effective utilization of the applications like Vehicle Ad-Hoc Network (VANET) applications. If one or more nodes in the wireless Ad-hoc network have little or no energy, then data transmission will be temporarily or permanently interrupted which might create a serious havoc in the Ad-hoc network especially during vital information transferred. This will, in turn, affect the performance of the entire network. Therefore transmission power control is one of the important research topics that needs to be focused in the wireless ad-hoc network in order to ensure effective energy consumption. Recently, we proposed a Dynamic Transmission Power algorithm to maintain network connectivity by adapting node's transmission power based on the distance between the vehicles in VANET. Our research aims to design a dynamic transmission power that can minimize the rate of energy consumption. Hence, in order to develop the proposed method, prerequisite experiment need to be done. This paper investigates the energy saving efficiency of dynamic and static transmission range in static mobility node wireless ad-hoc network which is prerequisite experiments before further experiment on VANET can be carried on. The simulation results prove that dynamic transmission range gives better energy consumption compared to static transmission range, so it is worth it to carry out the subsequent experiments on VANET.

1 Introduction

An ad hoc network is a collection of wireless mobile nodes dynamically forming a temporary network without the use of any existing network infrastructure or centralized administration [1]. In many ad hoc networks, two hosts that want to communicate may not be necessary within the wireless transmission range of each other. However, they could communicate with each other if the other hosts who are also participates in the ad hoc network are willing to forward packets for them. An advantage of the system is robustness, flexibility and mobility.

The ad hoc networks can be classified according to their application as Mobile Ad hoc Network (MANET) which is a self-arranging infrastructure-less network of mobile devices communicated through wireless link. Vehicular Ad hoc Network (VANET) uses travelling cars as nodes in a network to create a mobile network. Wireless Sensor Network (WSN) consists of autonomous sensors to control the environmental actions [2]. The nodes in the all of this application use an ad-hoc network to communicate with each other and they did not need any based station to communicate.

Since energy sources have a limited lifetime, power availability is one of the most important constraints for the operation of the ad hoc network [3]. If one or more nodes in wireless Ad-hoc network have little or no energy, then data transmission will be interrupted temporarily or permanently which might create a serious havoc in the Ad-hoc network especially when vital information is in transmission and requires an urgent response. This will in turn affect the performance of the entire network. Moreover, nodes involved in Wireless Ad-hoc network act as host and intermediate node to forward data of the neighbouring nodes and receiving node. Therefore, a lot of energy are needed to perform the task. The nodes require to use energy efficiently in order to avoid energy wasted, hence enable the nodes to perform all the given tasks. Therefore transmission power control is one of the important research topics that needs to be focused in Ad-hoc network application in order to ensure the effectiveness of energy consumption.

Recently we proposed Dynamic Transmission Power algorithm which is an algorithm to maintain network connectivity by adapting node's transmission power based on the distance between the vehicles in one of the ad hoc applications, which is VANET. The aim of the research is to design a dynamic transmission power that can minimize the rate of energy consumption. However, prerequisite experiment need to be done before the proposed method can be proceed. So, this paper investigates the energy saving efficiency of dynamic transmission range and static transmission

* Corresponding author: ilahyah_92@yahoo.com

DOI: 10.1051/matecconf/201714001018

range in static mobility node wireless ad-hoc network which is prerequisite experiments before further experiment on VANET can be carry on.

The rest of the paper is organized as follows. Explanation about the methodology of this experiment are presented in section 2. The simulation setup are explained in section 2.1. In section 3, the evaluation scenario for this experiment are discussed. Then, the selected simulation metric are presented in section 3.1. Section 4 discusses about the collected result and the final section concludes the paper.

2 Methodology

In this paper, OMNeT++ [4], was used to design the simulation model and perform all the experiments to evaluate and analyze the effects of using different packet sizes and packet rates on the energy consumption of wireless ad-hoc network.

2.1 Simulation Setup

The purpose of this phase is to prove that by using dynamic transmission range, the energy consumption can be further optimized. The significant parameters need to be determined in order to model dynamic transmission power algorithm in VANET environment. The significant parameters that have been decided to be in this experiment are shown in Table 1. The parameters are used for both dynamic and fixed transmission range.

Table 1. Parameter Setup for Fixed and Dynamic Transmission Range

Parameters Names	Values
Simulation Area	1000m * 1000m
Simulation Time	500s
Routing Protocol	AODV
Transport Protocol	UDP
Packet Size	1, 2, 3, 4 and 5 KB
Packet Rate	2, 4, 6 and 8 Packet / sec.)

For transport protocol, User Datagram Protocol (UDP) has been chosen because it can be very fast, with low delay and are not affected by congestion on a connection basis. On other hand, reduces the overhead with the packet header size of just 8 bytes. UDP has comparatively fast speed and is thus used for games or applications that require fast transmission of data. UDP also transmits fixed sized datagrams and can be used for multicasting. For routing protocol, AODV protocol is used because it uses a broadcast route discovery mechanism and it is responsive to changes in network topology.

Next step is to design and verify the simulation model. For this purpose, a few simulations need to be run. The details of the design simulations are described in Figure 1 and Figure 2.

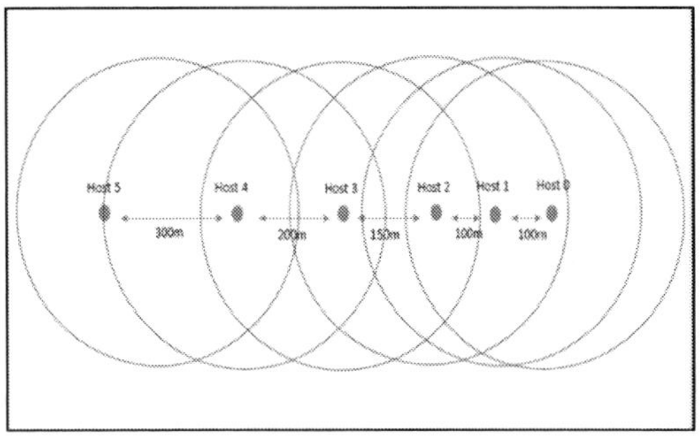

Fig.1. Fixed Transmission Range Illustration

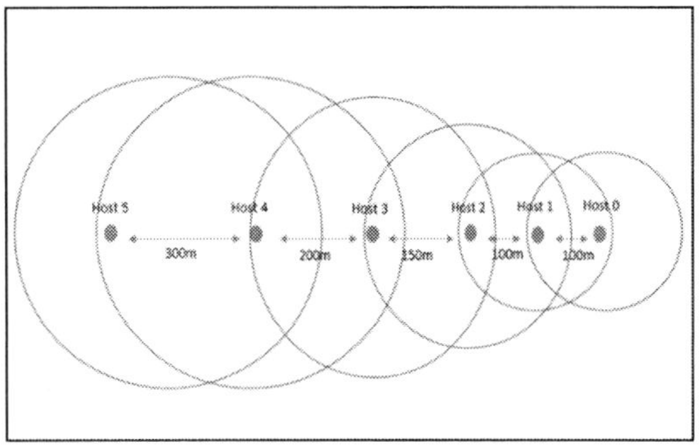

Fig.2. Dynamic Transmission Range Illustration

Figure 1 and Figure 2 shows the illustration of experiment setup for fixed transmission range and dynamic transmission range respectively. The simulation time is 500s for every experiment and all nodes are configured in static mobility. In this experiment, five sizes of packet data and packet rate are used. In Figure 1, all nodes in the simulation are configured with fixed transmission range, while in Figure 2, all the nodes are configured with dynamic transmission range. The transmission range for dynamic are assign based on the distance between the nodes. The details of the simulation are summarized in Table 2.

DOI: 10.1051/matecconf/201714001018

Table 2: Details of Simulation Setup

	Fixed Transmission Range	**Dynamic Transmission Range**
Distance Between Nodes	Host 0 to Host 1 = 50m Host 1 to Host 2 = 100m Host 2 to Host 3 = 150m Host 3 to Host 4 = 200m Host 4 to Host 5 = 250m	
Transmission Range	3 sets of experiments: 1st = 3mW 2nd = 5mW 3rd = 7mW	Host 0 = 0.3mW Host 1 = 0.6mW Host 2 = 0.9mW Host 3 = 1.5mW Host 4 = 2.2mW Host 5 = 2.5mW
Packet Size	1, 2, 3, 4 and 5 KB (Fixed packet rate = 1 packet/sec)	
Packet Rate	2, 4, 6 and 8 (Packet / sec.) (Fixed packet size = 1 KB)	

3 Evaluation Scenario

The experiments were evaluated at different network loads. The experiments were run by increasing the size and the rate of packet at using different transmission range.

3.1 Evaluation Metric

The evaluation metrics which are chosen to analyse the performance are energy consumption, PDR, throughput and transmission time delay. The evaluation metrics are explained in the Table 3.

Table 3: List of Evaluation Metric

Name	**Definition**
Energy consumption	Energy used by the node to transmit the data
Throughput	The average rate of successful packet that can be delivering to its destination over a communication channel per unit of time $Throughput = \frac{Received\ packet\ in\ bits}{Simulation\ time}$
Packet Delivery Ratio (PDR)	The number of packets that successfully received by the receiver to the number of packets sent by the source $PDR = \frac{Received\ Packets}{Sent\ Packets}$
Delay	The arrival time needed by the packet to its destination.

4 Result and Discussion

Different types of tests are performed to evaluate and investigate the effects of using different packet sizes and different packet rates on the energy consumption of static node wireless ad-hoc network. Each experiment was repeated ten times to get the accurate results. The evaluation and analysis was made by using the average result value of all the ten experiments.

4.1 Energy Consumption

The result of energy consumed by the network when using different packet size are illustrated as shown in Figure 3. The higher the packet size, the lower the energy used due to the fact of high packet collision and packet corrupted. Packet corruption rate increase as packet size increase due to the packet fragmentation. Packet fragmentation is a process that split the packet into smaller pieces, so that packets can pass through a link with a maximum transmission unit (MTU) compared to the original size. For example, if we transmit packet of 2048B and 5120B in size. For 5120B the packet will be split into 4 packets compared to 2048B, it only split into 2 packets. Since the MTU size is 1500B. So for higher packet size, more packets are produce compared to smaller packet size. Therefore, the probability of packet loss or drop due to packet collision and corrupted are high for larger packet size. In turn, the energy consumption will be low. From the graph also shown that dynamic transmission range give better energy saving compared to fixed transmission range.

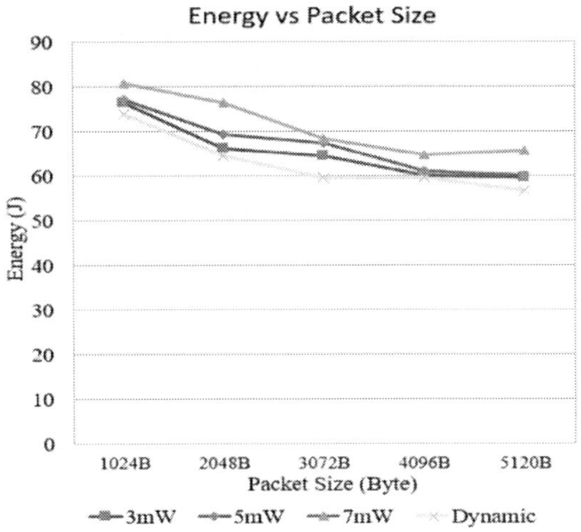

Fig. 3. Energy vs. Packet size

Figure 4 shows the graph of energy consumption over packet rate for different transmission range which is 3mW, 5mW, 7mW and dynamic transmission range. Packet rate is the amount of packet sent per unit of time during the communication process. In a simple word, it means how many packets per second (packet/sec). The higher the packet rate, the lower the energy used due to the fact that high probability of collision happen due to fast transmission rate and the network will be congested. When collision happened, the rate of successful packet arrived at the destination will decrease, giving rise to more number of packets being loss or drop. Furthermore, increasing the packet rate make the number of packet waiting in the MAC layer buffer to be sent also increase. The entire packets that exceed their waiting time limit are discarded by the MAC layer. When the number of packet loss is high, the energy used is low. The graph also shows that the energy consumption for dynamic transmission range is better than energy consumption for static transmission range.

From the Figure 3 and Figure 4, the graph shows that the higher the packet size and the higher the packet rate, the energy consumption will be low. This is due to more packet are loss due to high traffic generated. Since we used UDP that does not provide retransmission, the energy used will be low due to less number of packets were completed the transmission task. Although it is achieve our target to get lower energy consumption, but in term of PDR and throughput, it will be low. Hence to avoid this problem happen to our purposed method that will be implemented in VANET, an optimization method will be used. The optimization method are used to find the most suitable packet size and packet rate that can give higher PDR and throughput and at the same time the energy consumption will be lower compared to the existing method.

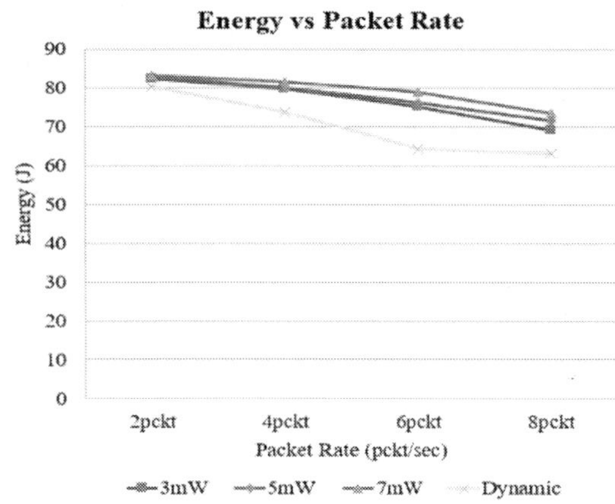

Fig. 4. Energy vs Packet Rate

4.2 Throughput and PDR

The result of throughput (bps) versus different number of packet size are shown in Figure 5 and the result of throughput versus different packet rate are shown in Figure 6. Both of the figure, which is Figure 5 and Figure 6 depicts that an increasing number of packet rates and packet size, increase the amount of data injected into the network. This data injection leads to an increase in throughput. The larger the packet rate, the larger the throughput and the larger the packet size, the larger the throughput.

Throughput and packet delivery ratio (PDR) are directly proportional to each other. When the PDR increase, the throughput also increase. Figure 7 demonstrate the result of PDR versus packet size and Figure 8 show the result of PDR over packet rate. It shows that, when packet size or packet rate increase, the PDR also increases.

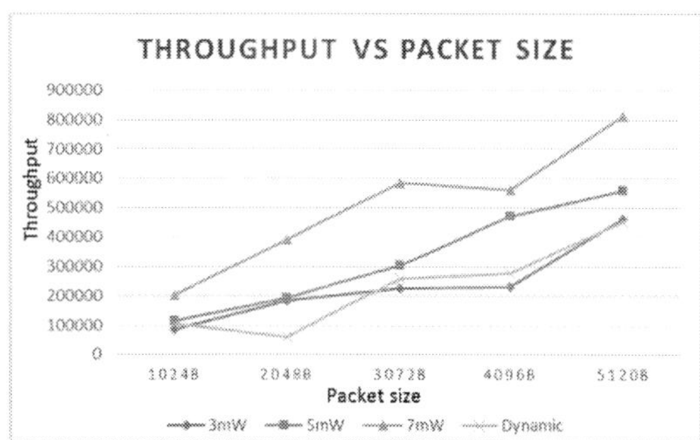

Fig.5. Throughput vs. Packet Size

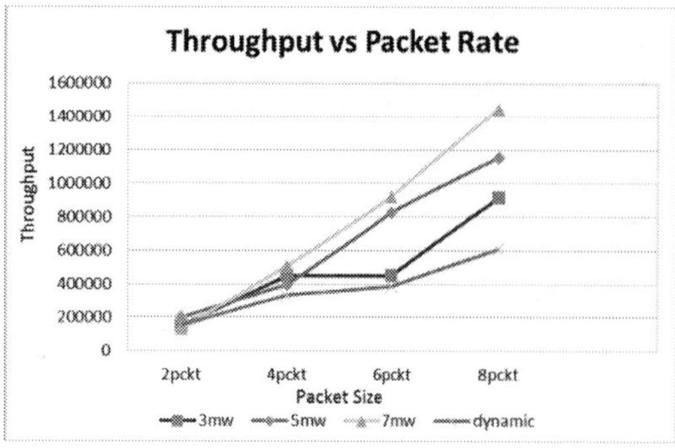

Fig.6. Throughput vs. Packet Rate

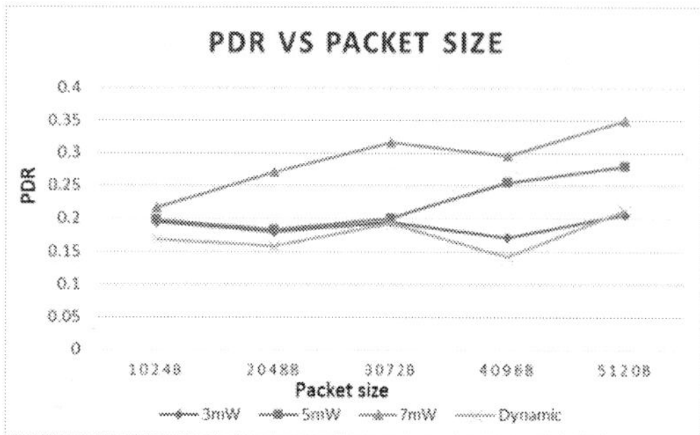

Fig.7. Packet Delivery Ratio (PDR) vs. Packet Size

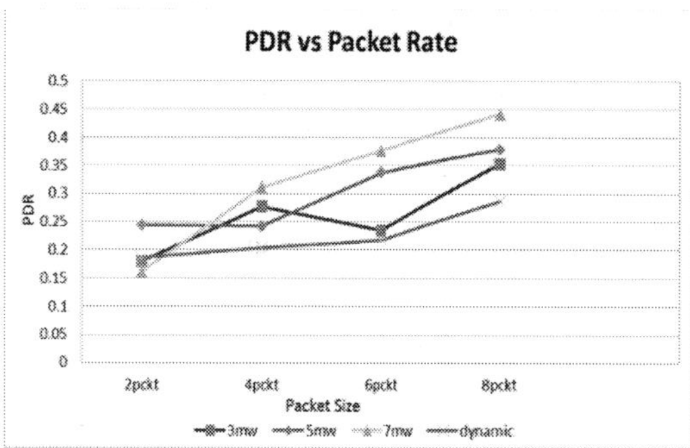

Fig.8. Packet Delivery Ratio (PDR) vs. Packet Rate

4.3 Transmission Time Delay

The transmission time delay for larger packet size is higher than the transmission time delay for smaller packet size as shown in Figure 9. The higher the packet size, the higher the delay. For example, delay for 5120B packet size is higher compare to 4096B packet size as shown in Figure 9. This is because the transport layer of each node needs more time to transmit a larger packet size compare to smaller packet size.

Moreover, increasing the packet rate increase the number of packets waiting in the Mac layer buffer to be sent, this in turn leads to an increasing in the transmission time delay. Increasing the packet rate also make the transport layer of each node need more time to send the data. This is the reason why transmission time delay in Figure 10 increase with the increasing number of packet rate. This is true for all cases.

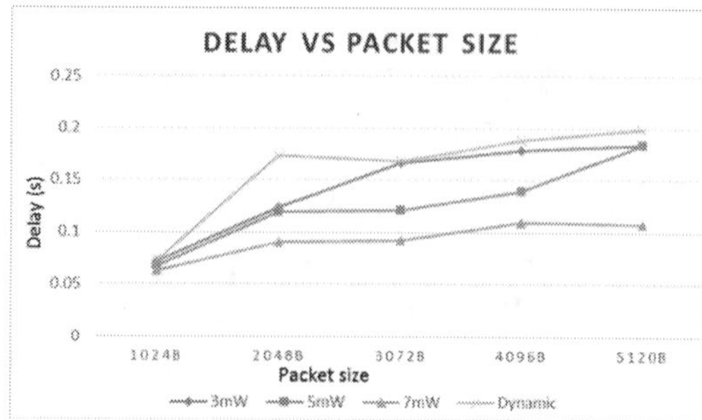

Fig.9. Transmission Time Delay vs. Packet size

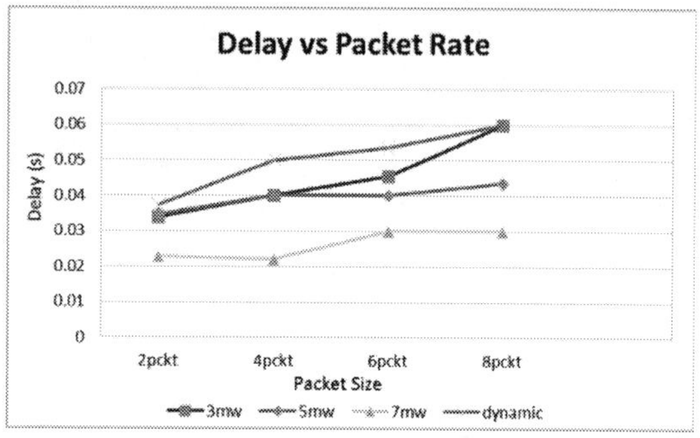

Fig.10. Transmission Time Delay vs. Packet Rate

Conclusion

This paper investigates the energy saving efficiency of dynamic transmission range and static transmission range in static mobility node wireless ad-hoc network which is prerequisite experiments before further experiment on VANET can be carried on. The simulation results prove that dynamic transmission range gives better energy consumption compared to static transmission range, so it is worth it to carry out the subsequent experiments on VANET. Although the result of PDR, throughput and delay from this experiments shows that dynamic transmission range is lower compared to static transmission range, this reasons is not a backstop to the subsequent experiments in VANET. This is due to the fact that been proved by the authors in the paper [5][6][7][8][9], which is the performance of dynamic transmission range in VANET is better compared to the performance of static transmission range. This is because, the application of VANET use a travelling cars as nodes in a network to create a mobile network. That means, the node in VANET used a high speed mobility node and the location of the nodes frequently change. This maybe the factor why the dynamic transmission range in this experiment give lower performance. In VANET the information may be directly sent to the destination since the nodes frequently change the location. The nodes did not need to use the intermediate node to send or receive the information. Hence less packet maybe drop or loss during the communication process compared to this experiments. This is because by using dynamic transmission range on static mobility ad-hoc network in this experiments, the transmitted data need to use intermediate node to reach the destination. The transmission range of dynamic transmission range are assign based on the distance between the nodes. Hence, more data might be dropped and lost during the transmission process since the data need to travel from one node to another node to reach the destination. This is the reason why the PDR, throughput and delay for the dynamic transmission range in this experiment give lower result compared to static transmission range. Since the objective of this experiment was to investigate and compare the energy consumption between the dynamic and fixed transmission range, therefore the result of PDR, throughput and delay of this experiment were not take into consideration.

Acknowledgment

The research reported in this paper is supported by Research Acculturation Grant Scheme (RAGS) [Grant number: 9018-00082]. The author would also like to express gratitude to the Malaysia Ministry of Higher Education (MOHE) and University Malaysia Perlis for the facilities provided.

References

1 S. S. . S, "A Survey on Wireless Ad Hoc Network Issues and Implementation," *2016 Int. Conf. Emerg. Trends Eng. Technol. Sci.*, pp. 1–6., 2016.

2 D. Helen and D. Arivazhagan, "Applications , Advantages and Challenges of Ad Hoc Networks," vol. 2, no. 8, pp. 453–457, 2014.

3 S. Verma and P. Singh, "Energy Efficient Routing in MANET : A Survey," vol. 3, no. 2, pp. 3971–3977, 2014.

4 "OMNeT++ Discrete Event Simulator - OMNeT++ Discrete Event Simulator," 2014. [Online]. Available: https://omnetpp.org/component/jdownloads/download/32-release-older-versions/2290-omnet-4-6-source-ide-tgz. [Accessed: 11-Nov-2016].

5 R. Chai, B. Yang, L. Li, X. Sun, Q. Chen, Rong Chai, Bin Yang, Lifan Li, Xiao Sun, and Qianbin Chen, "Clustering-based data transmission algorithms for VANET," *2013 Int. Conf. Wirel. Commun. Signal Process.*, no. 61102063, pp. 1–6, 2013.

6 R. hen, H. Yang, W. Jin, and A. Regan, "Dynamic Transmission Range in Inter-Vehicle Communication with Stop-and-Go Traffic," *2010 IEEE Intell. Veh. Symp.*, pp. 1–6, 2010.

7 J. M-Y. Lim, Y. C. Chang, J. Loo, and M. Y. Alias, "Improving VANET Performance with Heuristic and Adaptive Fuzzy Logic Scheme," *Wirel. Pers. Commun.*, vol. 83, no. 3, pp. 1779–1800, 2015.

8 D. B. Rawat, D. C. Popescu, and S. Member, "Enhancing VANET Performance by Joint Adaptation of Transmission Power and Contention Window Size," vol. 22, no. 9, pp. 1528–1535, 2011.

9 M. M. Artimy, "Local Density Estimation and Dynamic Transmission-Range Assignment in Vehicular Ad Hoc Networks," *IEEE Trans. Intell. Transp. Syst.*, vol. 8, no. 3, pp. 400–412, 2007.

Dual Band Planar Inverted F Antenna (PIFA) with L-Shape Configuration

Mohamad khlouf Munzer *Saidatul Norlyana* Azemi, Azremi *Abdullah Al-Hadi*, Ping Jack *Soh*, Mohd Faizal *Jamlos*

Advanced Communication Engineering Centre (ACE) CoE, School of Computer and Communication Engineering, Universiti Malaysia Perlis (UniMAP), Pauh Putra, Arau 02600, Malaysia.

Abstract. One of the most used antennas in mobile devices is planar inverted F antenna (PIFA). PIFA can be design in dual band frequencies due to the coverage of the wireless service in a mobile device that requires a multiple frequencies. However, the consideration of technical operation has to be combined with an evaluation of the antenna radiation impact on the users. A procedure of PIFA work in GSM (867-960MHz) and GSM (1710-1899MHz) is done using CST Software. The dual band frequency response is obtained by means of an insertion of an L-shaped slot, which is use to tune the operation frequencies. The prototype of the antenna is fabricated as model by CST Software and evaluated. It is found out that the PIFA antenna has a good efficiency, bandwidth as well as produce a maximum gain for the antenna. A key and innovative research is still underway to broaden performance parameters of the antenna.

1 Introduction

Most common built in antenna which becomes very promising candidates for applications in mobile handsets is Planar Inverted F Antenna (PIFA) [1]. Planar Inverted-F antennas (PIFAs) are among the most generally utilized receiving wires in portable terminals [2]. The PIFA have some unique characteristic that makes it suitable to use in portable wireless device especially on mobile handsets. It has many advantages compared to other microstrip antennas, where it has a low profile, small size and can be located in a place such as at the back of the mobile phone [3]. On the other hand, the other major antennas are easy to fabricate, sample structure, low manufacture cost and low Specific Absorption Rate (SAR) value where it has a small backward radiation toward the used head [4].

There are two type of antenna used for mobile phone which are external and internal. At first, the use of the external antenna had been used widely with the previous generation of the mobile such as Helical Antenna mobile, however due to today's market trends the demand has been placed on reducing the handset size where the antenna must be within size and performance as required compared to the hand-phone [5-6]. The size of the mobile have been reducing in size and weight over time, where an internal antenna must be used instead of external antenna. Therefore, there have been many evolutions of antenna structure due to the decreasing in mobile size and high performance required.

Due to the new era where human depend on mobiles and the size of the phone become slimmer and the performance become higher, hence it is essential to develop an antenna that can be able to accommodate to all the requirement with maximum performance. But, antenna size is a major factor that limits device miniaturization. Hence, the objective of this project is to develop low profile and realizable antenna that is capable to operate at GSM frequency band.

2 Antenna development

The fundamental parameters of PIFA design are the longitude of the patch (L and W), the height of the structure h, the width of the shorting wall w, and the distance from the feeding to the wall. The dimensions of the L, W and the h are chosen to achieve a certain resonant frequency as follow:

$$(L+h)\sqrt{\varepsilon_r} \cong \lambda 4 \quad (1)$$

If the size of shorting pin is reducing, the electromagnetic waves will travel a longer distance, for which the condition of resonance is transformed into:

$$(L+W+h)\sqrt{\varepsilon_r} \approx \lambda 4 \quad (2)$$

Several considerations obtained from diverse parametric studies will be taken to design a PIFA: the bandwidth of the PIFA increases with the h (as in the case of the conventional patch antenna), and the dimension of the patch determine the resonant frequency (bigger patch, lower frequency). Also, the variations of parameters have been carried out (position of the shorting wall, position of the feeding, height of the layer of air) for tuning the structure. At the beginning, a patch that operates at GSM 900 band was designed, and the second frequency is excited with the insertion of the L-shaped slot with a length around λ/4 for GSM 1800 band.

$$f(GHz) \approx 300 / [4(L(mm) + W(mm) + h(mm))\sqrt{\varepsilon_{eff}}] \quad (3)$$

The following equation is used to obtain the upper frequency band

$$f(GHz) \approx 300 / [4(L2(mm)+L3(mm))\sqrt{\varepsilon_{eff}}] \quad (4)$$

DOI: 10.1051/matecconf/201714001019

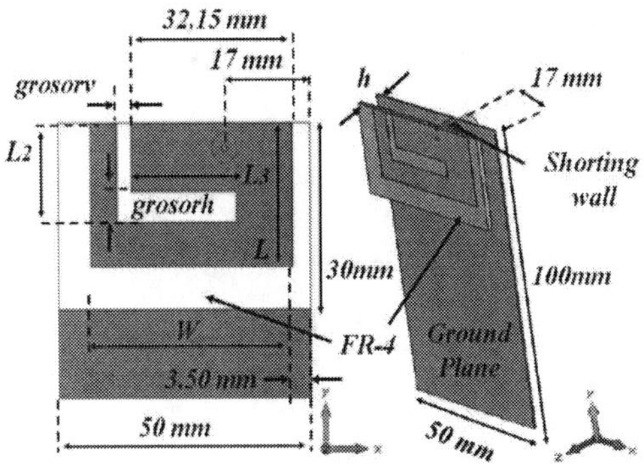

Fig. 1. Configuration of the PIFA antenna with L-shaped slot.

In order to maintain the first resonant frequency (the lower) the dimension of the patch is adjusted. The configuration of the proposed antenna is presented in Figure 1, and it can be seen that there is an air separation between the ground plane and the substrate that hold the patch. For this case an effective dielectric constant is used in the above equations

$$\varepsilon_{eff} = \varepsilon_r.(h_s + h_{air})\varepsilon_r.\ h_{air} + \varepsilon_{air}.\ h_s \quad (5)$$

The feeding and the shorting wall are in line in Figure 1, separated by 4 mm from the edge where is located the shorting wall to the centre of the coaxial probe feed.

Table 2. Parameter of PIFA antenna with L-shaped slot

Parameters	Value
L2	16 mm
L3	21 mm
W	40 mm
L	23 mm
h (height of the shorting wall)	7.5 mm
hs(height of the FR substrate-4)	1.58 mm
hair (height of the separation of air)	5.92 mm
hpt (height of the copper ground plane)	0.6 mm
hc (height of the copper microstrip antenna)	0.035 mm
Total height (*hpt + h + hc)*	8.135 mm
grosorv (width of the vertical slot of the L)	2.7 mm
grosorh (width of the horizontal slot of the L)	5 mm
εr (dielectric constant of the FR-4 substrate)	4.3
*tan*δ (loss tangent of the FR-4)	0.025
width of the shorting wall	0.6 x 1 mm

3 Result and Discussion

3.1 S_{11} measurement Vs simulation

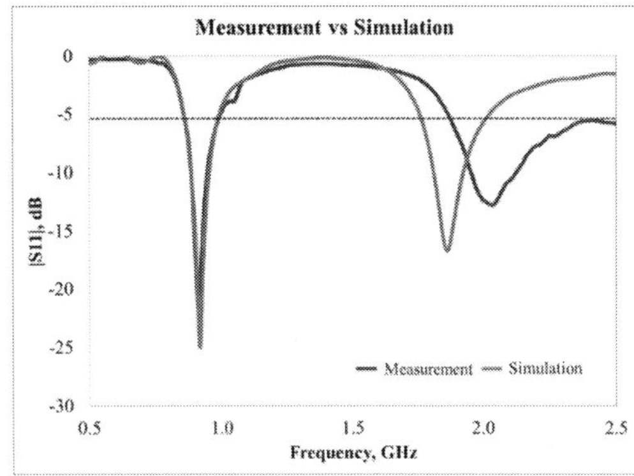

Fig. 2. S_{11} Measurement Vs Simulation Result

The simulation and measured result are Figure 2 where both result simulation and measurement are approximately the same. Measurement result drops at 0.9 GHz, whilst simulation result drops at 0.9 GHz. The antenna copper wire angle of the design hardware will affect the result. These may happen because of manual fabrication and designing hardware where the angle is very difficult to make it 90° which is like a sharp angle as same in simulation design. The length of copper wire of antenna played a significant role because of the curved angle not sharply 90° and possibility length of copper wire is more than the simulation design. From the analysis, slightly different of the length and the curved angle antenna will affect the actual result. Nevertheless, the different is almost neglect able. Return loss value of the measured result shows that it is in good value for the radiation signal where the insertion loss value is less than -10.

3.2 VSWR measurement Vs simulation result.

Shown in figure 3 is VSWR measurement versus simulation result. It is shows that, both of the results are approximately the same. VSWR is defined as the ratio of the maximum voltage to the minimum voltage in standing wave pattern along the length of a transmission line structure. It varies from 1 to (plus) infinity and is always positive. From the result, all the value is positive at the targeted frequency and it is approaching to 1.

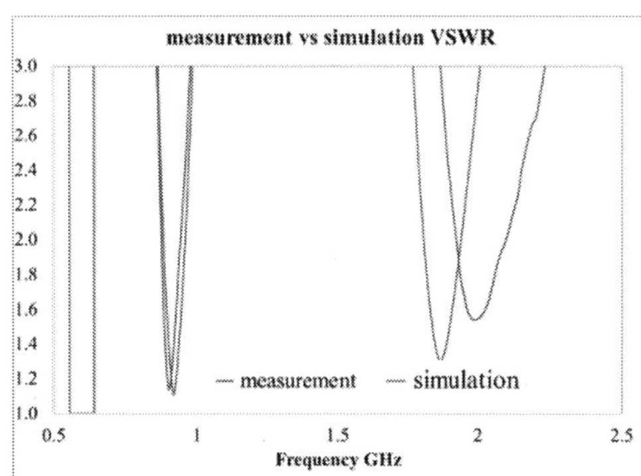

Fig. 3. VSWR Measurement Vs Simulation Result.

3.3 Antenna Gain

Antenna gain is the main performance which is combined the antennas directivity and electrical efficiency. The gain of an antenna is defined as the amount of energy radiated in a particular direction compared to the amount of energy directed in the same direction of an isotropic antenna. For this PIFA antenna gain value are 1.79 dB at 0.9GHz and 4.02 dB at 1.9GHz, Since the gain is sufficiently low, The low gain value will not affect the human body.

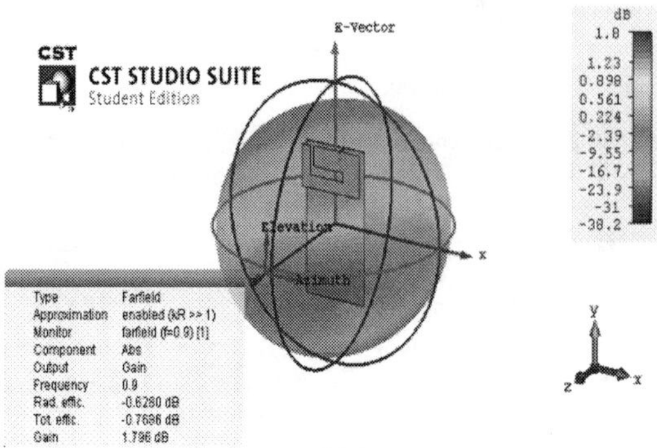

(a)

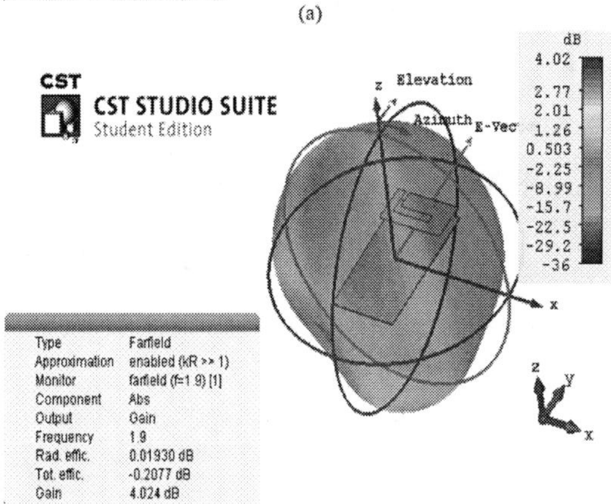

(b)

Fig. 4. PIFA Antenna Radiation Pattern a) at 0.9 GHz, b) at 1.9 GHz.

3.4 Surface current distribution

Figure 5 shows the parts where more current intensity flows at each frequency. For the lower frequency of 0.9 GHz, in the outer part circulates the majority of the current, meanwhile for the superior frequency of 1.9 GHz is in the inner rectangular part. In Figure 6, it is observed the influence of the L-shaped slot in the surface current distribution, where it is more intense the current near the slot for 1.9 GHz.

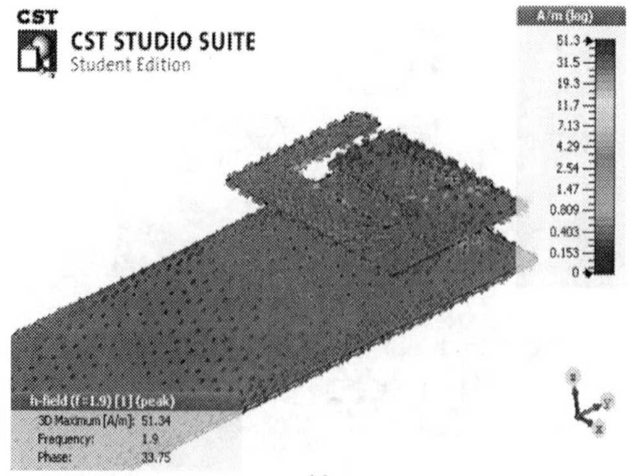

(a)

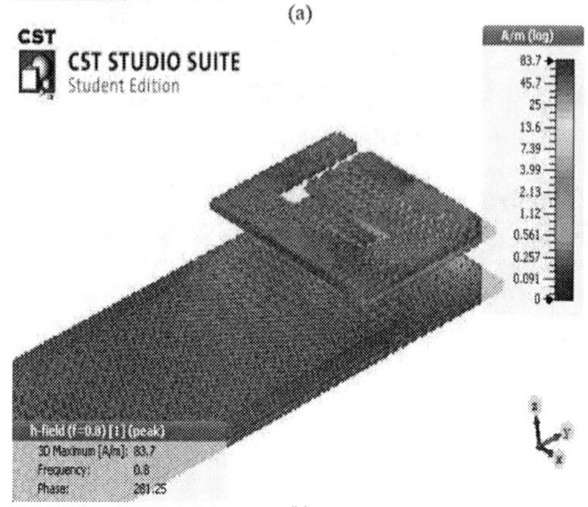

(b)

Fig.5. Surface Current Distribution In The Structure. a) at 1.9 GHz, b) at 0.8 GHz.

3.4 Antenna Efficiency

Efficiency is measured by integrating the total power flux or gain measured over the whole spherical surface. It is measure the radio antenna which is converts the radio frequency power into radiated power. Efficiency is the ratio of power actually radiated to the power put into the antenna terminals. The good percentage efficiency of antenna is must be more than 90% for the better radiation of antenna. This PIFA antenna was in good efficiency with the percentage of efficiency around 90% as shown in Figure 6.

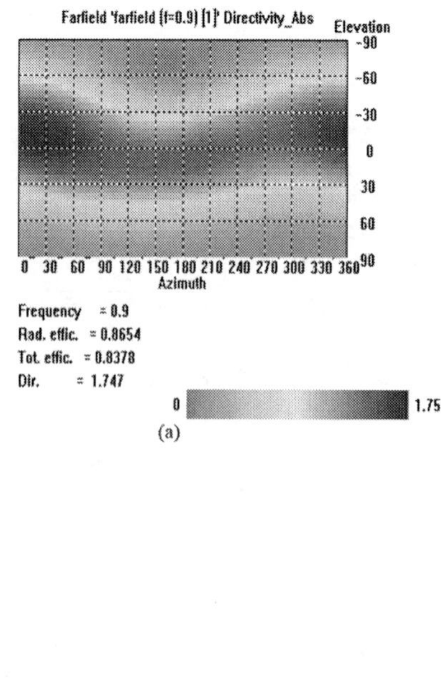

(a)

DOI: 10.1051/matecconf/201714001019

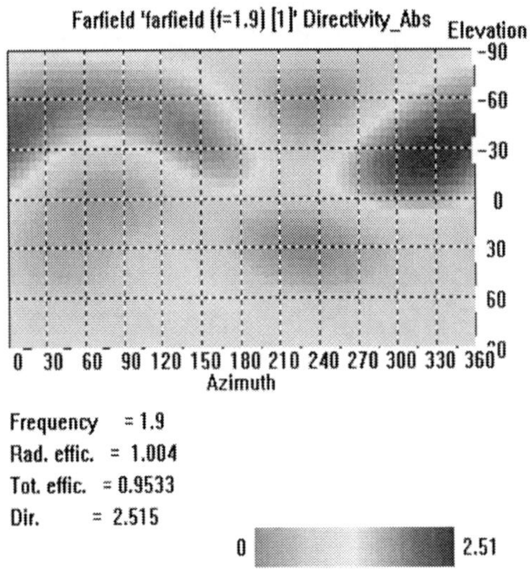

(b)

Fig. 6. PIFA antenna efficiency. (a) at 0.9 GHz, (b) at 1.9 GHz

4 Conclusion

This research is to develop a PIFA antenna that covers the frequency range for mobile phone applications. The antenna is a modification of traditional design of PIFA by creating an L-slot that covers GSM (876-960MHz) and (1710-1889 MHz) frequency band. The antenna was tested for suitability in mobile telephony applications, analysis of radiation patterns were made and gain and efficiency. It was observed that the radiation patterns were omnidirectional in the two planes and thus makes the antenna very suitable for mobile phone applications. Vertical and horizontal polarization effects were analysed due to the need of the phone working correctly regardless of the antenna's orientation. The radiation patterns indicated a high front to back ratio, this could point to the fact that the antenna produces a low Specific Absorption Rate value.

References

1. P. Lindberg, "Wideband active and passive antenna solutions for handheld terminals", Uppsala University, 2007.
2. N. Misran, M. M. Yunus and M.T. Islam, " Small Dual-Band Planar Antenna with Folded Patch Feed", Journal of Applied Sciences Research, 6(12), 2010.
3. K. QuLuhaibboa, S. Luhaib and B. Abaoy, "Design and Simulation Dual-Band PIFA Antenna for GSM Systems", International Multi-Conference on Systems, Sygnals & Devices, pp. 1-4, 2012.
4. D. G. Kang and Y. Sung, "Compact hexabandpifa antenna for mobile handset applications," Antennas and Wireless Propagation Letters, IEEE, vol. 9, pp. 1127-1130, 2010.
5. Y. Nie and L. Song, "A Compact Triband Fractal PIFA Antenna for Mobile Handset Applications", 2013.
6. J. Chen, L. Ying and Y. Ban, "Compact PIFA using capacitive coupled-fed for LTE/GSM/UMTS WWAN operation in the mobile application", vol. 1, pp. 414 - 417, 2011.

MATEC Web of Conferences **140**, 01020 (2017) DOI: 10.1051/matecconf/201714001020
ICEESI 2017

Optical Generation of 80 GHz Downstream Data in Radio over Fiber System Based on Stimulated Brillouin Scattering

M.M.H Husaini, Mohd Rashidi Che Beson, M.S Anuar, R. Endut and Aljunid Syed A.*

Advanced Communication Engineering, Center of Excellence, School of Computer and Communication Engineering, Universiti Malaysia Perlis (UNIMAP), Perlis, Malaysia

Abstract. In this work, we proposed Stimulated Brillouin Scattering (SBS) in an optical generation of millimeter waves (mm-waves) for radio over fiber (RoF) system with the transmission of 2.5 Gbps data rate. SBS method is achieved by imposing continuous wave (CW) laser in a single mode optical fiber (SMF) via Mach-Zehnder Modulator (MZM) with Erbium-Doped Fiber Amplifier (EDFA) and two pump lasers. Mm-waves signal of 80 GHz was generated from a 40 GHz radio frequency (RF) synthesizer. Simulation result and the analysis comparison of effective light power are investigated with different SMF loop length from 20 km until 120 km. The performance of SBS system is analysed and the generated of millimeter waves signal can be transmit with length up to 120 km based on eye pattern and Bit Error Rate (BER) curve.

1. Introduction

Radio-over-Fiber (RoF) is a fundamental property of communication which light propagates by radio frequency (RF) and transmitted over an optical fiber link [1]. RoF are promising significant advantages such as immune to electromagnetic interference, offers large bandwidth, low attenuation and low cost which these system can be used in various types of application such as Wi-Fi, sensor networks, radar and so on[2]. Optical generation of mm-waves signals is one of the method that would execute the RoF communication system[3]. Mm-waves can be named electromagnetic spectrum that ranges between 30 GHz to 300 GHz which capable to overcome the issue of spectral congestion at lower frequency ranges[4] .

Over the past few years, many optical generation of mm-waves signal approaches have been employed. These approaches comes with advantages and drawbacks as depicted in table 1.

From the table 1, one of the simplest technique based on external modulator has been widely used since it is low cost and high reliability but there are several problem occurs with chromatic dispersion of mm-waves generation signal[5]. Single sideband (SSB), double sideband (DSB) and optical carrier suppression (OCS) are external modulator schemes that have been demonstrated, DSB modulation bear with fading effect which is not suitable for long transmission distance while SSB and OCS requires dual arm MZM to drive optical source. In [6], generate OCS mm-waves signal based on two parallel Dual-Drive MZMs is proposed. However, complicated setup and generated mm-waves produces high BER at 60 km transmission distance.

Table 1. Comparison of technique in mm-waves signal generation

Method	Advantages	Drawbacks
Direct modulation [7]	Low cost and simplest scheme.	Low frequency response and limited by the laser chirp.
External modulation [8][9][10]	High frequency response and scalability	Requires optical filter and larger insertion loss
Optical heterodyne [11]	High spectral purity	Quality of the mm-waves relies on the coherence of the two laser lightwaves.

Thus, an alternative method is proposed which is Stimulated Brillouin Scattering (SBS) to overcome above mention problem. Stimulated Brillouin Scattering (SBS)

* M.M.H Husaini: mhhusaini15@gmail.com

DOI: 10.1051/matecconf/201714001020

has generally been perceived as one of the causes degrading framework execution in fiber optic systems because of the impact that the signal energy is exchanged to the back scattering signal and SBS has low threshold. In any case, SBS has advantages qualities, for example, frequency selective amplification that can be boost up to microwave and mm-waves photonics applications[12].

1.1 Principle of SBS

SBS is a nonlinear impact due to the measure of light backscattered and the measure of light transmitted by the fiber does not depend directly on the power input to the fiber. In depth, low input power of the backscattering is influence by straightforward Brillouin and Rayleigh scattering which are direct and vary from each other by brillouin shift. When the power increased, the brillouin scattered light is amplified by the output process. The backscattered light slightly increased with the increasing input power until it initiate most of the input light at SBS threshold also known as power level. Hence, the SBS threshold can lie in the span of 6-10 dBm for single mode fiber (SMF) with transmission distance greater than 10 km. Over this threshold, the input power is not independent of the insertion loss of the fiber.

1.2 Proposed of SBS system Model

The optical generation of mm-waves utilizes SBS technique is shown in fig. 1. A Continuous Wave (CW) laser at 1550 nm is intensity-modulated via MZM driven by the electrical sine generator, F_{rf} = 40 GHz. Dual series of MZMs component is used which MZM 1 modulates the optical source and MZM 2 is modulated via pseudo-random bit sequence (PRBS) at 2.5 Gbps NRZ electrical signal to generate baseband data signal. The voltage that is connected on the MZM is sufficiently high so that the laser wave is regulated nonlinearly with the applied frequency of the electrical generator.

A few optical sidebands isolated by frequency of electrical generator from the optical carrier are created by MZM non linearity. The transmission distance are connected into up to 120 km long of Single Mode Fiber (SMF) loop. The optical signal is then amplified using an erbium-doped fiber amplifier (EDFA) before being launched into the optical fiber loop, EDFA act as control power of signal wave which generated sidebands by nonlinear modulation from CW laser will be amplified by SBS in an optical fiber, whereas the rest will be attenuated due to natural attenuation in the fiber[13].

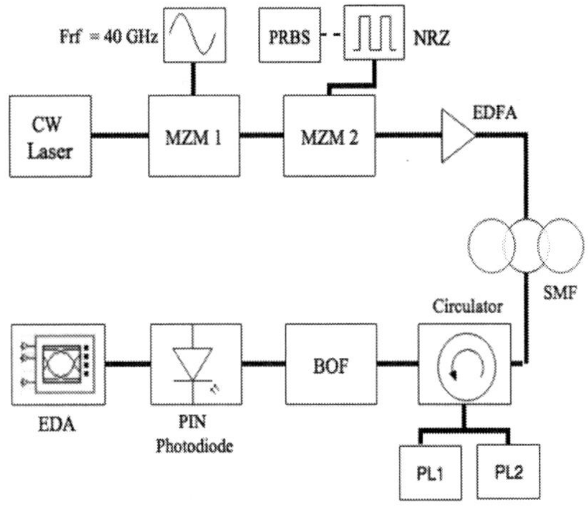

Fig. 1. Experimental setup for optical mm-waves signal generation based on SBS technique.
(CW: Continuous Wave, MZM: Mach-Zehnder Modulator, EDFA: Er.doped Fiber Amplifier, SMF: Single Mode Fiber, PL1 & PL2: Pump Laser, BOF: Bessel Optical Filter, EDA: Eye Diagram Analyzer)

SBS effects gain in the SMF is creating from combined output signal of two pump lasers which pump source driven via an optical circulator that propagates modulate signals through opposite direction. The wavelength of each pump laser is set to 980 nm. Hence, generated mm-waves output
is detected by a photodiode. The execution of this RoF system are utilized by eye diagram analyzer and Bit Error Rate (BER) curve.

Theoretically, the generation of mm-waves frequencies rely on the RF of the electrical generator and on the sidebands that were picked for amplification. The generated mm-waves can be expressed as :

$$f_{mm} = 2nf \quad [14]$$

where n as the number of the sideband used and f as the RF of the electrical generator. If the f =5 Ghz, generated frequencies will be 10,20,30 and so on can be deploy.

2. Result and Discussion

In this section, the generated optical mm-waves for SBS system is validate by using OptiSystem software package. All necessary simulation setup value is shown in the following table.

DOI: 10.1051/matecconf/201714001020

Table 2. General parameter for simulation

Parameters	Value
Freq. of electrical generator	40 GHz
Reference wavelength of CW laser	1550 nm
Effective light power of CW laser	0,-0.5,5 dBm
Linewidth of CW laser	1 MHz
Samples per bit	64
Bit rate	2.5 Gbps
EDFA	5 m
Optical fiber length	50 KM
Attenuation of Optical Fiber	0.2 db/km
Dispersion of Optical fiber	16.75ps/nm/km
Sensitivity	-100 dBm
Resolution	0.1 nm

The model of SBS system is constructed as shown in table 2, whereas these output signal is measured by the eye diagram analyser that indicate which is best performance of the system can be determine by the wide eye opening.

The demodulation analysis of 2.5 Gbps of the 80 Ghz mm-waves signal are presented when vary SMF loop length within range 20 km until 100 km with EDFA and without EDFA. As can be seen in fig. 2(a-f), for transmitted signal over 20 km, the eye pattern shows both techniques are in good performance but when optical signal mm-waves propagating over 50 km to 100 km distances, the outline of the simulated eye diagram gradually decreases without optical amplifier technique. Hence, the optical amplifier and SBS effect excites the power occur at 20 km then performance degraded over longer fiber transmission.

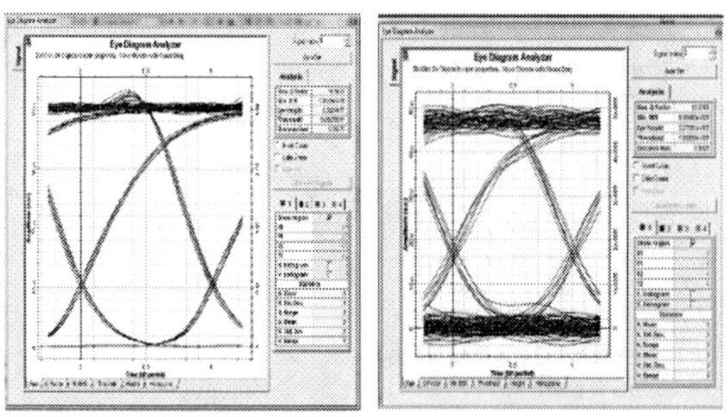

a) b)

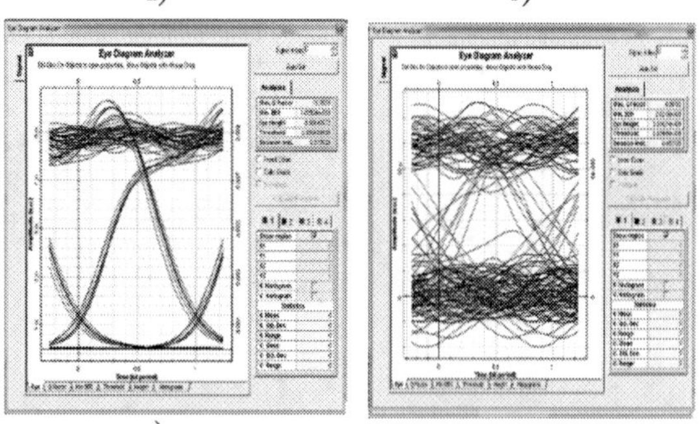

c) d)

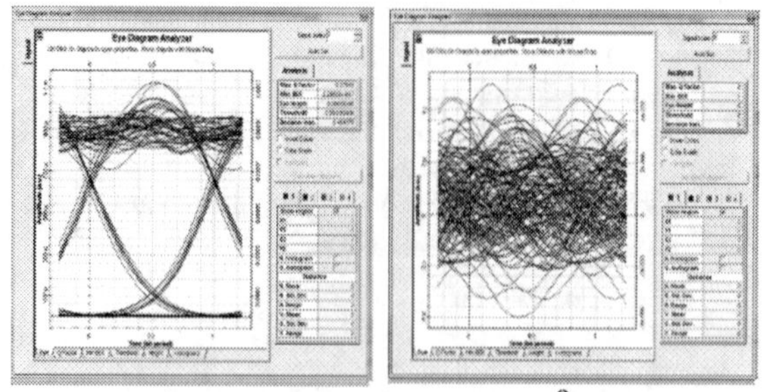

e) f)

Fig. 2(a-f). Simulated eye diagram at receiver for 2.5 Gbps data rate with different SMF transmission distances a) 20-km with amplifier b) 20-km without amplifier c) 50-km with amplifier d) 50-km without amplifier e) 100-km with amplifier f) 100-km without amplifier.

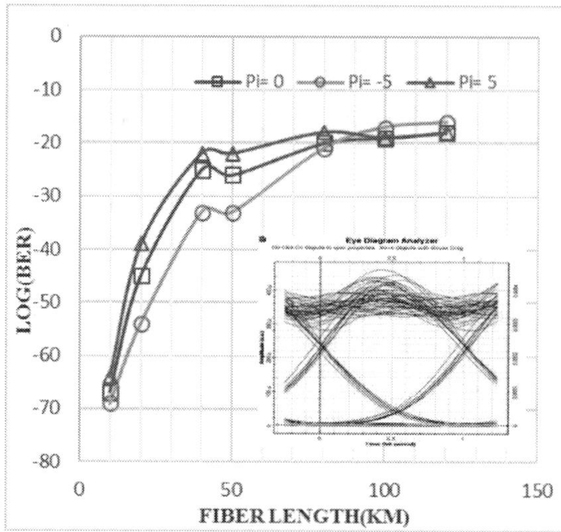

Fig. 3. Measured BER versus fiber length with different transmission distance (inset shows the eye diagram at 120 km for -0.5 dBm of effective light input power).

Fig. 3 compares the experimental data on the effective light power, P_i when varies 0, -0.5 and 5 dBm of the downstream 2.5 Gbps signal with different optical fiber transmission distance. From the graph above, we can see that, the value of BER displays an increasing pattern over longer fiber transmission distance due to attenuation. BER is approximately at (10^{-33}) at 50 km transmission distances with effective light power, P_i= -0.5 dBm. When the effective light power varies to 0 dBm and 0.5 dBm, the system presents high BER with (10^{-25}) and (10^{-20}) respectively. This SBS method capable to perform the system until 120 km transmission distance with the BER is (10^{-19}). Thus, a larger dispersion and attenuation will occur over the longer fiber length as well as increasing the BER.

3. Conclusion

In this paper, we have demonstrated 80 Ghz optical mm-waves signal and transmission of a 2.5 Gbps downlink stream in RoF system based on SBS in SMF. The proposed method of 80 Ghz signal were generated from a 40 Ghz RF oscillator corresponding to theoretical equation. The analysis of the system performance have been made with eye diagram visualizer which eye pattern decreased over long range optical fiber length. This scheme successfully propagates over 120 km in optical fiber length with low BER value at (10^{-19}) from -0.5 dBm of light power. Thus, it is strongly proven that the proposed technique is essential for long range wireless communication networks system.

References

[1] M. S. Transmission and P. C. Systems, "Millimeter-Wave Subcarrier Transmission," no. c, pp. 555–558, 1992.

[2] M. R. C. Beson, S. A. Aljunid, F. Ghani, M. S. Anuar, and H. A. Fadhil, "Alleviation noise effect using flexible cross correlation code in spectral amplitude coding optical code division multiple access systems," *J. Commun.*, vol. **8**, no. 2, pp. 101–109, 2013.

[3] T. A. Chowdhury, M. A. Ghuri, and R. Kibria, "Millimeter Wave Generation Based on Optical Frequency Multiplication in Radio over Fiber Systems," pp. 94–97, 2016.

[4] C. B. M. Rashidi, S. A. Aljunid, F. Ghani, H. A. Fadhil, and M. S. Anuar, "Cardinality enhancement using Flexible Cross Correlation (FCC) code for Spectral Amplitude Coding Optical Code Division Multiple Access systems," *J. Appl. Sci. Res.*, vol. **8**, no. 12, pp. 5614–5626, 2012.

[5] G. H. Smith, "Overcoming chromatic-dispersion effects in fiber-wireless systems incorporating external modulators," *IEEE Trans. Microw. Theory Tech.*, vol. 45, no. 8 PART 2, pp. 1410–1415, 1997.

[6] N. A. Mohammed, S. I. S. Hassan, F. Malek, R. Ngah, and S. A. Abbas, "Optical generation of 60-GHz signal for millimeter wave wireless communication," *RFM 2013 - 2013 IEEE Int. RF Microw. Conf. Proc.*, vol. **0**, no. 1, pp. 437–440, 2013.

[7] H. Shams and J. Zhao, "First Investigation of Fast OFDM Signals at 60GHz using Direct Laser Modulation," *CLEO/Europe-EQEC 2013*, vol. **487**, no. 318415, p. 318415, 2013.

[8] J. Ma, J. Yu, C. Yu, X. Xin, J. Zeng, and L. Chen, "Fiber dispersion influence on transmission of the optical millimeter-waves generated using LN-MZM intensity modulation," *J. Light. Technol.*, vol. **25**, no. 11, pp. 3244–3256, 2007.

[9] N. A. Al-shareefi, S. I. S. Hassan, F. Malek, R. Ngah, and S. A. Abbas, "Optical Generation of 60 GHz Downstream Data in Radio over Fiber Systems Based on Two Parallel Dual-Drive MZMs," vol. 6, no. **2**, pp. 579–587, 2014.

[10] C. Jiang, X. Chen, and Z. Liu, "Effect of Modulation Index of Mach-Zehnder Modulator on Single Sideband Radio over Fiber System," *Proc. 2012 2nd Int. Conf. Comput. Inf. Appl. (ICCIA 2012)*, no. Iccia, pp. 574–577, 2012.

[11] G. J. Simonis and K. G. Purchase, "Optical generation, distribution, and control of microwaves using laser heterodyne," *IEEE Trans. Microw. Theory Tech.*, vol. **38**, no. 5, pp. 667–669, 1990.

[12] N. M. Nawawi and S. M. Idrus, "Investigation of stimulated brillouin scattering for the generation of millimeter waves for radio over fiber system," *Proc. IEEE 2008 6th Natl. Conf. Telecommun. Technol. IEEE 2008 2nd Malaysia Conf. Photonics, NCTT-MCP 2008*, no. August, pp. 33–36, 2008.

[13] M. Junker, T. Schneider, M. J. Ammanno, and A. T. Schwarzbacher, "Carrier Generation in the Millimetre Wave Range based on Stimulated Brillouin Scattering for Radio over Fibre Downlink Systems," pp. 191–195, 2006.

[14] T. Schneider, D. Hannover, and M. Junker, "Investigation of brillouin scattering in optical fibers for the generation of millimeter waves," *J. Light. Technol.*, vol. **24**, no. 1, pp. 295–304, 2006.

 DOI: 10.1051/matecconf/201714001021

Comparison between Conventional OCDMA and Subcarrier Multiplexing SAC OCDMA System Based on Single Photodiode Detection

N. A. A Ahmad, *M.* N Junita, Syed Alwi Aljunid, Mohd Rashidi Che Beson, Rosdisham Endut

Advanced Communication Engineering,
Centre of Excellence,
School of Computer and Communication Engineering,
Universiti Malaysia Perlis,
Arau, 01000, Perlis
Malaysia.
n.amirahamniahmad@gmail.com

Abstract. This paper demonstrates the comparison between conventional OCDMA system and subcarrier multiplexing (SCM) SAC-OCDMA system by applying Recursive Combinatorial (RC) code based on single photodiode detection (SPD). SPD is used in the receiver part to reduce the effect of multiple access interference (MAI) which contributes as a dominant noise in incoherent SAC-OCDMA systems. From this analysis, the performance of SCM OCDMA network could be improved by using lower data rates and higher received power. The hybrid SCM OCDMA system shows better performance compare to conventional OCDMA system although the number of users involved is very high. This is because, for hybrid SCM OCDMA system, the number of users can be increased by increasing the number of subcarriers without affect the number of code length and optical codes. Increasing the number of subcarriers will enhance the power consumption by applying hybrid SCM system in OCDMA compared to the conventional OCDMA system. This is because increasing the number of users for hybrid SCM system does not affects the number of code length and the number of optical codes but only increase the number of subcarriers. Thus, hybrid SCM OCDMA system has to increase spectral efficiency and produce better performance compared to conventional of OCDMA system.

1 Introduction

Generally, there are three types of multiplexing techniques in optical system such as Optical code division multiple access (OCDMA), Wavelength Division Multiplexing (WDM) and optical Time Division Multiplexing (TDM). OCDMA technique is a method of sharing the bandwidth of optical fiber among the number of users in the local area network [1] meanwhile WDM technique and TDM technique are applicable for long-haul networks. In addition, both of TDMA and WDMA techniques required deterministic wavelength assignment and strict synchronous time-slot control, respectively[2]. Besides that, OCDMA becomes attention because of their benefits such as it capable to uphold asynchronous access network, dynamic bandwidth assignment, all optical processing, and the ability to support multimedia services. These benefits are too important for implementation of local area network.

Basically, detection is a vital part of the system as the design of transmitter and receivers depends on the type of detection. The detections are divided by two types which are coherent and incoherent [3] but mostly system applied incoherent detection. This is because incoherent detection has numerous benefits compared to coherent. For instance, the hardware complexity of incoherent detection is less rather than coherent detection. Besides that, the phase information knowledge is needed when coherent detection sending the signal while for incoherent detection does not need the phase information knowledge. In addition, for coding operation, coherent detection performed in bipolar approach meanwhile incoherent detection performed in unipolar approach [4]. Coherent technique is difficult to apply compared to incoherent technique. Thus, in this research, we applied incoherent detection.

However, the cross correlation function is always generated in the incoherent code words which give rise to the multiple access interference (MAI) in the system. Then, in order to reduce the effects of MAI in the OCDMA system, the detection technique is proposed. There are many types of detection techniques but the

DOI: 10.1051/matecconf/201714001021

well-known detection techniques are AND subtraction detection technique, direct detection technique, spectral direct detection technique (SDD), single photodiode detection technique and NAND subtraction technique. In this paper, we applied SPD technique to reduce the effects of MAI in the hybrid SCM system.

The effects of MAI can limit the performance of hybrid SCM OCDMA system especially when employing large number of users. So, to reduce the effects of MAI, SAC-OCDMA system needs to employ code sequence with fixed in-phase cross correlation. SAC-OCDMA systems are receiving more attention because of their ability to completely reduce multi access interference. There are numerous of code families have been developed for SAC-OCDMA such as modified quadratic congruence codes, M-sequence codes, double weight codes, etc but in this word we applied recursive combinatorial code. A part from that, the SCM scheme is capable of improving the channel data rate of OCDMA systems. This hybrid scheme is proposed for the benefit of combining the advantages of both schemes. The use of hybrid Radio Frequency technique over SCM-spectral amplitude coding (SAC) Optical Code Division Multiple Access (OCDMA) is proposed to carry the high data over fiber link and to increase the number of users. This enables the reduction of Multi Access Interference (MAI) with high data rate, leading to the increase in the number of active users. Consequently, the hybrid system is robust against interference and is much more spectrally efficient.

We carried out this research by using a hybrid SCM technique by employing SAC OCDMA code sequences. By employing SCM technique, it will increase the spectral efficiency of the OCDMA system. In addition, the number of users can be increase by increasing the number subcarrier without effects the number of code length and number of SAC-OCDMA codes.

2 Recursive Combinatorial Code

Recursive Combinatorial (RC) code was introduced and developed by M. N Junita [5] in 2004. The advantages of RC codes are, it has minimum cross correlation, which is one and minimum code lengths compare to other codes. Besides that, RC code is flexibility in choosing code parameter. It can be shown when adding the number of users in hybrid system, it will not disturb the code words given for existing users, since new code sequence for additional users are added at the next available bandwidth. A part from that, increasing the number of weight will not affect other parameter such as the number of users, K. Thus, it will not affect the code sequences for every existing user; i.e. increasing the code weight does not affect the chips positioning

3 Single Photodiode Detection (SPD)

In OCDMA systems, the detection schemes affect the design of transmitter and receivers. The types of detection are divided by two which are coherent system and incoherent system. Incoherent system is unipolar sequence for the code sequence whereas coherent system is bipolar sequence for the code words. In our system, we applied incoherent system since the coherent system is phase sensitive, which would be practically difficult to implement in this system.

Figure 1 illustrates the implementation of SCM OCDMA structure at the transmitter part. Here, data from users are modulated into electrical signal within different specific frequencies. These RF frequencies are then combined using electrical adder before being modulated into optical signal codes via a Mach Zhender Modulator. The optical RC codes are executed by splitting the broadband optical sources into smaller wavelengths, and encoded it back in the form of codes using an encoder. Optical combiner is then used to combine the optical codes before being transmitted via a single mode fiber.

Figure 2 shows the implementation of SPD detection at the receiver. The received optical signal is decoded by the decoder, which has an identical spectral response to the intended encoder for the data to be received. The remainder of the signal from the decoder is then transmitted to the subtractive decoder (s-Decoder) to cancel out signals with mismatched interferers. The s-Decoder contains only frequency bins from different interferers. The output from the s-Decoder is either zero power unit for active user or cross-correlation power unit for interferers.

After optical subtraction, the output is either code weight power unit for active user or zero power units for interferers. This implies that the interference signals are suppressed in the optical domain before the conversion of the signals to the electrical domain, as a result, the proposed SPD scheme alleviates both PIIN and MAI in the optical domain. Moreover, the two interference signals at the optical subtractor are assumed to be equal and cancel each other out. However, practically, the interference signals differ slightly at the optical subtractor and results in a small amount of optical power to reach the photodiode.

The main advantage of using the SPD is that the cancellation of the interference signals in the optical domain allows the use of only a single photodiode rather than two photodiodes as in typical subtraction detection schemes. This reduces the amount of optical-to-electrical conversion and shot noise generated at the receiver part [1].

At the photodetector, the photo diode only detects and filters the clean chips or the non-overlapping chips. Then, the effects of MAI are successfully reduced in the optical domain because there is only the required signal

 DOI: 10.1051/matecconf/201714001021

spectra are filtered. Then, there is no subtraction process in SPD technique. It is because; the photo detector only detects the clean optical pulses and converted into electrical domain. The electrically modulated signal is then passed through a Band Pass Filter (BPF) to remove the unwanted frequency. In order to remove the higher frequency, a low pass filter (LPF) is assigned. The original data signal is detected at the output of LPF using a BER analyser.

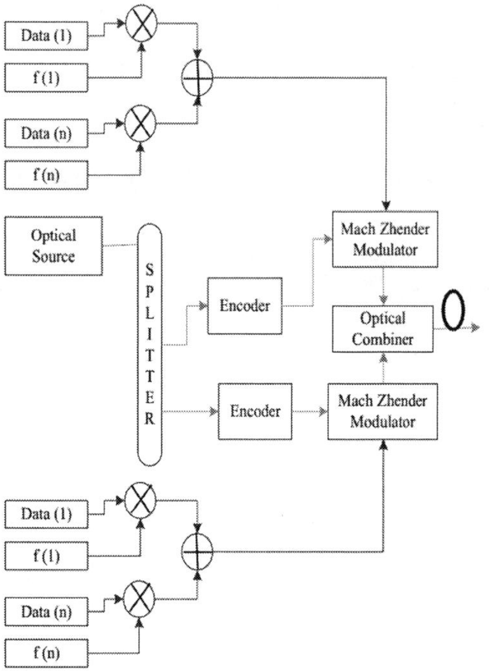

Fig. 1 : Transmitter part of hybrid SCM SAC- OCDMA system

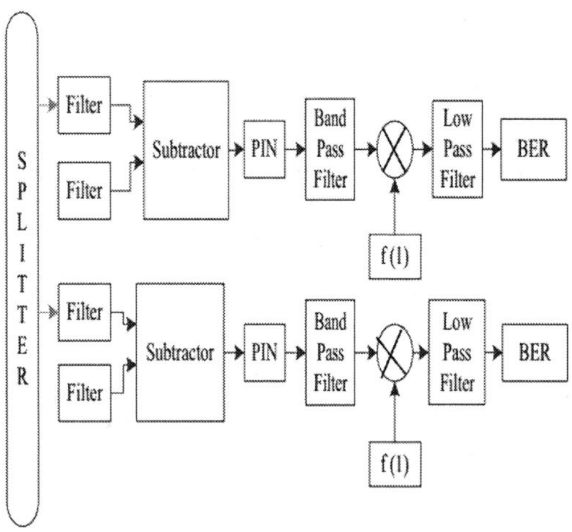

Fig. 2 : The receiver part hybrid SCM SAC-OCDMA system based on SPD detection technique

4 Theoretical Development of Hybrid SCM OCDMA System Based on SPD Technique

The performance of RC code in hybrid system can be analysed by describing the properties of RC code which are the cross correlation and auto correlation function of the code take into account. In equation 1, $C_d(i)$ represents the ith element of Ath RC code sequence. So that, the properties of RC code [5] based on SPD method can be illustrated as:

$$\sum_{i=1}^{L} C_d(i)C_e(i) = \begin{cases} W; for\ d=e \\ 1; for\ d \neq e \end{cases} \quad (1)$$

Where C_d and C_e depict the RC code sequences, meanwhile L depicts the code length and W depicts the number of the weight. At the receiver part, the photo detector only detects and decodes the non-overlapping chips. Therefore, the total incident power during one period at the input of the photo detector *PIN 1* for the SPD detection technique can be expressed by:

$$\int_0^\infty G_{dd}(v)dv = \int_0^\infty \frac{P_{sr}}{\Delta v}\sum_{k=1}^{K} d_k \sum_{i=1}^{L} C_d(i)\, C_e(i) u\left[\frac{\Delta v}{L}\right] dv \quad (2)$$

$$= \frac{P_{sr}}{\Delta v}\left[\frac{\Delta v}{L}\right]\sum_{k=1}^{K} d_k \sum_{i=1}^{l} C_d(i)C_e(i) \quad (3)$$

$$= \frac{P_{sr}}{L}(w-1) \quad (4)$$

The photodiode current, I indicate the signal from the desired user and can be displays as:

$$I = I_{dd} = \Re\int_0^\infty G_{dd}(v)dv \quad (5)$$

Then, substitute the equation (4) in (5), the equation becomes:

$$I_{dd} = \Re\frac{P_{sr}}{L}(w-1) \quad (6)$$

Based on equation (6), (W-1) represents the power spectral density of the overlapping chips or cross correlation are subtracted at the received power signal. Thereby, the photodiode only detects and filters the PSD of the non-overlapping chips.

For the noise variances, there are three types of noises are considered which are shot noise, thermal noise and intermodulation noise while PIIN noise is neglected. The PIIN noise is neglected because it only appearance when the optical signal is incident more than one on the photo detector surface. The shot noise can be expressed as:

$$I_{shot}^2 = 2eB\, I_{dd}$$

$$I_{shot}^2 = 2eB\, \Re \frac{P_{sr}(w-1)}{L} \quad (7)$$

The thermal noise [6] can be expressed as:

$$I_{thermal}^2 = \frac{4K_b T_N B}{R_L} \quad (8)$$

The inter-modulation noise [7] can be expressed as:

$$I_{IMD}^2 = P_{sr}^2\, \Re^2 m_{n,k}^6 \left[\frac{D_{111}}{32} + \frac{D_{21}}{64}\right] \quad (9)$$

The Signal Noise Ratio or SNR equation of the hybrid SCM OCDMA using RC code using SPD method can be expressed as:

$$SNR = \frac{\left(\frac{\Re P_{sr}(W-1)}{L}\right)^2 m_{n,k}^2}{eB\Re\frac{P_{sr}(W-1)}{L} + \frac{4K_b T_n B}{R_L} + P_{sr}^2 \Re^2 m_{n,k}^6 \left[\frac{D_{11}}{32} + \frac{D_{21}}{64}\right]} \quad (10)$$

Based on equation (10), the formula to calculate the bit error rate (BER) can be written as:

$$\text{BER} = \frac{1}{2}\, erfc\left(\sqrt{\frac{SNR}{8}}\right) \quad (11)$$

Here, $erfc = \frac{2}{\sqrt{\pi}} \int_x^{\infty} \exp(-z^2)\, dz$ (12)

Table 1 : List of parameter used in the analysis

Symbol	Parameter	Value
η	The efficiency of photodetector quantum	0.6
Δv	The source of Linewidth broadband	3.75THz
λ_0	The wavelength of operating	1550 nm
B	Electrical bandwidth	80/331 MHz
R_b	Data Bit Rate	155/622 Mbps
T_n	The temperature of Receiver noise	300K
R_L	Receiver load resistor	1030 Ω
E	Electron charge	1.6 x 10^{-19} C
K_b	Boltzmann's constant	1.38 x 10^{-23} J/K

5 Analysis of Results and Discussion

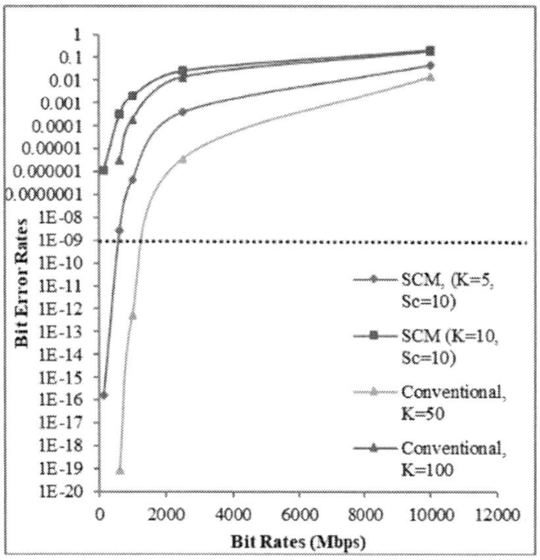

Fig. 3: The performance of bit rates versus BER at number of users, K=50 and K=100

From Figure 3, it indicates the comparison of performance of conventional OCDMA system and hybrid SCM SAC OCDMA system with respect to the different number of bit rates for active users equal to 50 and 100. The bit rates are varied from 155Mbps to 10 Gbps. As observed in Figure 3, increasing the number of bit rate will degrades the BER performance for the both system. When higher the number of bit rate, the pulse width will be lower, thereby the signal are more sensitive to the noise. Generally, increasing the number of bit rates will generate greater electrical bandwidth and then can results in greater noise. However, hybrid SCM OCDMA system indicates better performance compared to conventional OCDMA system although the number of users involved is very high. This is because, for hybrid SCM OCDMA system, the number of users can be increase by increase the number of subcarriers without effect the number of code length and optical codes.

DOI: 10.1051/matecconf/201714001021

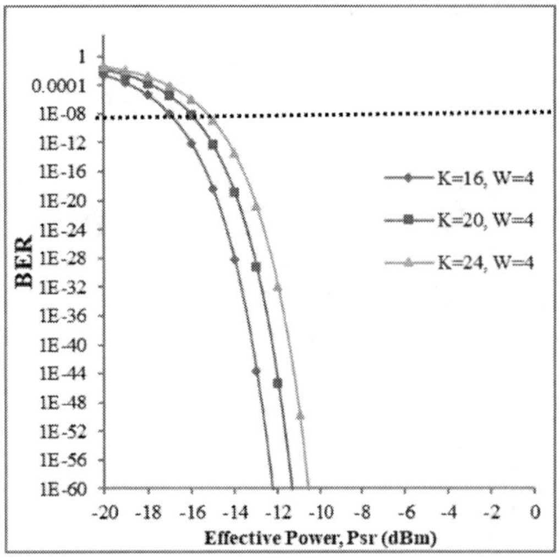

Fig. 4 : Performance of hybrid system for BER versus effective transmitted power, Psr for conventional OCDMA system.

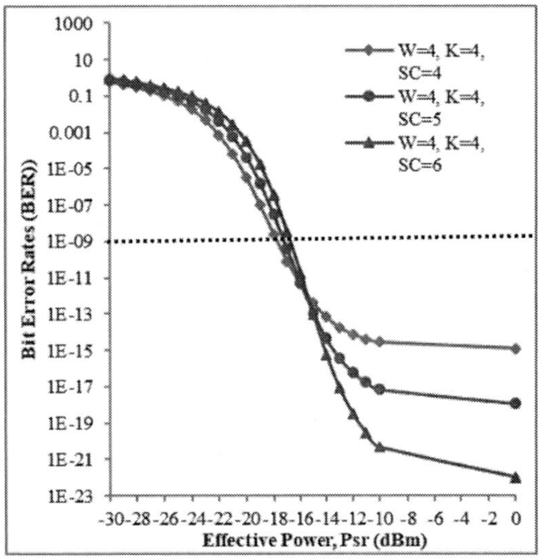

Fig. 5 : The performance of hybrid SCM OCDMA system with respect to effective power at the different number of subcarriers.

From Figure 4 and Figure 5, it depicts the effect of effective power at the photo detector versus BER. Figure 4 presents the effects of received power with varying the number of optical channel meanwhile Figure 5 indicates the effects of received power with varying the number of subcarriers. From Figure 4 and 5, it can be conclude that, the performance of BER value degrades as the number of received power decrease. The BER can degrade by some order of magnitude eventhough a tiny fall happened in the optical signal power.

Based on Figure 4, it demonstrates the performance of conventional OCDMA system with SPD detection. At the threshold BER value of 10^{-9}, the effective power for 16 optical is about -17 dBm, for 20 optical is approximately -16 dBm and for 24 optical is about -14 dBm. So, the power different for conventional OCDMA system in order to get from 16 channels to 24 channels is about 3 dBm

Besides that, Figure 5 demonstrates hybrid SCM OCDMA system by varying the number of subcarriers and constant the number of channels. For four number of subcarriers channel, the received power is about -19 dBm, meanwhile for five subcarriers is about -18 dBm, for five subcarrier channel is about -17 dBm and for six subcarriers, the received power is about -16 dBm respectively. So, the power different is about 2 dBm in order to get from 16 channel to 24 channels by utilized hybrid system. From this analysis, it can be conclude that the power range for Figure 5 is smaller than power range in Figure 4. Therefore, it is clearly presents, increasing the number of subcarriers will enhanced the power consumption by applying hybrid SCM system in OCDMA compared to the conventional all optical system in OCDMA. This is because, increasing the number of users for hybrid SCM system does not effects the number of code length and the number of optical codes but only increase the number of subcarriers. Thus, hybrid SCM OCDMA system has increase spectral efficiency compared to conventional of OCDMA system.

6 CONCLUSION

This paper demonstrates the comparison of conventional OCDMA system between hybrids SCM OCDMA system by applying SPD technique. The implementation of SPD technique can evaluate the performance of hybrid SCM system by employing Recursive Combinatorial code (RC). The results revealed that, it will get better performances if the number of bit rates is lower and the number of received power is higher. Besides that, hybrid SCM OCDMA system shows good performance compare to conventional OCDMA system. As the bit rates increase, it will decrease the bandwidth of the system, but for hybrid SCM OCDMA system, it still shows better performance compare to conventional OCDMA system. For the receives power, the power consumption for hybrid SCM OCDMA system also lower compared to conventional OCDMA system. This is because hybrid SCM OCDMA system can supports large number of users by increasing the number of subcarrier without effects the number of code length and optical codes. In addition, its bandwidth becomes more efficient. Thus, it can be conclude that, hybrid SCM OCDMA system shows better performance compared to conventional OCDMA system.

DOI: 10.1051/matecconf/201714001021

ACKNOWLEDGEMENT

The authors would like to thank the Ministry of Higher Education (MOHE) Malaysia for the funding under the Fundamental Research Grant Scheme (FRGS).

REFERENCES

[1] S. G. Abdulqader, S. A. Aljunid, H. M. R. Al-khafaji, and H. A. Fadhil, "Enhanced performance of SAC-OCDMA system based on SPD detection utilizing EDFA for access networks," *J. Commun.*, vol. 9, no. 2, pp. 99–106, 2014.

[2] N. A. M. Ariff, A. Abdullah, and M. F. Nasrudin, "Advanced Computer and Communication Engineering Technology," *Lect. Notes Electr. Eng.*, vol. 362, pp. 1209–1217, 2016.

[3] I. Rao and D. Rana, "Comparison of Different Detection Techniques Based on Enhanced Double Weight Code in Optical Code Division Multiple Access System," pp. 126–132, 2015.

[4] R. Sahbudin, S. Aljunid, M. Abdullah, M. B. A. Samad, M. Mahdi, and M. Ismail, "Comparative Performance of Hybrid SCM SAC- OCDMA System Using Complementary and AND Subtraction Detection Techniques," *Iajit*, vol. 5, no. 1, pp. 61–65, 2006.

[5] J. M. Nordin, S. A. Aljunid, A. M. Safar, A. R. Arief, R. A. Rahim, and R. B. Ahmad, "Design Construction and Performance of Recursive Combinatorial Code for Enhancement in Optical Code Division Multiple Access Systems," *Aust. J. Basic Appl. Sci.*, vol. 8, no. 1, pp. 343–350, 2014.

[6] Somaya A. Abd E, Mottaleb, Heba A. Fayed, Ahmed Abd El Aziz, and Moustafa H.Aly, "SAC- OCDMA System Using Different Detection Techniques," *IOSR J. Electron. Commun. Eng.*, vol. 9, no. 2, pp. 55–60, 2014.

[7] T. H. Abd, S. A. Aljunid, H. A. Fadhil, I. F. Radhi, R. B. Ahmad, and M. A. Rashid, "Performance improvement of hybrid SCM SAC-OCDMA networks using multi-diagonal csode," *Sci. Res. Essays*, vol. 7, no. 11, pp. 1262–1272, 2012.

Substrate Integrated Waveguide (SIW) Coupler on Green Material Substrate for Internet of Things (IoT) Applications

Nurehansafwanah Khalid[1*], *Siti Zuraidah* Ibrahim[1], and *Fwen Hoon* Wee[1]

[1] Advanced Communication Engineering Centre (ACE) School of Computer and Communication Engineering, Universiti Malaysia Perlis, Arau, Perlis, Malaysia

Abstract. This paper shows designed coupler on paper-substrate with Substrate Integrated Waveguide (SIW) techniques. Types of paper-substrate that used are photographic paper. The rectangular coupler is presented with Substrate Integrated Waveguide (SIW) which metallic via on paper-substrate. The structures of the coupler are designed and analysed using Computer Simulation Tools (CST) Studio Suite 2014 Software. This designed coupler operating within frequency of 3.8-5 GHz. The paper-based substrates are permits the implementation of green materials (Eco-friendly) technology. The design of the coupler and its simulated results are reported in this paper.

1 Introduction

Nowadays, there are huge interest with a new emerged technology known as Substrate Integrated Waveguide (SIW) in communication systems [1]. Substrate Integrated Waveguide (SIW) functions as a bridge technology which combines metallic waveguide and planar circuits. Integrated Waveguide Subject Technology (SIW) is also called as laminate-waveguide where it has the advantage to prevent leakage of electromagnetic waves by using metallic via holes [2,3].

The rectangular waveguide is widely used in communications systems due to increases of Q-factor and increases of power. However, the key challenges of this issues is that they are not eco-friendly and environmentally friendly to be implemented in microwave and communications systems especially for IoT applications and life-quality [2,4,5]. The paper substrate is the ideal choice as it is easily accessible, flexible and requires a simple fabrication process [6]. The implementation of paper-based (organic substrate) in electronic components is carried out without the need for etching and the use of chemical acids as opposed to inorganic substances that are hazardous to humans [7]. Paper substrate is suitable for wireless on body network and biomedical applications [7]. Moreover, the basic microstrip lines are easy to design and implement but the drawback it has a high loss radiation [5–8].

The conventional LTCC and PCB technology has increased the pollution impact due to the process of chemical etching and metallization on both surfaces [9],[4]. This substance is significantly unsuitable for the environment because the fabrication process uses acids, heavy metals and non-recyclable substrates [10,11].

Thus, the Substrate Integrated Waveguide (SIW) supports the frequency in Transmission Lines (TL) which provides better frequency performance on low substrate dielectric permittivity.

Then, the geometry of Substrate Integrated Waveguide (SIW) structure is not affected and influenced by the dielectric loss. In fact, energy leakage occurs because of the distance (gap) between metallic via [11][13]–[16]. In addition, the advantages of Substrate Integrated Waveguide (SIW) to the performance of electronic system is that it produces compact, flexible and small-sized designs, reducing radiation loss, enhancing power capability and Q-factor system.

The review of previous works of Substrate Integrated Waveguide (SIW) using paper substrate was studied. For example in [5], a half-mode Substrate Integrated Waveguide (SIW) transmission line implemented on paper substrate was presented, indicating that the simulated and measured results of the insertion loss are 2.8 dB at 4 GHz. In turn in [7], the designed bandpass filter was demonstrated at 5 GHz with a insertion loss of 5 dB.

In this paper, Substrate Integrated Waveguide (SIW) using paper substrate is investigated within operating frequency range of 3.8-5 GHz. There are few types of paper substrates that are commercially available in the market including watercolour paper, raw paper and photographic paper. A watercolour paper substrate is commercially available at a thickness of 0.30 mm, with a tangent loss of 0.045 and the dielectric constant of 2.09. A raw paper substrate is commercially available at a thickness of 0.35 mm, has a tangent loss of 0.037 and dielectric constant of 2.11. The photographic paper substrate is commercially available at a thickness of 0.69

* Corresponding author: nurehansafwanah@yahoo.com

DOI: 10.1051/matecconf/201714001022

mm, with a tangent loss of 0.044 and a dielectric constant of 2.35. In this paper, the photographic paper substrate is used in the coupler design because the substrate loss in SIW is dependent on the substrate thickness; the thicker the substrate, the better the insertion loss. The design is accomplished using Computer Simulation Tools (CST) software.

2 Design Techniques

2.1 The configuration of Substrate Integrated Waveguide (SIW) cavity

The configuration of SIW transmission lines is illustrated in Fig. 1.

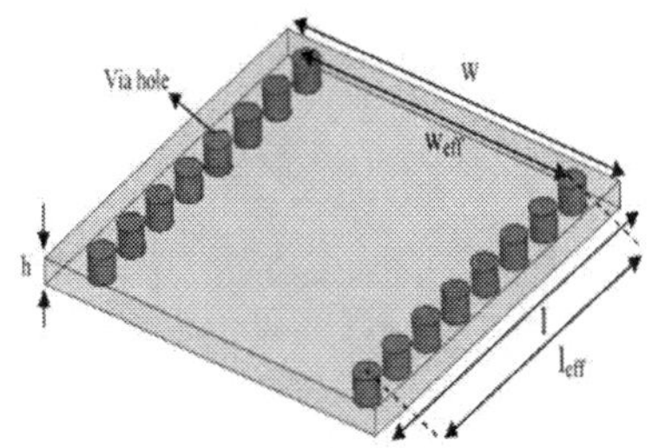

Fig. 1. Basic Structure of Substrate Integrated Waveguide (SIW) [14]

The waveguide substrate are covered with two metal (high conductivity) surfaces on top and bottom to let the propagation of all TE_{n0} modes [15]. The two silver epoxy holes are placed on the side line of the rectangular waveguide that functions to prevent and reduce signal leakage. Both metal posts are set symmetrically in order to control the signal flow. Two extra silver epoxy posts that set in every port act as reflection cancelling elements by varying the sections of Substrate Integrated Waveguide (SIW) [14]. Via hole must be shorted to both planes in order to provide vertical current paths, otherwise the propagation characteristics of SIW will be significantly degraded [14].

2.2 SIW Coupler's Design

The configuration and parameter dimensions of SIW coupler are shown in Fig. 2.

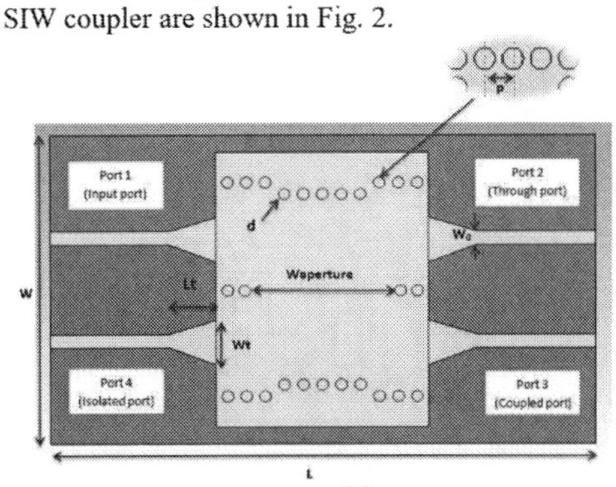

(a)

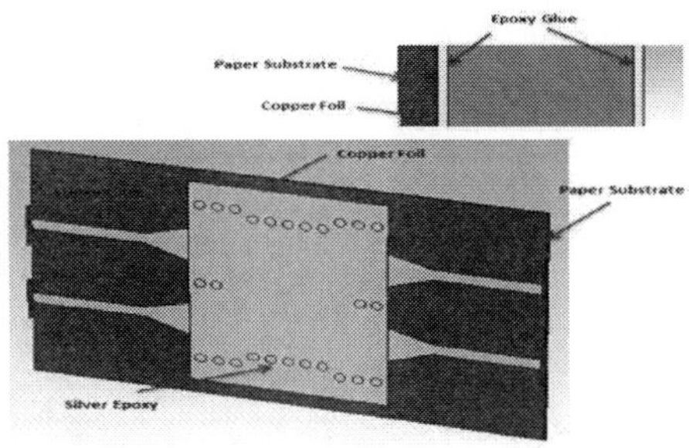

(b)

Fig. 2. Configuration of SIW Coupler (a) Top view (b) Side view

As depicted in Fig. 2, the coupling is obtained by two narrow apertures in the common broadside wall of two adjacent SIWs. To accomplish such coupling ratio, Port 1 is defined as the input port, Port 2 as the through port, Port 3 as the coupled port and Port 4 as the isolated port. The transition between SIWs and microstrip line is realized using microstrip taper to match both electrical and magnetic field distributions between the two medias [14].

In this paper, the proposed coupler is designed on photographic paper substrate materials where the metal (conductive) used is copper as a ground plane and metallic via holes. The relative dielectric constant (ε_r) of the photographic paper is 2.35, thickness (h) is 0.69 mm and loss tangent (tan δ) is 0.044.

The effective width (W_{eff}) and effective length (L_{eff}) of the Substrate Integrated Waveguide (SIW) cavity shown in Fig. 1 can be determined using expression (1) and (2).

$$W_{eff} = W - \frac{d^2}{0.95s} \qquad (1)$$

$$L_{eff} = L - \frac{d^2}{0.95s} \qquad (2)$$

Where,
W and L are the width and length of Substrate Integrated Waveguide (SIW) cavity.

Fig. 3 illustrates the two of SIW cavity design main parameters, diameter of metallic via holes (d) and the gap distance between via holes (s).

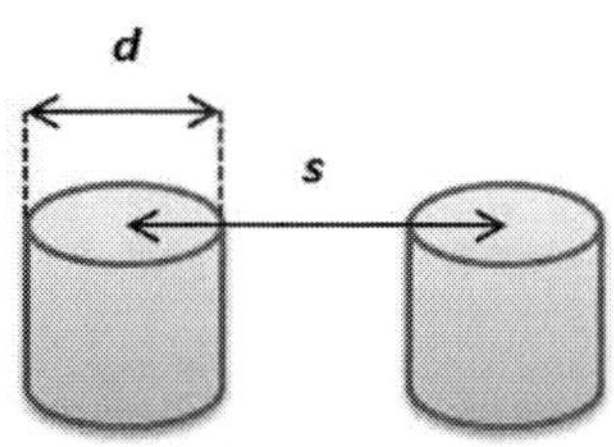

Fig. 3. Via holes SIW cavity's design parameters[11]

For SIW cavity designs, conditions expressed in (3) and (4) must be followed.

$$d < \frac{\lambda_g}{5} \quad (3)$$

$$s \leq 4d \quad (4)$$

Where,

$$\lambda_g = \frac{2\pi}{\sqrt{\frac{\varepsilon_r (2\pi f)^2}{c^2} - (\frac{\pi}{w})^2}} \quad (5)$$

f is the frequency
c is the speed of light

The ratio of d/p is important in order to inhibits leakage loss in SIW, which applicable range is $0.5 < d/p < 0.8$ [17]. Then, the width of SIW can be defined by expression (6) and (7) [14]:

$$W_{siw} = W_{equi} + 1.08(\frac{d^2}{s}) - 0.1(\frac{d^2}{W_{siw}}) \quad (6)$$

$$W_{equi} = \frac{c}{2f_c\sqrt{\varepsilon_r}} \quad (7)$$

The configuration of SIW to microstrip transition is shown in Fig. 4.

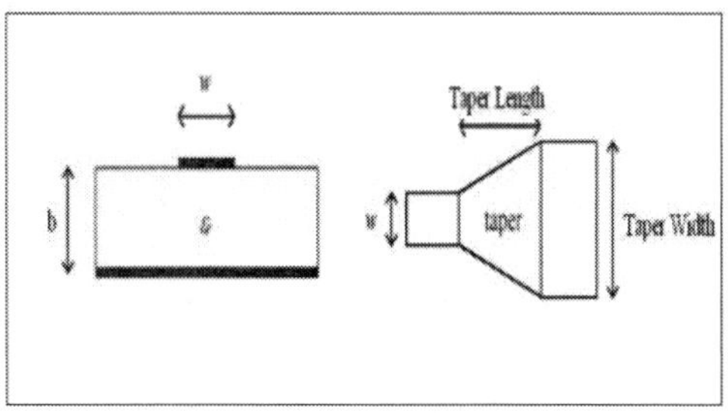

Fig. 4. Basic Geometry of microstrip line and tapered microstrip

The transition between SIW and microstrip basically comprises of tapered microstrip line interconnected with microstrip-waveguide with different width [4]. A tapered microstrip line is important to match the SIW with 50Ω transmission lines impedance. By following the guidelines given in [18][19], the initial value of tapered width (W_t) and tapered length (Lt) shown in Fig. 4 are determined [9][19] using expression (8) to (11).

$$\frac{W_t}{A_g} \approx 0.4 \quad (8)$$

$$(\frac{\lambda_g}{2}) < L_t < \lambda_g \quad (9)$$

Where,

$$\lambda_g = \lambda_c / \sqrt{\varepsilon_r} \quad (10)$$

$$L_t = \frac{n\lambda_g}{2}, n = 1,2,3,4\ldots \quad (11)$$

Table 1 shows the optimized dimensions of the SIW coupler.

Table 1. Dimensions of Coupler

Parameters	Descriptions	Dimensions (mm)
L	The length of substrate	150.00
W	The width of substrate	40.00
d	Diameter of circular via	3.175
p	Pitch of circular via	5.175
Wt	Width of microstrip taper	13
Lt	Length of microstrip taper	13.6
$W_{aperture}$	Width of aperture	42.39
W_{siw}	Width of siw	27.52
W_o	Width of microstrip line	4

3 Results

The final overall dimensions of the SIW coupler are 150 mm x 40 mm. Fig. 5 shows the simulated results of the proposed SIW coupler. The reflection coefficient, S11 and isolation, S41 are better than 10dB for the frequency range of 4 – 4.8 GHz. The flat coupling of 3 dB±1.5 dB can be observed within frequency range of 4.06 – 4.8 GHz. Excellent $90^0 \pm 3^\circ$ phase difference between S_{21} and S_{31} is observed in Fig. 5(b) with the same frequency range. Considering all the S-parameters performance, the proposed SIW coupler using photographic paper shows acceptable performances and suitable for IoT applications.

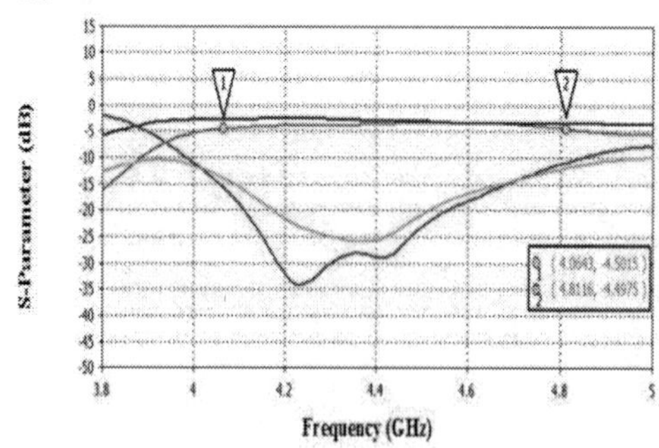

(a)

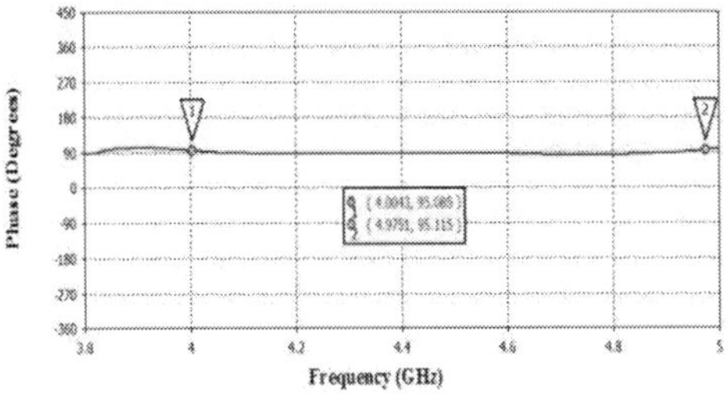

(b)

Fig. 5. Simulated S-parameters of SIW Coupler (a) magnitude in dB (b) Phase difference between output port

4 Conclusions

The design of SIW coupler using photographic paper as a substrate has been presented. The SIW coupler design has been accomplished using electromagnetic software tool, CST Microwave Studio. The simulated results of the coupler show acceptable performances for IoT applications such as for wireless on body network and biomedical applications. The paper-based substrates are permits the implementation of green materials (Eco-friendly) technology.

Acknowledgement

The author would like to thank the Ministry of Higher Education of Malaysia (MOHE) for the financial support for this research work in terms of FRGS grant of 9003-00555.

References

1. N. Keltouma, F. Mohammed, A. Saghir, "Design and characterization of tapered transition and inductive window filter based on Substrate Integrated Waveguide technology (SIW)," **vol. 8**, no. 6, pp. 135–138, (2011).
2. M. Bozzi, "Substrate integrated waveguide (SIW): An emerging technology for wireless systems," *2012 Asia Pacific Microw. Conf. Proc.*, pp. 788–790, (2012).
3. S. Agneessens, M. Bozzi, R. Moro, H. Rogier, "Wearable textile antenna in substrate integrated waveguide technology," *Electron. Lett.*, **vol. 48**, no. 16, pp. 985–987, (2012).
4. S. Moscato, R. Moro, M. Pasian, M. Bozzi, L. Perregrini, "Innovative manufacturing approach for paper-based substrate integrated waveguide components and antennas," **vol. 10**, pp. 256–263, (2016).
5. S. Moscato, N. Delmonte, L. Silvestri, M. Pasian, M. Bozzi, L. Perregrini, "Compact Substrate Integrated Waveguide (SIW) Components on Paper Substrate," pp. 24–27, (2015).
6. R. Moro, S. Agneessens, H. Rogier, A. Dierck, M. Bozzi, "Textile microwave components in substrate integrated waveguide technology," *IEEE Trans. Microw. Theory Tech.*, **vol. 63**, no. 2, pp. 422–432, (2015).
7. R. Moro, M. Bozzi, "Novel Inkjet-Printed Substrate Integrated Waveguide (SIW) Structures on Low-Cost Materials for Wearable Applications," pp. 72–75, (2012).
8. F. Parment, A. Ghiotto, T. P. Vuong, J. M. Duchamp, K. Wu, "Air-filled substrate integrated waveguide for low-loss and high power-handling millimeter-wave substrate integrated circuits," *IEEE Trans. Microw. Theory Tech.*, **vol. 63**, no. 4, pp. 1228–1238, (2015).
9. H. Mansor, R. Abdul-rahman, "Optimal Transition in Air-filled Substrate Integrated Waveguide," pp. 172–176, (2015).
10. A. Nasri, H. Zairi, A. Gharsallah, "Design of a Novel Structure SIW 90° Coupler," *Am. J. Appl. Sci.*, **vol. 13**, no. 3, pp. 276–280, (2016).
11. A. Kumar, G. Saini, S. Singh, A. Kumar, G. Saini, S. Singh, "Electrical & Electronic Engineering | Review Article A Review On Future Planar Transmission Line Electrical & Electronic Engineering | Review Article," *Cogent Eng.*, **vol. 48**, no. 1, pp. 1–12, (2016).
12. H. A. Ali, R. Bedira, H. Trabelsi, A. Gharsallah, "Cost-effective and Green Manufacturing Substrate Integrated Waveguide (SIW) BPF for Wireless Sensor Network Applications," **vol. 7**, no. 6, pp. 150–154, (2016).
13. S. Kumari, S. Srivastava, "Losses in Waveguide and Substrate Integrated Waveguide (SIW) For Ku Band : A Comparison," **vol. 3**, no. 1, pp. 53–57, (2013).
14. B. H. Ahmad, S. S. Sabri, A. R. Othman, "Design of a Compact X-Band Substrate Integrated Waveguide Directional Coupler," **vol. 5**, no. 2, pp. 1905–1911, (2013).
15. N. A. Smith, "Substrate Integrated Waveguide Circuits and Systems," no. May, (2010).
16. B. Rahali and M. Feham, "Divider , Circulator and Coupler in [10-15]."
17. Z. Kordiboroujeni, J. Bornemann, "Designing the width of substrate integrated waveguide structures," *IEEE Microw. Wirel. Components Lett.*, **vol. 23**, no. 10, pp. 518–520, (2013).
18. B. Rahali, M. Feham, "Substrate Integrated Waveguide Power Divider , Circulator And Coupler In [10-15] GHz Band," **vol. 4**, no. 1, pp. 1–11, (2014).
19. D. Deslandes, "Design equations for tapered microstrip-to-Substrate Integrated Waveguide transitions," *IEEE MTT-S Int. Microw. Symp. Dig.*, pp. 704–707, (2010).

DOI: 10.1051/matecconf/201714001023

Review Article of Radio Frequency and Microwave Heating Treatment to Disinfest *Kutu Beras* in Malaysian Rice

Siti Zulaika Abdul Nyzam[1], *Rosemizi* Abd. Rahim[1,*], *and Hasliza* A. Rahim @ Samsuddin[1]

[1]Biolectromagnetic Research Group (BioEM), School of Computer and Communication Engineering, Universiti Malaysia Perlis, Arau, Perlis, Malaysia.

Abstract—This article presents the basic principle of radio frequency (RF) and microwave heating and its application in agricultural sector specifically to control rice insects for storage purpose. Experiment findings of the past few years are highlighted according to the heating mechanism which influence the response of rice and insect exposed to the microwave field.

1 Introduction

Microwave heating has been considered to be one of the methods to control stored grain insects since early 20s[1]. The production of rice in Malaysia is seasonal. It is important to ensure the storage of stock meet the continuous demand and avoid the problem of shortage of supply. According to a study conducted by two partners of Malaysia in rice industry, Malaysian Agricultural Research and Development Institute (MARDI) and Bernas, near 50% of the rice storage is damage caused by the insect i.e. rice weevils, also known as *bubuk* or *kutu beras* in Malaysia. It is one of the most serious stored grain pests and originated in India but has been spread worldwide by trade[2]. The current insects controlling methods involve the use of physical, biological and chemical methods but they come with several disadvantages such as it takes time to present the results, low precision in ability to control disinfestation of large volume of grain, consequently much damage occur before the treatment is working and drawback of insecticide pests used in chemical method has grows attention in developing treatment method by using radio frequency (RF) and microwave heating method. This method has never been practically used in the industry and this paper will highlight the trial of experimenting this method in order to bring it into the industry for commercial purpose especially in controlling rice infestation.

2 Methods of Controlling Rice Insects

The process of paddy rice has to go through a lot of stages before it is marketed to the consumers. Inevitably, the major loss of paddy rice during the processes has been estimated caused by the rice weevils. This insects also act as the transportation of storage fungi which encouraging the infestation in rice[3]. Current insects controlling methods involve the use of physical, biological, chemical methods and a few developing technology in electrical methods. However, these methods have several disadvantages and some safety concern due to their effects on the rice nutrition and consumers.

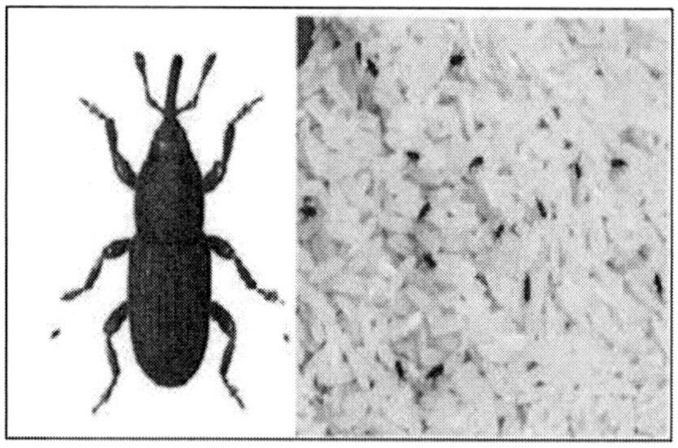

Fig. 1. Rice Weevil

1.1 Physical Method

The physical methods were the main methods in controlling stored grain before the use of chemical substances. It involves the manipulation of the physical environment or applying physical treatment to the rice such as manipulation of the physical environment, mechanical impact, physical removal, abrasive and inert dusts[4].

* Corresponding author: rosemizi@unimap.edu.my

In manipulating the physical environment, the temperature control is usually used to slow down degradation of produce caused by the insects. Both low and high temperature can be effective depending on rate of temperature change and the period of exposure. Next, the use of inert dusts are effective due to their low mammalian toxicity[4]. They slowly effect several mechanism that lead to dehydration especially absorption of cuticular lipid and abrasion. In mechanical aspect, polishing techniques is used by rubbing off the pericarp of rice grains[4]. Result experiment conducted by Luca et al shows that polishing to a weight loss of 11% of rice grains causes 40% severe mortality in rice weevil, Sitophilus Oryzae, eggs, followed by another 40% mortality remaining to the poor suitability of polished rice grains[5].

Nevertheless, physical control method could take a long time to show a satisfied result and it needs highly precise actions to control insects within a large volume of grain.

1.2 Biological Method

Biological control is a purposeful manipulation of population to living useful organisms to control the population of pests[6]. This useful organisms are called as natural enemies which attack the stored grain insects. The three major classes of natural enemies are predators, parasite and pathogens[7]. The major advantage of biological enemies that the natural enemies only need to be release at only one location[8]. They will keep reproducing their species and spread through the wide area of stored grain.

Specific species of natural enemies are needed to kill each of the insect pests. These natural enemies are released in the stored grain and they will attack the insects. This method is considered safe to the consumer and environment as there is no chemical substance involved.

The disadvantage of this method is a lot of precise information and perfect timing are needed to release the natural enemies. The release has to be early enough before the insect pests outnumber them[8]. Most of the natural enemies are specific host, which make it necessary to ensure the right pests species that causing the problem[8]. Also, this method is considered slow in action and taking risk of consequently damage may occur before the treatment is effective.

1.3 Chemical Method

Chemical methods are used widely due to low cost, the essential for the efficient production, preservation of food product and ease of use. Two classes of the chemical used to control the insects in the bulk stored grains are contact insecticide and fumigants.

The contact insecticide is sprayed directly to the rice grains and kill the insects when they contact with the treated surface. The popular insecticides are malathion, chlorypyrifosmethyl and pirimiphosmethyl. However, some of the insect have become immune to these chemicals substances due to wide spread resistance in insects population[3]. A survey of stored product insects shows 87% of 505 strain of the red flour beetle, Trobolium castaneum, from 78 countries were resistant to malathion and resistance to phosphine is great in Australia and India[9].

Fumigants are gaseous insecticides types applied to control the insects in the grains. It is a better chemical method as it can reach the inaccessible area by the contact insecticides type. The fumigants gaseous are methyl bromide (MeBr) and phosphine. However, it has been found that methyl bromide is a risk to the Earth's ozone layer[10]. It also creates damage to the environment and dangerous for human's health as it may still remain in the rice even though the gaseous is applied limitedly. Also, the fumigants is proven to kill larvae and mature insects but it does not guaranteed to successfully disinfect the survived eggs inside the kernels[11]. Therefore, the interest in discovering covering the alternative method with chemical-free has drawn the attention to the application of electromagnetic wave to control stored grain insects.

1.4 Electrical Methods

Conventional heating is considered as an electrical method. Heat disinfestation is an alternative way to replace chemical method. In conventional heating, heat is diffused from the surface into the interior of the material by convection or conduction in the particles. Many application of conventional heating use hot air or hot water to control the insect infestation by increasing the temperature. The advantages of this method are that the process is easy to conduct, no chemical involve and it leaves no residue. It was reported that ideal temperature to kill all stages of insects is between 25°C to 35°C[3]. However, this method take a huge amount of treatment time because of the slow rate of heat transfer within the material. It took 23 min and 6 hours for the centre of the fruit to reach the desired maximum temperature for cherries and apple respectively[12], [13]. This disadvantage lead to another cause which is the alteration of flavour compound.

Dielectric heating and microwave heating also have been considered to replace most of the current methods in controlling stored grain insect. Dielectric heating is also known as RF heating. Both RF and microwave heating have the same concept but each of them use different electromagnetic spectrums. Microwave heating commonly use frequency of 915 MHz and 2450 MHz while RF use 13, 27 and 40 MHz. In RF, the most famous design used in conducting the heating is two parallel plates electrodes with the material in between them. The power dissipated in the material can be expressed as,

$$P=E^2\sigma=55.63\times10^{-12}fE^2\varepsilon'' \quad (1)$$

where: P = the power dissipated,
f = the frequency used,
E = the rms electric field intensity

DOI: 10.1051/matecconf/201714001023

ε'' = the dielectric loss factor.

3 General Principles

The permittivity of a material is a complex quantity consists of real part and imaginary part. Relative to free space, permittivity can be expressed as $\varepsilon^* = \varepsilon' - j\varepsilon''$. The real part, ε', is called as the dielectric constant representing the capability in storing the energy in the electric field in the material. The imaginary part, ε'', is called as dielectric loss factor representing the dissipation of electric energy in the material which transform the electric energy to heat.

Microwave heating is the use of the electromagnetic waves and it involves heat transfer to beam up the material exposed to its radiation. Its frequency range is from 300 MHz to 300 GHz and it lies between radio frequencies and infrared radiation[14]. The rapidly varying electric and magnetic fields lead to the occurrence of heating. Electric field induced by the microwave field will cause dipolar molecules such as water and fat to oscillate back and forth and lead to heat dissipation in the material exposed[15]. The conversion of microwave energy to heat can be expressed as,

$$P = 2\pi E^2 f \varepsilon'' \varepsilon_o V \quad (2)$$

where: P = power
E = the electric field intensity
f = frequency
ε'' = dielectric loss factor
ε_o = the permittivity of free space
V = volume of the material.

By comparing the penetration depth between RF and microwave heating, RF allow the deeper depth than in microwave heating according to the relationship of the frequency and the wavelength. Also, the two parallel plate electrodes design in RF can achieve a uniform field intensity is the material is homogeneous[1].

The major advantage of microwave heating is it heats the material volumetrically where material under the field directly absorb microwave energy and transform it to heat. Unlike the conventional heating, by using microwave energy, the heat is absorbed directly and internally by the material. Hence, microwave heating is considered efficient and reliable in overcoming not only the problem of long time treatment but also in getting rid of the insects both inside and outside the kernels.

4 Experimental Findings

This paper discusses several experiments related to RF and microwave heating treatment. The important element of this method is the measurement of dielectric properties of the targeted commodities and the specific host insects to determine the how the two subjects respond towards the electromagnetic wave. Dielectric properties of a material is influenced by temperature, frequency and moisture content[16]. Hence, it needs to be measured as a function of temperature and also important to choose the optimal frequency[17]. An attempt to measure data values of permittivity of adult Sitophilus Oryzae has been done by Stuart Nelson et al at 1997 at wide frequency of 0.2 GHz to 20GHz[18]. The measurement was taken at four different temperature which were 25, 40, 50 and 60°C with the bulk density of 0.613 g/cm^3. Due to the nearness between the size of the insect and the end of the coaxial probe, the variability in the taken measurement may be caused by the differences of the effective density of the sample and the mean sample bulk density calculated. It was resolved that the dielectric loss factor of insects does not show good respond as presented for dielectric heating of the insects at range frequencies up to 20 GHz[18]. It increased as the temperature rose and appear significant at temperature higher than 50°C.

Manjur et al in 2011, carried out the measurement of permittivitties values of Rhd and Malaysian paddy MR219. The data recorded the higher values of permittivities for insects compared to the paddy. The reading of permittivities for paddy was basically independent of frequency while for Rhd, ε' degrade as the frequency rise and ε'' increased and remain constant from 7.5 GHz to 20 GHz[17]. The result of permittivities of ε'' in the function temperature at 23-65°C under specific frequency showed close interaction between 2.45 GHz up to 5 GHz as shown in Figure 2. Therefore, it concluded as the operating frequency of 2.45 GHz for industrial microwave has important influence to reach complete mortality of Rhd.

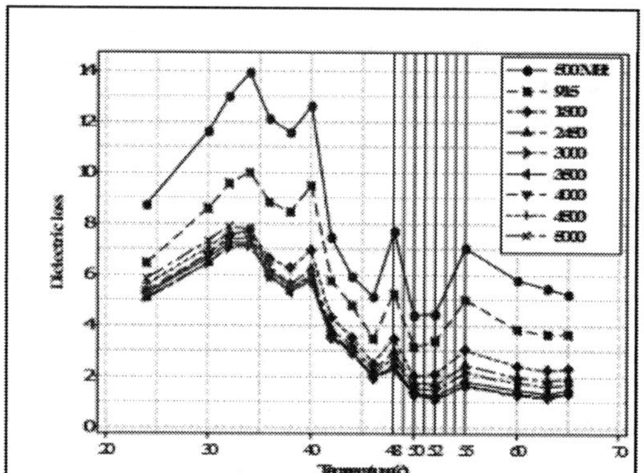

Fig. 2. Dielectric loss factor of Rhyzopertha Dominica (Rhd) in a function of temperature at specific frequencies

A new model was developed to directly obtain the values of ε' and ε'' of Rhd under the frequency and temperature as described in equation 3 and 4[17],

$$\varepsilon' = 52.272 - 1.407 \times 1.159 \times T + 0.017 \times F^2 + 0.006 \times T^2 + 0.017 \times F \times T \quad (3)$$

$$\varepsilon'' = 19.1867 - 0.100 \times F - 0.446 \times T + 0.012 \times F^2 + 0.03 \times T^2 + 0.005 \times F \times T \quad (4)$$

where F = frequency from 200 MHz to 20 GHz
T = temperature from 23 to 66°C.

Two famous insect pests often found in rice are Sitophilus Oryzae and Rhyzopertha Dominica (Rhd). Insect at adult stage is exposed the most under the electromagnetic field rather than insects at the immature stage and young adult are found more resistant to the treatment compared to the older ones[16]. Robert in 1974 recorded, the mature larva or pupae are more resistant to the microwave treatment compared to the immature larva or eggs[19]. Jose Franco in 2013, insects in at immature stages which are larva and pupae inside the rice are easier to be killed as the treatment time needed was only 100 seconds to achieve 100% mortaility while the egg stage needed a longer exposure which was 130 seconds[20]. The study of RF treatment in 2005, 100% mortality was reported in 25 adult Rhd inside and outside of the rice seed under the RF treatment at 27.12 MHz with temperature range from 70°C to 85°C for 180 seconds by Pattaya Janhang et al. 70°C was suggested to be the minimum temperature which provide complete mortality and maintain the seed viability at its maximum level at the same time[21]. With the same use of frequency, Wornwimol Wangspa led several experiments to determine the most tolerant stage of Sitophilus Oryzae and studied the number of progeny of this species after the treatment process. Under condition of 50°C for 180 seconds, it was found that eggs stage suffered complete mortality while the mortality of larva, pupa and adults are 79.1±5.1, 80.9±7.3 and 96.5±2.3% respectively[22]. It was shown the treatment sufficiently kill the insects compared to the untreated rice where the mortality of eggs, larva, purva and adult were 6.50±1.5d, 9.25±1.5d, 7.00±1.4d and 6.50±0.6d respectively. The survivor insects were used after 6 weeks of storage in succeeding experiment to observe the hatching eggs and reproduction capability and the number of progeny was proven to be less compared to the controlled rice sample. In another experiment, adult rice weevil was used to find the combination of temperature and time was experimented with temperature range of 50 to 70°C for 90 to 180 seconds and the result is as shown in Figure 3[22].

Time (sec)	Mortality of rice weevil (%)± SE at various temperature				
	50	55	60	65	70
Untreated sample	9.25±0.9d	4.25±1.2d	2.50±0.6d	3.00±1.1d	5.00±0.7d
90	58.22±1.9c	93.03±2.9a	99.11±0.8a	99.17±0.8a	99.04±0.9a
120	67.00±14.7b	98.08±1.9a	99.08±0.9a	100.00a	100.00a
150	69.72±7.1b	99.14±0.8a	100.00a	100.00a	100.00a
180	91.2±2.6a	100.00a	100.00a	100.00a	100.00a

Fig. 3. Average mortality of adult rice weevil for temperature range of 50 to 70°C for 90 to 180 seconds.

Microwave heating treatment was applied by Siming Zhao et al with Sitophilus Oryzae was subjected to the operating frequency of 2.45 GHz in 2006. The result obtained showed the adults was killed at 55°C and the eggs was killed at 54°C[23]. A lethal model was developed to investigate the precise temperature at which it kills the insects from eggs and adults stage(thermal death kinetic) and can be described as in equation 5[23],

$$\begin{cases} M = 1 - A\left(\frac{T_1 - T_0}{T - T_0}\right)^m, & (T \geq T_1) \\ M = e, & (T_0 < T < T_1) \\ M = 0, & (T \leq T_0) \end{cases} \quad (5)$$

where: T_1= initial temperature of rice,
T_0= the final temperature recorded when complete mortality is reached
A= the susceptibility factor which indicate how easy the easy are killed.
m=coefficient fitted to the model

Manjur et al conducted similar exploration with insect species of Rhd. It recorded mortality for time treatment from 40 to 60 seconds and the complete mortality was reached at the maximum treatment time[17]. With condition of 78°C for 60 seconds treatment, 300 gm of Rhd was studied on three different holding time to observe the mortality rate and the effect of holding time. The result as in tabulated in Table 1. The effect of holding time is need significant consideration in designing the treatment to reach lethal temperature in order to obtain high mortality rate[24].

Table 1. Holding time effect on mortality rate of Rhd under microwave power of 1kW.

Temperature and time	Holding Time (min)	Mortality (%)
78°C / 60 seconds	3	98.4
	7	99.2
	10	100

In term of heating rate and energy, it can be concluded that higher heating rate provide low mortality and it needs longer holding time to reach high mortality as low heating rate can provide[24]. Siming Zhao reported complete disinfestation above energy consumption of 0.017 and 0.016 kWh/ks for adults and eggs respectively as the rice temperature from 55 to 62°C[23]. Wormwimol Wangspa recorded 0.391kWh under treatment of 65°C for 120 seconds and 0.670kWh at 70°C for 180 seconds[22]. Both of the treatment caused 100% mortality. The energy consumption required to reach 100% mortality of adults is higher compared to eggs stage. Hence, eggs stage is easier to be killed and support the result obtained by Siming Zhao.

Halverson in 1999 has investigated the lethal temperature for rice insects in wheat by using extremely high frequency (EHF) and super high frequency (SHF) and found that eggs were more thermo tolerant compared to other insect stages[25]. However, the penetration depth of the electromagnetic waves should be considered as it decreases as the frequency increases.

In 1974, Rai et al claimed that heat can reduce the quality of sperm cells and ovarian tissue of insects[1]. Consequently, Pattaya Janhang tested the rice viability by doing tetrazolium test to observe the rice viability and the results showed the viability was decreased to 90, 82, 64 and 39% as the temperature rose from 70, 75, 80 and 85°C respectively[21]. Seed germination was affectedly drop to 89, 86, 57 and 27 % correspondingly to the rising temperature. Microwave specification for disinfestation purpose with high power, long time treatment and high energy consumption can provide high mortality rate of insects but it induces the rice to heat excessively. Siming Zhao suggested that 50°C as the minimum treatment temperature under controlled microwave power, treatment time and energy utilization to kill the insects and maintain the rice quality as high as possible[23].

5 Conclusion

This paper discusses several research and experiments related to the use of RF and microwave heating treatment to disinfest rice weevil for long time storage in the industry. The current methods used to overcome insect pest are physical, biological and chemical methods. However these methods result a few disadvantages as they take a long time to get rid of the insects, the needing of precise information of time and specific host insect to be killed and chemical methods is proven cannot reach the eggs stage inside the kernel and dangerous to the consumers and our environment. By using the electromagnetic wave energy, it can reduce the treatment time, ensure the complete mortality inside and outside the kernel and free chemical residue.

From the findings, it tells the importance of accurate data permittivities of the insects and commodities in order to investigate the suitable operating frequency, temperature and time treatment. Also, the data can help to develop the specification and design of application based on radio frequency and microwave heating. The permittivities values indicate the interaction between the rice and insect towards the radiation. High values of dielectric constant show strong interaction.

It also can be shown in most of the experiments above that the higher temperature caused higher mortality both in RF and microwave heating treatment. Pattaya Janhang recorded 70°C in the minimum temperature that caused completed mortality for 180 seconds in treating Rhyzopertha Dominica (Rhd) in rice seeds. Siming Zhao suggested 55°C to reach the complete mortality and maintain the maximum rice quality as well. Manjur reported in his experiment that 100% mortality of Rhd at 60 seconds with 1kW power. The most tolerant stage of insect was mostly found in eggs both in RF and microwave treatment. This suggests that the effect of temperature on pest is high.

Las but not least, unavoidably, this method causes damage to the rice and under condition of extremely high temperature and time can lead the rice to heat excessively. Therefore, it is importance to precisely determine the combination of temperature, time and power at the minimum level to maintain the quality of the rice and healthy to the consumers.

Acknowledgement

The authors highly appreciate the support and would like to thanks Ministry of High Education (MoHE) for the financial funding under Fundamental Research Grant Scheme (FRGS 2016, 9003-00597).

References

1. S. O. Nelson, Trans. Am. Soc. Agric. Eng., **39** 1475–1484, (1996)
2. P. G. Koehler, Univ. Florida IFAS Extension., 2–3, (2008)
3. I. Das, G. Kumar, N. G. Shah, Int. J. Food Sci., **2013** 1–114, 2013.
4. F. F.-L. Vincent Charles, Guy Hallman, Bernand Panneton, *Management of Agricultural Insects With Physical Control Methods* (Annual Review of Neuroscience, Quebec, Canada, 2003)
5. É. Lucas, J. Riudavets, Stored-Product Quar. Entomol., **93** 1837–1841, (2000)
6. D. Mahr, P. Whitaker, N. Ridgway, *Biological control of insects and mites: an introduction to beneficial natural enemies and their use in pest management* (Universiti of Wisconsin, Wisconsin, 2008)
7. M. a Altieri, C. I. Nicholls, M. a Fritz, *Manage Insects on Your Farm: A Guide to Ecological Strategie* (Sustainable Agriculture Research and Education(SARE), Maryland, 2005)
8. P. W. Flinn, M. Schöller, Stored Prod. Prot., 1–10, (1996)
9. R. Vadivambal, Can. Soc. Eng. Agric. Food Biol. Syst., 1–11, (2005)
10. J. G. Leesch, G. F. Knapp, B. E. Mackey, Stored Prod. Res., **36** 65–74, (2000)
11. S. J. Langlinais, ASAE Tech. Pap., **46** 16, (1989)
12. S. Wang, J. Tang, R. P. Cavalieri, Postharvest Biol. Technol., **22** 257–270, (2001)
13. S. Wang, J. Tang, R. P. Cavalieri, Postharvest Biol. Technol., **22** 29–38, (2001)
14. S. Jones, "Comparing Microwave to Conventional Heating and Drying Systems Mechanics Advantages Economics."
15. Y. Kueon, Int. Conf. Control. Autom. Syst. ICCAS 2008, 1966–1971, (2008)
16. S. O. Nelson, "Frequency dependence of the dielectric properties of wheat and the rice weevil," Iowa State University, 1972.
17. M. Ahmed, F. Malek, R. B. Ahmad, M. Rahman, K.

M. Juni, World Acad Sci Eng Technol, **81** 1–6, (2011)
18. S. O. Nelson, P. G. Bartley, K. C. Lawrence, IEEE Trans. Instrum. Meas., **46** 941–946, (1997)
19. R. L. Kirkpatrick, *Proc.* Int. Work. Conf. Stored Prod. Entomol., 431–437, (1974)
20. J. G. Franco, S. S. H. Franco, C. H. Franco, P. B. Arthur, V. Arthur, Ina. 2013 Int. Nucl. Atl. Conf., 1–23, (2013)
21. P. Krutmuang, S. Mekchay, Conf. Int. Agric. Res. Dev. Stuttgart-Hohenheim, 1–8, (2005)
22. W. Wangspa, Y. Chanbang, S. Vearasilp, Chiang Mai Univ. J. Nat. Sci., **14** 189–197, (2015)
23. S. Zhao, C. Qiu, S. Xiong, X. Cheng, J. Stored Prod. Res., **43** 430–434, (2007)
24. S. Wang, J. N. Ikediala, J. Tang, J. D. Hansen, J. Stored Prod. Res., **38** 441–453, (2002)
25. S. L. Halverson, T. W. Phillips, T. S. Bigelow, G. N. Mbata, Mark E. Payton, Atmos. Environ., **30** 3001–3009, (1996)

DOI: 10.1051/matecconf/201714001024

The Performance Analysis of K-Nearest Neighbors (K-NN) Algorithm for Motor Imagery Classification Based on EEG Signal

Nurul E'zzati Md Isa[1,*], *Amiza* Amir[1], *Mohd Zaizu* Ilyas[1], and *Mohammad Shahrazel* Razalli[1]

[1]School of Computer and Communication Engineering, UniMAP, 02600 Arau, Perlis

Abstract. Most EEG–based motor imagery classification research focuses on the feature extraction phase of machine learning, neglecting the crucial part for accurate classification which is the classification. In contrast, this paper concentrates on the classifier development where it thoroughly studies the performance analysis of k-Nearest Neighbour (k-NN) classifier on EEG data. In the literature, the Euclidean distance metric is routinely applied for EEG data classification. However, no thorough study has been conducted to evaluate the effect of other distance metrics to the classification accuracy. Therefore, this paper studies the effectiveness of five distance metrics of k-NN: Manhattan, Euclidean, Minkowski, Chebychev and Hamming. The experiment shows that the distance computations that provides the highest classification accuracy is the Minkowski distance with 70.08%. Hence, this demonstrates the significant effect of distance metrics to the k-NN accuracy where the Minknowski distance gives higher accuracy compared to the Euclidean. Our result also shows that the accuracy of k-NN is comparable to Support Vector Machine (SVM) with lower complexity for EEG classification.

1 Introduction

Motor imagery classification is used to classify the imaginary task such as hand and foot movements by using brain signal which usually used in Brain-Computer Interfaces (BCI). BCI is an alternative communication pathway between the brain and computer which does not requires muscular movement or control [1] which has potential application value in wide range of the fields such as medical, military and entertainment [2].

The classification of brain signals is challenging as the signals are low signal to noise ratio, non-linearity and limited training data due to difficulties in collecting signals [3]. Electroencephalogram (EEG) is one of the techniques that usually been chosen by researchers to collect the signals instead of others such as computer tomography (CT), magnetic resonance imaging (MRI) and functional MRI (fMRI). The collection of signal using EEG is done by placing the electrode on the scalp to records electrical activity and brain waves.

Two main phases in data classification is feature extraction phase and training/ classification phase (classifier development). Unlike the existing researches [4-8] which focus on feature extraction, this paper focuses on the classifier development.

Classification task of EEG data can be done with many algorithms, one of them is k-Nearest Neighbor (k-NN). K-NN classifier is used in [9] to discriminate between seizure and non-seizure events for automated seizure detection using EEG signals. In [10], the k-NN is used as classifier for its classification task to compare three different signal decomposition methods for motor imagery BCI systems. K-NN classifier provides better classification accuracy and requires less training and testing times compared with the Multiple Layer Perceptron (MLP) and Support Vector Machine (SVM) [11]. In [12], the k-NN classifier obtained the best accuracy in classifying EEG signals to identify the engagement, enjoyment, frustration and difficulty compared to Bayes Network, Naïve Bayes, SVM, Multilayer Perceptron, Random Forest and J48. Besides that, [13] shows that the k-NN also provides better accuracy in eye state classification compared to Multilayer Perceptron Neural Networks.

This research study the classification technique which using k-NN algorithm. K-NN is simple supervised learning algorithm [14] that classified a sample by a majority vote of its neighbors which means the sample is allocated to the class based on the most common class among its k closest neighbors. In k-NN, distance metrics is used to calculate distance between a new samples and existing samples in dataset. The literature is strongly influenced by a commonplace in using the Euclidean distance metric [10-13][15-18]. In fact, we are not aware of any studies with a focus on a performance comparison among various different distance metrics for k-NN. This paper aims to thoroughly investigate the effect of distance metrics and parameter k to the k-NN performance. In this paper, we study five distance metrics of k-NN: Manhattan, Euclidean, Minkowski, Chebychev and Hamming and vary k value in the

* Corresponding author: nurulezzati@studentmail.unimap.edu.my

DOI: 10.1051/matecconf/201714001024

experiments. Additionally, the paper also verifies whether the k-NN algorithm is suitable or not to use for classification method in BCI system as in [15] states that k-NN is one of the most important non-parametric algorithms in BCI implementation and compare it with the performance of SVM which commonly used for EEG data classification.

This paper is study the performance analysis of k-NN algorithms by using EEG data for BCI motor imaginary classification. The results helps in choosing better k value and a proper metric that makes a good classification performance. The rest of this paper is organized as follows: Section 2 gives the introduction of k-NN and its parameter. Section 3 contains the EEG processing using k-NN. In section 4, the results and discussion is presented. Lastly, section 5 is conclusion.

2 K-Nearest Neighbors

K-NN classification is a non-parametric model that is described as instance-based learning which the model are characterized by memorizing the training dataset. K-NN also a typical example of a lazy learner. It is called lazy not because of its apparent simplicity, but because it does not learn a discriminative function from the training data but memorized the training dataset instead. Lazy learning is a special case of instance-based learning that is associated with zero cost during the learning process [19].

The k-NN algorithm is suitable to classify EEG data as it is a robust technique for large noisy data [20]. The samples which is the data is classified by the majority vote of its neighbor's class. In order to determine the class, this algorithm requires training data and predefined k value as it will search through the training sample space for the k-most similar samples based on a similarity measure a distance function [21]. The value of k and distance metric will affect the result of classification [20].

Fig. 1 illustrates the concept of k-NN algorithm when applied the distance metric to determine the appropriate class of new data with k = 9. The data to be classified is at point (0.6, 0.45), which is shown with "X". The big circle with dot line is represented the distance metric using Euclidean distance computation. It has two possible classes which is circle class with six instances and triangle class with three instances. The algorithm will classify mark "X" to the circle class as the circle class have the majority of data within the radius.

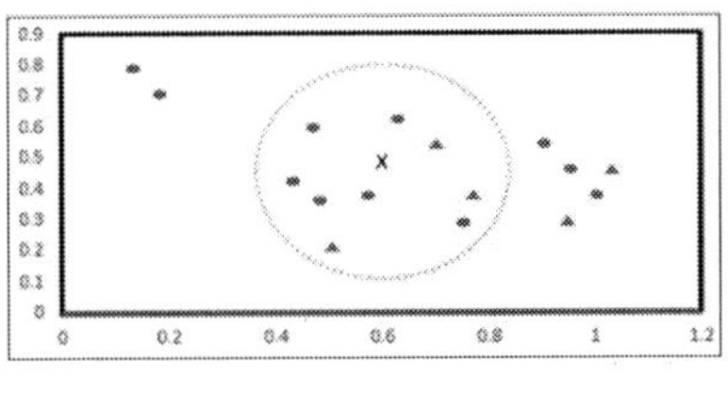

Fig. 1. The classification of K-NN algorithm with using Euclidean distance and k=9

2.1. Value of k

The value of k is usually small and integer with positive value. If k = 5, class allocation of the sample is based on the nearest five neighbors within a certain distance. In this paper, the value of k is use from 1 to 15.

2.2. Distance metric

Distance metrics are a method to find distance between a new data point and existing training dataset [22]. There are 5 distance metrics used in this research which can be explained as follows:

2.2.1 Manhattan / City Block Distance

The Manhattan distance or also be called as city block distance is the distance between two points (x,y) which the sum of the absolute difference of Cartesian coordinates, defined by:

$$d(x,y)=\sum_{i=1}^{k}|x_i-y_i| \tag{1}$$

2.2.2 Euclidean Distance

The Euclidean distance is a measure to find distance between two points. In Cartesian coordinates, if x and y are two points in Euclidean k-space, then the distance (d) from x to y or from y to x is defined by the Pythagoras's theorem. The formula is as follow:

$$d(x,y)=\sqrt{\sum_{i=1}^{k}(x_i-y_i)^2} \tag{2}$$

2.2.3 Minkowski Distance

The Minkowski distance is a method to find distance between two points (x,y) in a normed vector space with various value of p. The distance is defined as follow:

$$d(x,y)=\left(\sum_{i=1}^{k}(|x_i-y_i|)^p\right)^{1/p} \tag{3}$$

For the special case of Minkowski distance: When p = 1, the Minkowski metric gives the Manhattan distance; when p = 2, the Minkowski metric gives the Euclidean distance; and when p = ∞, the Minkowski metric gives the Chebychev distance.

2.2.4 Chebychev Distance

The Chebychev distance is a measure to find the distance between two vectors or points (x,y) which have the greatest of their differences along any coordinates dimension, defined by:

$$d(x,y)=\lim_{p=\infty}\left(\sum_{i=1}^{k}|x_i-y_i|^p\right)^{1/p} \tag{4}$$

DOI: 10.1051/matecconf/201714001024

2.2.5 Hamming Distance

The Hamming distance, which is the percentage of coordinates that differ, can be defined by:

$$d(x,y)=\left(\frac{\#(x_i \neq y_i)}{n}\right) \quad (5)$$

3 K-NN for EEG Classification

3.1. Data description

The dataset used is Dataset 1 from BCI Competition IV [23] which contain recorded EEG data for motor imagery task perform by four healthy human subjects (a, b, f and g). In the experiment, each subject was asked to select two mental tasks to perform out of three tasks which are left hand, right hand or foot movements. The arrows pointing left, right or down is presented as visual cues on a computer screen as represented for left, right and foot movement respectively. The cues were displayed for a period of 4s which interleaved with 2s of blank screen and 2s with a fixation cross shown in the center of the screen. As in each run have 50 trials of each of the chosen two classes, give the total of 200 trials for each subject [24]. Eleven channels are used in this research, named as FC5, FCz, FC6, C5, C3, Cz, C4, C6, CP5, CPz and CP6.

3.2. Feature Extraction

The extraction of the wanted signals from the raw EEG signals and removing the irrelevant signals is called as feature extraction process as shown in Fig.2. The EEG data should be process in term of frequency as it is been classified based on frequency bands which are delta, theta, alpha, beta and gamma band. Therefore, the Fast Fourier Transform (FFT) is used to transform the time domain to frequency domain. The FFT features are then feed into k-NN classifier for classification process.

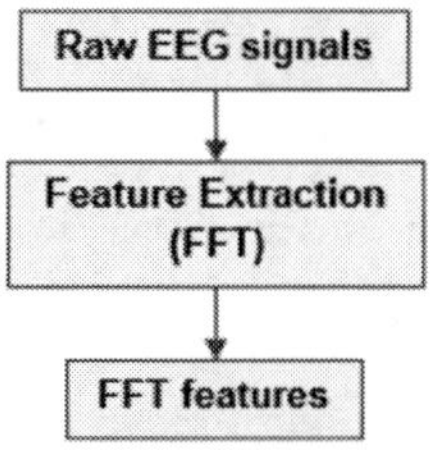

Fig. 2 Block diagram of the feature extraction process

As signals in time domain can be split into a group of sinusoids, the EEG signals which are lengthy and noisy can be easily convert into frequency domain. The hidden features can become visible after the conversion. The original signal can be restored by adding all the sinusoids up after FFT. Therefore, no information is lost.

3.3. Classification

In machine learning, classification is a prediction process to predict the categories or classes of the samples in the dataset by applying classification algorithm. The classifier used in this paper is k-NN classifier as shown in Fig. 3. This process is performed in python programming.

Five distance metrics: Manhattan distance, Euclidean distance, Minkowski distance, Chebyshev distance and Hamming distance are used to determine the best value of k to maximize the classification performance. The k-value is searched from 1 to 15. The classification is performed for all subjects and the result are expressed in percentage of accuracy.

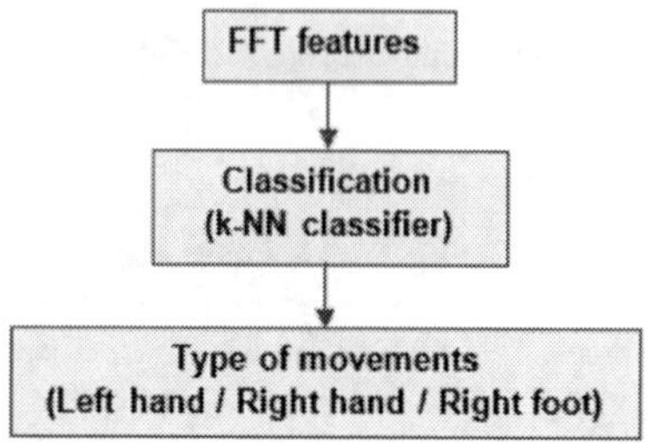

Fig. 3 Block diagram of the classification process

SVM is usually used in BCI research as classifier for this dataset [4-8]. Thus, in this paper, the k-NN is investigated and compared with SVM.

4 Result and Discussion

The classification is between two classes which either motor imagery of right hand and left hand or left hand and right foot. In this section, the result of EEG data classification by applying the k-NN classifier with various value of k and distance metrics are tabulated in tables and illustrated in graphs. The results have been analyzed for the performance analysis of k-NN classifier and verified by using 10-fold cross validation technique.

The results have been analyzed for the performance analysis of k-NN classifier. In order to prove the reliability of the classification results obtained, the results is verified using 10-fold cross validation technique. This process randomly divides the data into 10 folds for tested 10 times. The testing process is done on one of the folds while the remaining folds undergoes the training process. The process was repeated until all folds are used for testing and training the classifier. The accuracy of classification was calculated from the average of 10-fold cross validation.

4.1. K-NN Accuracy with Varying Parameters

The data classification has been performed with the same classifier which is k-NN with various value of k and

DOI: 10.1051/matecconf/201714001024

distance measurement. Table 1 shows the result of accuracy for all 4 subjects.

Table 1. Result for Subject a, f, b and g

Metric Distance	Value of k	Accuracy (%)			
		Left hand & Right foot		Left hand & Right hand	
		a	f	b	g
Manhattan Distance	1	53.52	51.21	48.15	55.62
	2	54.96	54.28	47.63	57.41
	3	56.98	55.17	48.04	57.85
	4	57.16	57.63	47.54	60.80
	5	57.23	59.02	47.47	61.75
	6	58.34	59.26	47.74	62.19
	7	59.42	60.37	48.01	63.46
	8	60.47	60.92	45.78	63.24
	9	60.96	61.03	45.75	63.86
	10	62.06	61.50	46.33	64.03
	11	62.57	61.90	47.79	64.74
	12	63.08	62.71	48.44	65.14
	13	62.78	63.13	47.88	64.57
	14	62.15	62.87	48.21	65.58
	15	62.46	63.13	48.79	66.37
Euclidean Distance	1	54.73	55.10	48.33	57.52
	2	56.24	53.28	45.76	58.79
	3	56.64	57.17	46.84	61.20
	4	58.24	58.05	45.37	63.77
	5	60.17	58.60	45.74	64.04
	6	61.73	59.98	47.82	63.64
	7	61.37	60.25	50.97	64.40
	8	62.51	60.33	51.74	63.93
	9	61.37	60.31	51.61	65.36
	10	61.10	61.31	53.00	65.55
	11	61.76	60.84	51.87	66.33
	12	62.54	60.99	52.46	67.80
	13	63.10	61.63	52.43	68.40
	14	64.03	61.96	52.05	68.28
	15	64.57	62.26	52.72	68.33
Minkowski Distance (p=11)	1	54.51	54.31	51.37	59.40
	2	55.49	53.71	52.11	59.32
	3	58.39	55.25	51.80	62.73
	4	60.67	56.98	51.59	65.30
	5	61.21	58.67	50.39	65.70
	6	61.22	59.53	50.44	65.70
	7	62.82	61.08	50.14	66.16
	8	63.35	60.80	50.44	67.45
	9	64.65	60.86	52.32	68.03
	10	65.61	61.60	52.09	68.32
	11	65.60	61.24	52.49	68.20
	12	66.15	61.74	51.98	68.54
	13	66.99	61.78	54.13	68.93
	14	67.67	62.07	53.87	69.71
	15	67.66	62.95	52.95	70.08
Chebyshev Distance	1	56.50	51.99	51.52	57.27
	2	56.80	52.90	52.64	60.99
	3	56.85	55.89	52.79	63.92
	4	59.03	56.46	52.99	67.07
	5	59.74	58.21	52.89	67.10
	6	60.79	58.94	51.95	66.69
	7	62.73	61.44	53.33	66.17
	8	64.22	61.91	54.92	65.79
	9	65.39	61.13	54.15	66.81
	10	66.03	61.22	53.39	67.75
	11	66.70	60.62	52.96	67.74
	12	66.73	61.42	54.77	67.78
	13	67.33	61.29	55.26	67.47
	14	67.07	61.09	53.98	67.69
	15	67.11	61.16	53.85	67.78
Hamming Distance	1	50.46	51.05	48.53	50.51
	2	50.12	51.24	48.88	50.07
	3	49.85	50.14	47.81	50.44
	4	50.09	50.61	47.01	49.24
	5	49.65	50.91	47.87	50.38
	6	49.84	50.93	47.29	49.84
	7	49.80	51.31	46.34	50.15
	8	50.11	51.65	48.31	50.73
	9	49.71	51.67	49.01	49.63
	10	49.00	50.74	48.06	49.69
	11	50.45	50.31	50.01	50.28
	12	49.24	50.84	49.67	49.79
	13	48.82	52.20	49.00	49.41
	14	49.91	50.98	48.43	50.26
	15	50.50	51.47	47.21	49.79

The left hand and right foot imaginary movements classification is done by subject a and f. For subject a, the highest accuracy is 67.67% when using Minkowski distance at p = 11 and k = 14. For subject f, the highest accuracy is 63.13% when using Manhattan distance at k = 13,15. Subject b and g is used to classify the right hand and left hand imaginary movements. For subject b, the highest accuracy is 53.87% when using Minkowski distance at p = 11 and k = 14. For subject g, the best classification accuracy achieved is 70.08% when using Minkowski distance at p = 11 and k = 15.

The comparison accuracy results between different distance metrics and value of k for motor imaginary task classification of left hand vs right foot imaginary movement and right hand vs left hand imaginary movement are illustrated in Fig. 4. Based on the overall results, the Minkowski distance at p = 11 with the value of k between 13 and 15 give the highest accuracy compared to other distances which should be chosen to get a good performance for BCI motor imagery classification.

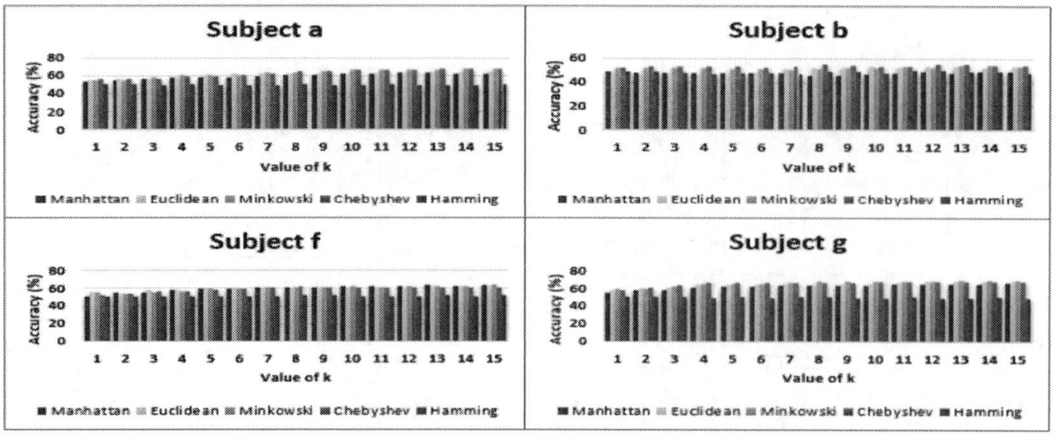

Fig. 4 Comparison of motor imaginary task classification for left hand vs right foot movement (Subject a and Subject f) and right hand vs left hand movement (Subject b and Subject g) by using different distance metrics and value of k

4.2 K-NN vs SVM

Although SVM classifier is commonly used for this dataset, the results shows that the k-NN outperformed most of the classification accuracy for subjects when the classification accuracy using SVM and k-NN is compared. The k-NN is very simple algorithm to implement and need least time compared to SVM.

For this comparison, Linear Discriminant Analysis (LDA) is applied before run the classifiers to reduce the data dimensionality in order to get optimum results. As the Minkowski distance at p =11 give the highest accuracy compared to other distances, this metric distance is used to be compare with SVM. Table 2 shows the result of classification accuracy of k-NN and SVM and Fig. 5 shows the graph of comparison between k-NN and SVM after applying the LDA.

Table 2. Results for classification accuracy of k-NN and SVM

Classifier	Accuracy (%)			
	Left hand & Right foot		Left hand & Right hand	
	a	f	b	g
k-NN	85.24	78.08	67.71	87.35
SVM	80.30	77.71	70.95	85.49

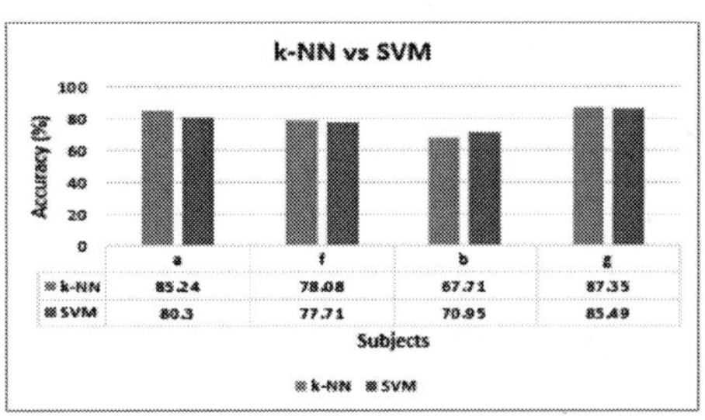

Fig. 5 Comparison classification accuracy between k-NN and SVM

* Corresponding author: nurulezzati@studentmail.unimap.edu.my

From the results obtained, it can be concluded that the k-NN algorithm is also suitable to use for classification method in BCI system as its result comparable to SVM.

5 Conclusion

The motor imagery classification of EEG signal for BCI applications is performed in this paper. Feature extraction of the signals is obtained using FFT, and the k-NN classifier is used to classify the data. The classification of motor imagery task using EEG data is performed to analyze performance analysis of k-NN. This paper demonstrates a detail the performance analysis of k-NN for EEG signal classification. The results of this research show that different accuracy of k-NN is obtained when applied with different k-value and distance metrics. This demonstrates that distance metrics and k-value affect the classifier performance and therefore worth to be considered in designing an EEG signal classifier. The results show that k-NN obtained comparable accuracy to SVM which commonly used for EEG signal classification. Given the expensive computational time of SVM, the lower complexity k-NN is promising for low-cost EEG classification. Since, the Hamming distance metrics perform badly on all datasets as having the lowest accuracy compared to others distance. Thus, Hamming distance not suitable for EEG data classification. In future work, more algorithms will be thoroughly investigated for EEG classification.

The Dataset 1 of BCI Competition IV is a property Berlin BCI group. We like to thank to Berlin BCI group by providing this dataset. In addition, this research is supported by Fundamental Research Grant Scheme (FRGS) from Ministry of Education, Malaysia. [Grant No: 9003-00525]

References

1. S. A. Sprague, M. T. McBee, E. W. Sellers, *The*

effects of working memory on brain-computer interface performance, Clin. Neurophysiol. **127**, 1331–1341 (2016)

2. X. Tang, N. Zhang, J. Zhou, Q. Liu, *Hidden-layer visible deep stacking network optimized by PSO for motor imagery EEG recognition*, Neurocomputing **234,** 1–10 (2016)
3. Denis R. de Moraes Piazentin and J. L. G. Rosa, *Motor imagery classification for Brain-Computer Interfaces through a chaotic neural network through a chaotic neural network*, Int. Jt. Conf. Neural Networks (2014)
4. M. E. A. Abdel-Hadi, R. A. El-Khoribi, M. I. Shoman, M. M. Refaey, *Classification of motor imagery tasks with LS-SVM in EEG-based self-paced BC,* 5th Int. Conf. Digit. Inf. Process. Commun. (CDIPC), 244–249 (2015)
5. M. Ming, L. Shaona, M. Haitao, M. Yuliang, G. Yunyuan, *Feature extraction method of motor imagery EEG based on DTCWT sample entropy*, 34th Chinese Control Conference (CCC), 3964–3968 (2015)
6. J. Hong, X. Qin, , J. Bai, P. Zhang, Y. Cheng, *A combined feature extraction method for left-right hand motor imagery in BCI*, IEEE Int. Conf. on Mechatronics and Automation (ICMA), 2621–2625 (2015)
7. Z. Qiu, J. Jin, H. K. Lam, Y. Zhang, X. Wang, A. Cichocki, *Improved SFFS method for channel selection in motor imagery based BCI*, Neurocomputing (2015)
8. B. Yang, H. Li, Q. Wang, Y. Zhang, *Subject-based feature extraction by using fisher WPD-CSP in brain-computer interfaces*, Comput. Methods Programs Biomed. **129,** 21–28 (2016)
9. J. Birjandtalab, M. B. Pouyan, D. Cogan, M. Nourani, J. Harvey, *Automated seizure detection using limited-channel EEG and non-linear dimension reduction*, Computers in Biology and Medicine ***82***, 49-58 (2017)
10. J. Kevric, A. Subasi, *Comparison of signal decomposition methods in classification of EEG signals for motor-imagery BCI system*, Biomedical Signal Processing and Control ***31***, 398-406 (2017)
11. T. Kayikcioglu, O. Aydemir, *A polynomial fitting and k-NN based approach for improving classification of motor imagery BCI data,* Pattern Recognit. Lett. **31,** 1207–1215 (2010)
12. J. Fan, J. W. Wade, D. Bian, A. P. Key, Z. E. Warren, L. C. Mion, N Sarkar, *A Step towards EEG-based brain computer interface for autism intervention,* Proceedings of the Annual Int. Conf. of the IEEE Engineering in Medicine and Biology Society (EMBS), 3767–3770 (2015)
13. K. Sabancı, M. Köklü, *The Classification of Eye State by Using kNN and MLP Classification Models According to the EEG Signals,* Int. J. Intell. Syst. Appl. Eng. **3,** 127 (2015)
14. A. Raheel, S. M. Anwar, M. Majid, B. Khan, Ehatisham-ul-Haq. *Real time text speller based on eye movement classification using wearable EEG sensors,* Proceedings of 2016 SAI Computing Conf., 161–164 (2016)
15. R. Aldea, M. Fira, A. Lazər, *Classifications of motor imagery tasks using k-nearest neighbors,* 12th Symposium on Neural Network Applications in Electrical Engineering, NEUREL 2014, 115–120 (2015)
16. F. A. Mousa, R. A. El-Khoribi, M. E. Shoman, *An integrated classification method for brain computer interface system*, 5th Int. Conf. on Digit. Inf. Process. and Commun., ICDIPC 2015, 141–146 (2015)
17. M. Maleki, T. Kayıkçıoğlu, *Classification of EEG signal during gaze on the different rotating vanes*, 24th Signal Processing and Communication Application Conference, SIU 2016 - Proceedings, 593–596 (2016)
18. J. T. Oliva, J. L. G. Rosa, *How an epileptic EEG segment, used as reference, can influence a cross-correlation classifier?*, Applied Intelligence, 1-19 (2017)
19. S. Raschka, *Python Machine Learning*. (Packt Publishing Ltd., (2016)
20. R. Bose, A. Khasnobish, S. Bhaduri, D. N. Tibarewala, *Performance Analysis of Left and Right Lower Limb Movement Classification from EEG*, 3rd Int. Conf. on Signal Processing and Integrated Networks (SPIN), 174–179 (2016)
21. M. Mohammadpour, M. Ghorbanian, S. Mozaffari, *Comparison of EEG Signal Features and Ensemble Learning Methods for Motor Imagery Classification*, 8th Int. Conf. on Information and Knowledge Technology (IKT), Hamedan, Iran, 288–292 (2016)
22. K. Chomboon, P. Chujai, P. Teerarassammee, K. Kerdprasop, N. Kerdprasop, *An Empirical Study of Distance Metrics for k-Nearest Neighbor Algorithm*, The Proceedings of the 2nd Int. Conf. on Industrial Application Engineering 2015, 280–285 (2015)
23. B. Blankertz, G. Dornhege, M. Krauledat, K. R. Müller, G. Curio, *The non-invasive Berlin Brain-Computer Interface: Fast acquisition of effective performance in untrained subjects,* Neuroimage **37,** 539–550 (2007)
24. M. Tangermann, K. R. Müller, A. Aertsen, N. Birbaumer, C. Braun, C. Brunner, R. Leeb, C. Mehring, K. J. Miller, G. R. Müller-Putz, G. Nolte, G. Pfurtsscheller, H. Preissl, G. Schalk, A. Schlögl, C. Vidaurre, S. Waldert, B. Blankertz, *Review of the BCI competition IV*. Frontiers in Neuroscience (2012)

MATEC Web of Conferences **140**, 01025 (2017) DOI: 10.1051/matecconf/201714001025
ICEESI 2017

A Computational Study on the Magnetic Resonance Coupling Technique for Wireless Power Transfer

*N.A.*Zakaria[1], *M.*Jusoh[1], *N.H.*Ghazali[1], *M.N.*Yasin[1], *T.*Sabapathy[1], *M.N.*Osman[1], *N.*Ahmad[1], M.Z.Zakaria[2]

[1]Bioelectromagnetic Research Group (BioEM), School of Computer and Communication Engineering Universiti Malaysia Perlis, 02600 Arau, Perlis

[2]School of Manufacturing Engineering, Universiti Malaysia Perlis, 02600 Arau, Perlis

Abstract. Non-radiative wireless power transfer (WPT) system using magnetic resonance coupling (MRC) technique has recently been a topic of discussion among researchers. This technique discussed more scenarios in mid-range field of wireless power transmission reflected to the distance and efficiency. The WPT system efficiency varies when the coupling distance between two coils involved changes. This could lead to a decisive issue of high efficient power transfer. This paper presents case studies on the relationship of operating range with the efficiency of the MRC technique. Demonstrative WPT system operates at two different frequencies are projected in order to verify performance. The resonance frequencies used are less than 100MHz within range of 10cm to 20cm.

1 Introduction

Nowadays, there are a lot of technologies applied in order to enhance our daily life such as transferring energy wirelessly which has been a recent topic of discussion. Wireless energy transfer has drawing more and more attention among researchers and in this sense a question on how the energies being transferred take place. Surveys found that there are a lot of methods for wireless energy transfer that exists either by using acoustic, electromagnetic, capacitive or inductive technique. These techniques are then classified into three, whether it is near field, mid-range field or far field depending on the distance between the transmitter and receiver [1].

An investigation to develop wireless power transfer system for wireless power charging application has attracted researchers, due to the need of more convenient way of power charging system [2]. The power requirement of each application to be recharged depends on the typical range from few microwatts to tens of milliwatts based on the battery power consumption [3].

In this paper, a case study of wireless non-radiative mid-range within 10cm-20cm power transfer using magnetic resonance coupling (MRC) method is proposed. The study is in between the transfer distance with the transmission efficiency. Based on this technique, the transmitter and receiver resonator are designed by following the Wheeler's formula parameter. In case study 1, resonance frequency of 73MHz is used. While in case study 2, the resonance frequency is 32MHz. It has been reported that in Tesla's work, the operating frequency for mid-range WPT ranging from 10kHz to almost 200MHz [4].

2 Related Theory

2.1. Magnetic Induction

It has been over 180 years since Michael Faraday figured out that power can be exchanged by magnetic induction through the air [5]. In 1914, Tesla first reported about the wireless power and transferring of data based on the magnetic coupling of two loops. The principle of mutual induction between two coils can be utilized in energy transfer for the exchange of electrical power which does not include any physical contact in between [6]. Transformer is the simplest example of how the mutual induction works. There is no physical contact between both primary and the secondary coils.

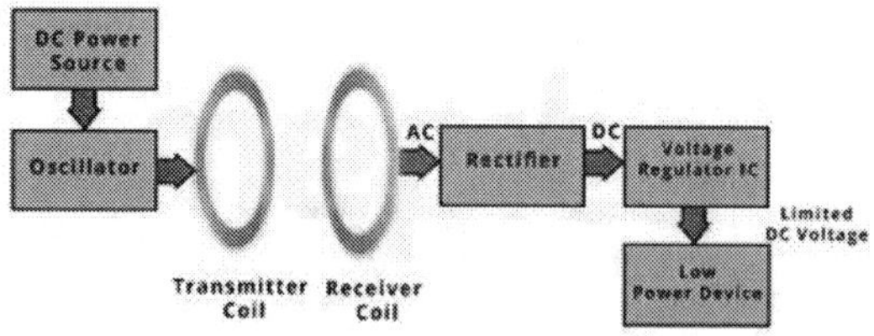

Fig. 1. Block diagram of inductive coupling

Fig. 1 shows the overview of inductive coupling where the technique of mutual induction is used to transfer power to the battery of any low power devices. The power transmission occurs when the transmitter coil transmit power to the receiver coil. In inductive power

* Corresponding author: atiqahzakaria92@gmail.com

DOI: 10.1051/matecconf/201714001025

transfer system, it has been common to adopt Tesla's principles which state that [4]

1) By using near-field which is non-radiative for magnetic coupling gives out magneto-inductive effects.
2) Resonances technique is applied for both transmitter and receiver designs.

However, inductive coupling can transfer power with high efficiency only in the range of a few centimeters (short-distance) [7]. Therefore, any devices used are required to get very close with the induction coil of transmitter so that charging process could take place.

2.1. Magnetic Resonance Coupling

Magnetic resonance coupling technique is a near-field magnetic inductive coupling methods considering the same resonance frequency at both transmitter and receiver which then known as the mid-range MRC.

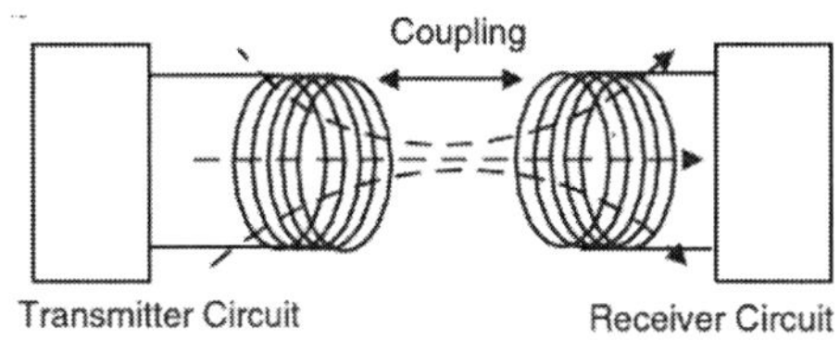

Fig. 2. Wireless charging system[6].

Fig. 2 shows the diagram of wireless charging system along with MRC. The transmission occurs due to the electromagnetic coupling between both coils without any change in the frequency used. Magnetic resonant coupling can transfer a huge amount of energy over an air gap at high efficiency [8]. The resonance frequency can be defined by using the given formula

$$f = \frac{1}{2\pi\sqrt{LC}} \quad (1)$$

Where, L and C are the inductance and capacitance of the coil respectively.

The coupling between two basic LC circuits with a certain distance is fragile. However, if both transmitter and receiver coils have the same natural resonance frequency, when they operate at the resonant frequency the two coils could generate resonance and produce strong coupling [6].

3 Coil Design

The coil designed in Fig. 3 is used in both case study 1 and case study 2 with different diameter sizes and compared with the same range of distances. The aim of these case studies is to prove that the relationship between distances and efficiencies are significant.

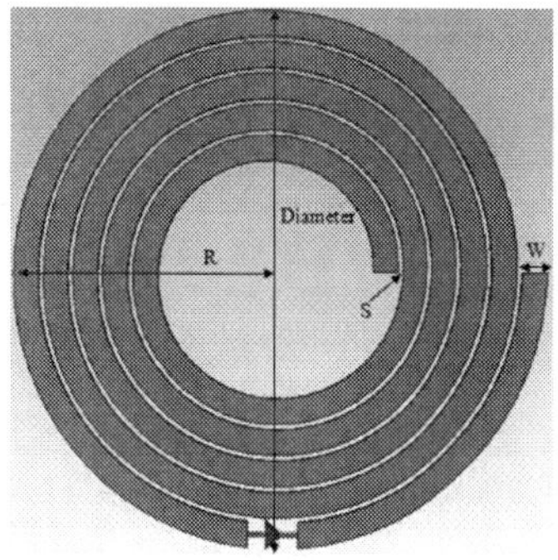

Fig. 3. Coil design parameter.

All the coil design and simulation is carried out in CST Studio Suite. While the parameter value involved is calculated by using classic Wheeler's formula for spiral coil given as [9][10]

$$L = \frac{(NR)^2}{2.54(8R + 11W)} \quad (2)$$

where,
L is inductance in micro-Henries
N is number of turns of wire
W is width of coil in cm
R is radius of coil in cm

3.1. Case Study 1

In case study 1, the size of coil used is 10cm x 10cm x 0.05cm with resonance frequency of 73MHz. All the transmitter and receiver coils parameters are listed in Table 1. Both coils use the same size and parameter values. The number of turns, N is 5, with radius 5cm and width of each turn of coil is 0.5cm.The height is 0.05cm while the spacing between coil's turn is fixed by 0.1 cm.

Table 1. Transmitting and receiving coil specification.

Parameters	Transmitter	Receiver
N (turns)	5	5
Radius (cm)	5	5
Width (cm)	0.5	0.5
Height (cm)	0.05	0.05
Spacing (cm)	0.1	0.1
Resonce frequency (MHz)	73	73

3.2. Case Study 2

In case study 2, the size of coil used is 20cm x 20cm x 0.05cm. All the parameter values involved are recorded in the Table 2. Both transmitter and receiver coil use the same parameter value. 5 turns of coil is designed with radius of 10cm and width of 0.5cm for each turn. The height is fixed likewise in case study 1 which is 0.05cm while gap between each turn is increased to 0.3cm.

Table 2. Transmitting and receiving coil specification.

Parameters	Transmitter	Receiver
N (turns)	5	5
Radius (cm)	10	10
Width (cm)	0.5	0.5
Height (cm)	0.05	0.05
Spacing (cm)	0.3	0.3
Resonance frequency (MHz)	32	32

4 Simulation Result

Simulation of coil design in CST studio suite gives out result of S_{11}(dB) and S_{21}(dB). Each graph S_{11} shows the transmission coefficient at a simulated resonance frequency of magnetic resonance coupling system. Graph S_{21} shows the efficiency of the system which is calculated by using formula

$$\text{Efficiency} = (S_{21})^2 \times 100\% \qquad (3)$$

Where, S_{21} value from graph is first converted from dB into numerical.

4.1 Discussion case Study 1

Fig. 4 shows that at a closed distance of 5cm, there are two resonance frequencies. This is due to the splitting frequency phenomenon between both coils where both coil are tightly coupled. The transmission coefficient S_{11} at 67.27MHz is -4.876dB with efficiency of 21.74%. While at 81.2MHz, the transmission coefficient is -6.572dB with efficiency of 35.56%.

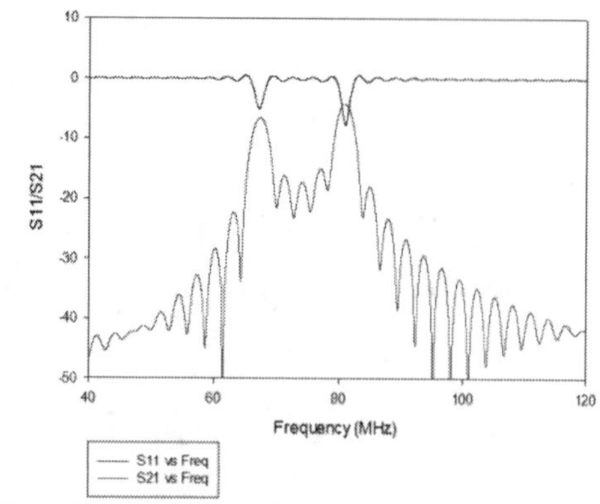

Fig. 4. Simulation result at 5cm

At a distance of 10cm, Fig. 5 shows the two resonance frequencies approach each other and become one at 73MHz. The transmission coefficient S11 is -5.2907dB with efficiency of 48.44%.

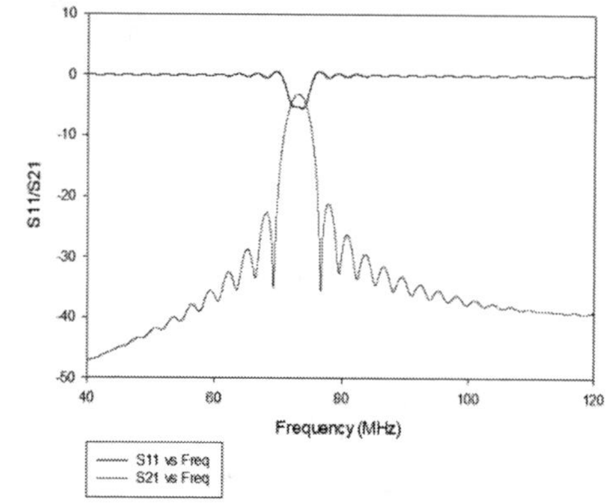

Fig. 5. Simulation result at 10cm.

As the distance becomes larger to 15cm, it can be observed that the transmission coefficient is drop to -21.07dB at 73MHz and the efficiency is 14.26% as in Fig. 6.

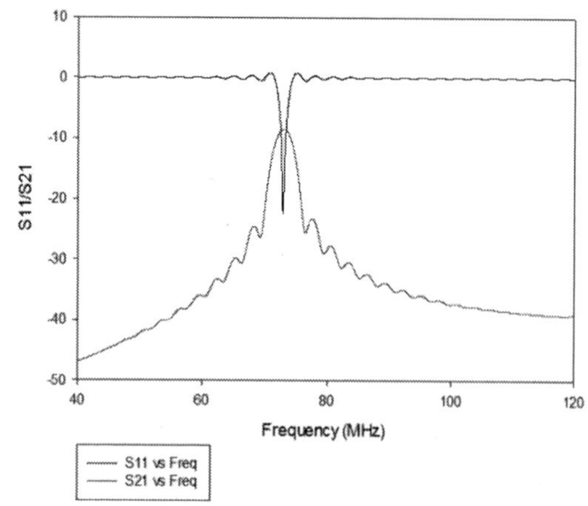

Fig. 6. Simulation result at 15cm.

When the distance is 20cm, Fig. 7 shows the transmission coefficient at 73MHz drop to -25.016dB and the efficiency is 0.172%.

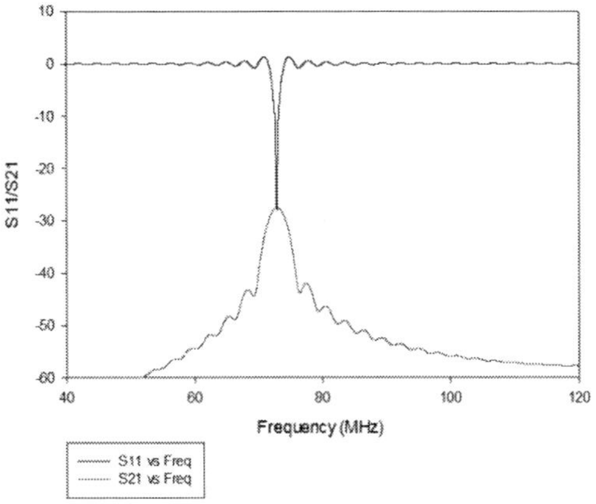

Fig. 7. Simulation result at 20cm.

4.2 Discussion case Study 2

In case study 2, the same observation likewise in the case study 1 is taken. In this case study, the observation is focused on the resonance frequency of 32MHz. At the distance of 5 cm, there are two resonance frequencies as shown in Fig. 8. At 29.1MHz, the transmission coefficient S_{11} is -2.5806dB with 6.34% efficiency. At 36.67MHz, the S11 is -4.359dB with 14.76% of efficiency.

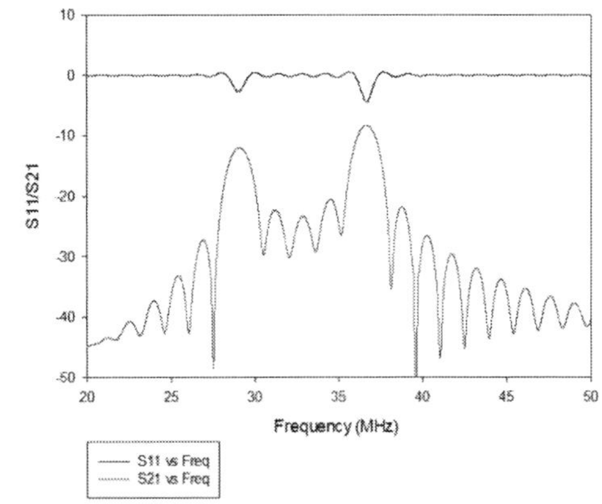

Fig. 8. Simulation result at 5cm.

When the distance is 10cm, there are also two resonance frequencies. Both frequencies become more close to each other. Fig. 9 shows the S_{11} is -3.089dB at 30.6MHz with 7.74% efficiency. At 34MHz, the value of S_{11} is -3.75dB with 10.83% efficiency.

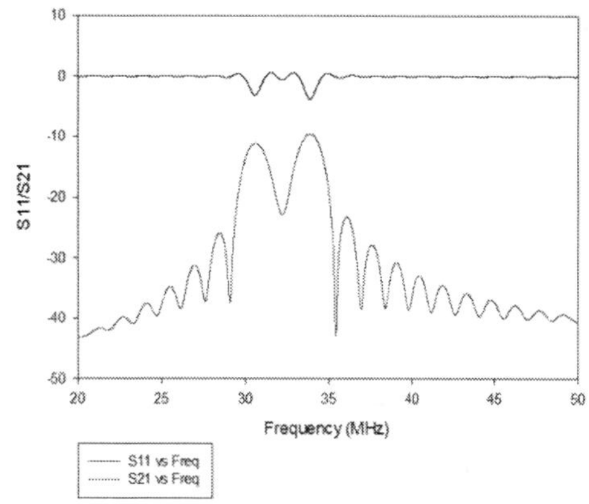

Fig. 9. Simulation result at 10cm.

The same observation is taken for a distance of 15cm. There are still two resonance frequencies at 31.2MHz with -3.43dB and 6.77%efficiency, and also at 33.05MHz with -3.7465dB and 8.35% efficiency as shown in Fig. 10.

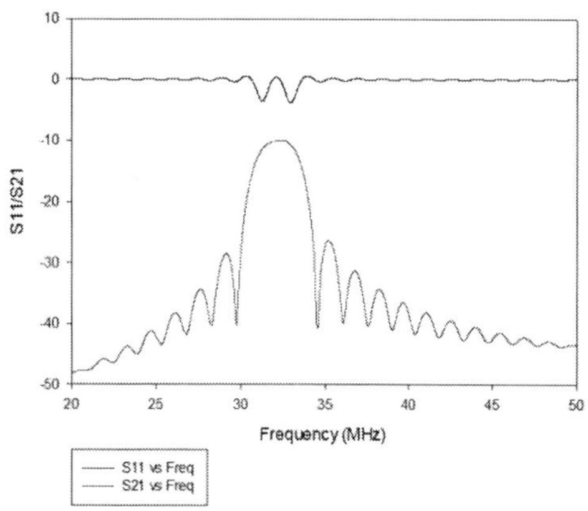

Fig. 10. Simulation result at 15cm.

Only when the distance is 20cm, the two resonance frequencies reach each other and become one at 32.33MHz with -3dB transmission coefficient and 17.72% efficiency as shown in Fig. 11.

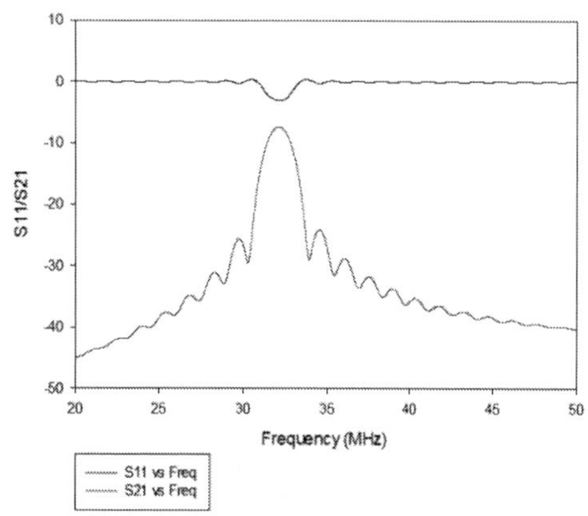

Fig. 11. Simulation result at 20cm.

5 Conclusion

From the discussion in case study 1 and case study 2, it is confirmed that the distance and the efficiency are related to each other. It can be concluded that the farther the distance, the lower the efficiency of the power transmission. Besides, the coil size and the gap between each turn also play a role in the change of resonance frequency. It seems like when the size of coil becomes larger, the resonance frequency becomes smaller. In this paper, simulation results verify the relationship between the distance and efficiency. Hence, it would be easier for further research to gain high efficiency with other design parameter.

The authors would like to thank the funding support from the Ministry of Higher Education for the MyBrain15 program. We also like to thank Dr Muzammil bin Jusoh for the FRGS grant number 9003-00494

References

1. K. Kalyan, S. Mohsin, A. Suresh, Transmission of Power through Wireless Systems. 394–398 (2013).
2. A. Munir, B. Ranum, T. Wireless Power Charging System for Mobile Device Based on Magnetic Resonance Coupling. 221–224 (2015).
3. O. Jonah, S. V. Georgakopoulos, Wireless Powering of Biomedical Device via Magnetic Resonance.
4. S. Hui, W. Zhong, C. Lee, A Critical Review of Recent Progress in Mid-Range Wireless Power Transfer. **29,** 4500–4511 (2014).
5. A. Amar, A. B. Kouki, H. Cao, Power Approaches for Implantable Medical Devices. 28889–28914 (2015). doi:10.3390/s151128889
6. B. Zhu, J. Li, W. Hu, X. Gao, Review of Magnetic Coupling Resonance Wireless Energy Transmission. **8,** 257–272 (2015).
7. A. K. Sah, Design of Wireless Power Transfer System via Magnetic Resonant Coupling at 13 . 56MHz. 202–210
8. Y. Yang, C. Wang, Wireless Rechargeable Sensor Networks. 1–7 (2015). doi:10.1007/978-3-319-17656-7
9. X. Lu, P. Wang, D. Niyato, D. I. Kim, Z. Han, Wireless Charging Technologies : Fundamentals , Standards , and Network Applications. 1–40 (2015).
10. J. Choi, J. Cho, C. Seo, IWPT-P-2 Analysis on Transmission Efficiency of Wireless Energy Transmission Resonator Based on Magnetic Resonance. 199–202 (2011).

DOI: 10.1051/matecconf/201714001026

Smart Rash Driver System via Internet of Things (IoT)

Kong Sheau Tong[1], *Mohd Natashah* Norizan[1,2,3,*], and *Ili Salwani* Mohamad[1,2,3]

[1]School of Microelectronic Engineering, Universiti Malaysia Perlis, Pauh Putra Campus, 02600 Arau, Perlis, Malaysia.
[2]Centre of Excellence Geopolymer and Green Technology (CeGeoGTech), Universiti Malaysia Perlis, Perlis, Malaysia.
[3]Advanced Multi-Disciplinary MEMS-Based Integrated NCER Centre of Excellence (AMBIENCE), Universiti Malaysia Perlis, Perlis, Malaysia.

Abstract. Nearly half a million accidents on Malaysians road occur in 2015. The aim of this research is to detect car speed, capture the photo of the speeding car and then transfer the data like car speed, date and time, location and lane number to an online database. A distance sensor is used to measure the distance range between two points on the road. The ESP8266 NodeMCU will be the control unit to process the data and calculate the speed with the formula of speed equal to distance over time. The ESP8266 NodeMCU is also a Wi-Fi module to help in transferring data via IoT to an online database. The Google spreadsheet acted as an online database and will receive all the data if detected a speeding car. In conclusion, the Smart Rash Driver System is successfully invented and able to detect vehicle speed, capture the photo of over speed vehicle and save it to the SD card and lastly transfer all data via IoT to the Google Spreadsheet. This invention will be able to help to decrease the road accident rate efficiently.

1 Introduction

In 2015, nearly half a million accidents on Malaysians road were recorded [1, 2]. Most of the cases happened were caused by the human error, faulty vehicle, driving under the influence of alcohol, drug effect and speeding [3-7]. Moreover, according to the cause of death record of World Health Ranking which is updated on 2014, road accident is rank 5 in Malaysia and rank 34 in worldwide [8-10].

The government has figured out a solution for this problem by placing Automated Enforcement System (AES) speed tracker near major roads in Malaysia [11]. According to Jabatan Pengangkutan Jalan Malaysia, after the first eight months since the AES was implemented, the accident rates at 14 locations nationwide have dropped by up to 30% [12]. Therefore, more speed trackers need to be installed to decrease the rate of road accident as mentioned by Malaysian Institute of RoadSafety (Miros) Chairman [13]. Road users might not be stopped from speeding in camera-free section, but they will be more aware on the speed limit all along the road. However, a huge cost needed to install those speed monitoring system such as AES in whole Malaysia, therefore human intervention as well is needed to detect the speed of moving vehicle [11]. The drawback of human intervention is human unable to work 24 hours and besides, human error might occur [14-17].

The objective of this project is to develop a low cost speed monitoring system, which can capture the photo for car that are speeding. Besides, this research also will provide real-time wireless reporting data speed, location, date and time, location and lane to the centralized database via Internet of Things (IoT). NodeMCU which is an open source IoT platform will be used for the data transfer as it promotes a breakthrough in wireless sensor networks [19] and very popular among various applications [20-23].

2 Proposed architecture

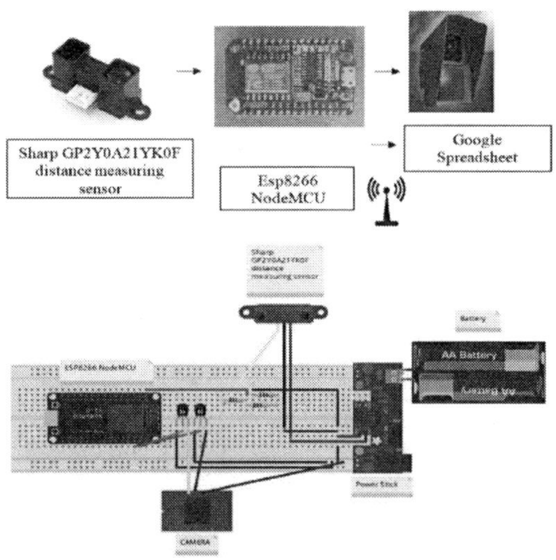

Fig. 1. Overview of system block design and schematic diagram.

The smart rash driver system is a speed monitoring system, which able to capture car speed, date, time, lane,

* Corresponding author: mohdnatashah@unimap.edu.my

location and then transfers the data to a database via IoT. Besides, when the system detected a speeding car, it will trigger the camera automatically to capture the car's photo. Therefore, the input of the ESP8266 NodeMCU is a Sharp GP2Y0A21YK0F distance measuring sensor and the output of the ESP8266 NodeMCU will be the camera and Google spreadsheet. Figure 1 shows the overview of the system block diagram and the schematic diagram of the design.

The main input for the speed monitoring system is the Sharp laser sensor and the main control unit will be the ESP8266 NodeMCU. ESP8266 NodeMCU will act as a control unit and Wi-Fi module at the same time. After processing the data, ESP8266 NodeMCU will send it via IoT to Google spreadsheet. Lastly, the authorities will able to observe the data from the Google spreadsheet.

Resistor in the circuit act as the voltage divider of the circuit. It can change the input voltage of the Sharp distance sensor from 5 to 3.3 V. Therefore, it will not burn out the ESP 8266 NodeMCU because of the high voltage. ESP 8266 NodeMCU only able to support the input voltage that less than 3.3 V. Moreover, the transistor used in the circuit is to push the input voltage to the maximum so that it can provide enough power to trigger the camera. The breadboard power stick can be powered by two 3.6 V batteries. Therefore, it makes the system become portable.

2.1. System design

Figure 2 shows the system design flowchart. At the beginning, the ESP8266 NodeMCU will check the Wi-Fi connection with the Internet. Then, the Sharp GP2Y0A21YK0F distance sensor will start to detect the distance range. The distance between two points will be calculated when the vehicle passed through the sensor. The system on road diagram in Figure 3 shows the measurement of the distance while speeding.

The system will calculate the speed with the distance range that is received from the sensor and the time range that provided by the Arduino IDE in the unit of a millisecond. The data of distance1, time1, distance2, and time2 are then collected. The formula of calculating the speed is:

$$Speed = \frac{Distance1 - Distance2}{Time2 - Time2} \quad (1)$$

After calculating the speed, the unit of the speed will be cm/ms, therefore a conversion needed for changing the unit to km/hour. For getting the timing, time will be provided from the Arduino IDE with the coding of "millis ()". The timing will be count in the unit of milliseconds. The reason of using millisecond instead of a second because it is the smaller unit of time and able to increase the accuracy of calculating the speed of the moving vehicle.

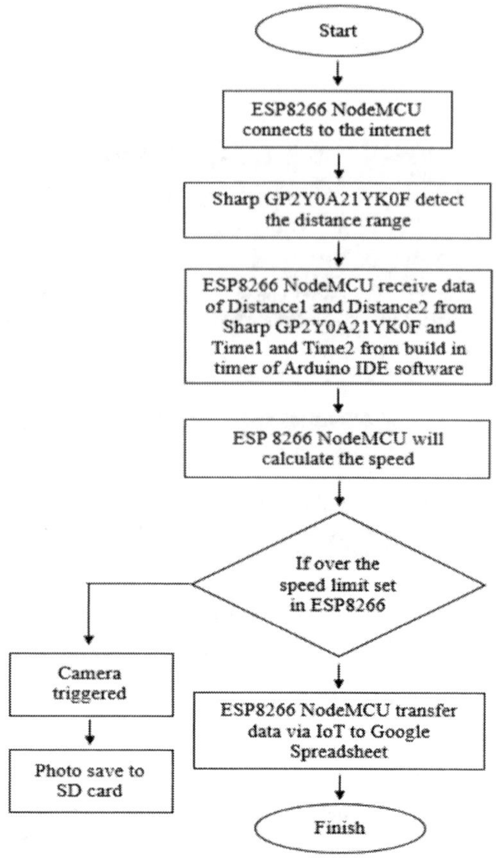

Fig. 2. System design flowchart.

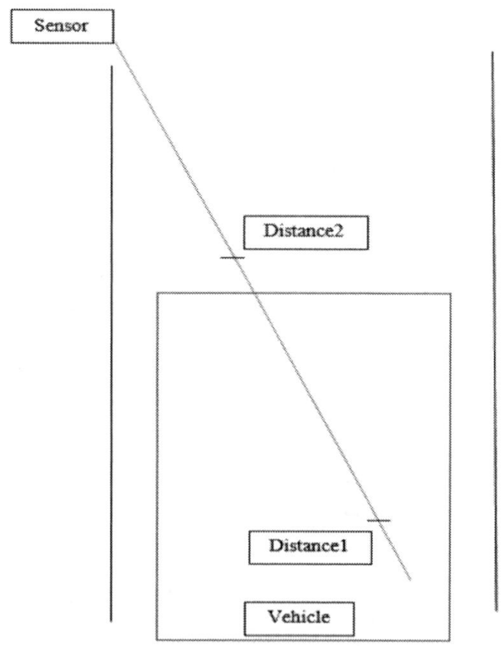

Fig. 3. System on road diagram.

After the speed is calculated, the ESP8266 NodeMCU will begin to process the data. If the speed of the vehicle are over the speed limit that being set, it will trigger the camera to capture the photo of the speeding vehicle. The photo captured will be saved on an SD card. Besides, the ESP8266 NodeMCU will also provide the Wi-Fi connection ability for sending the data wirelessly to an online database.

2.2. Internet of Things (IoT) design

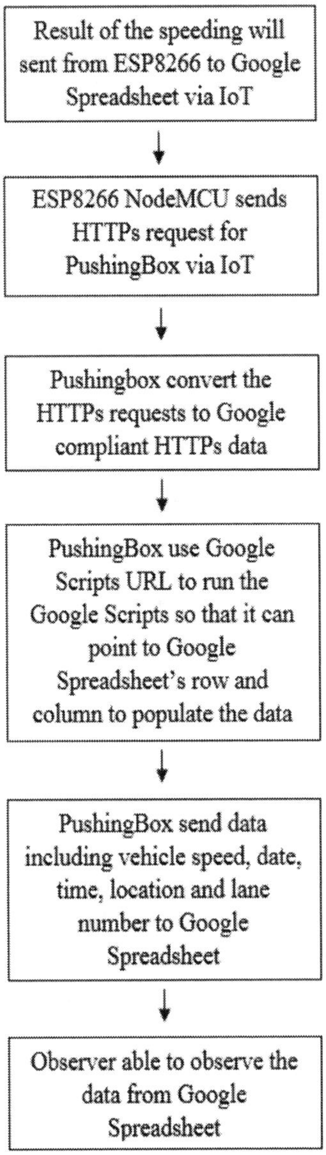

Fig. 4. IoT process flow chart.

Figure 4 shows the IoT process flow chart. Before uploading the data to the Google spreadsheet, there are a few steps need to be done. Firstly is creating and deploying a Google spreadsheet. Log into the free Google account and create a new "Google Sheet". This is the spreadsheet which will be populated with all the data of speed, location, date and time happened, lane number and location via Wi-Fi. After creating the Google spreadsheet, the spreadsheets URL key is copied and saved. This key is listed in the URL between the "/d/" and the "/edit" of the new spreadsheet. Lastly, the title of the spreadsheet is renamed. Figure 5 shows the example of Google spreadsheet.

Fig. 5. Google spreadsheet.

Next, a Google app script (similar to JavaScript) is created which will process the data and populate to the specific spreadsheet correctly. To insert the Google app script, first navigate from the spreadsheet to the script editor. Then, the coding that prepared for the Google spreadsheet is keyed in to Google script and then the Google spreadsheet URL key is inserted to make sure the data point to the right spreadsheet as shown in Figure 6. Users need to program the Google Scripts so that user able to program a build in date and time, all the data are then populated into specific rows and columns.

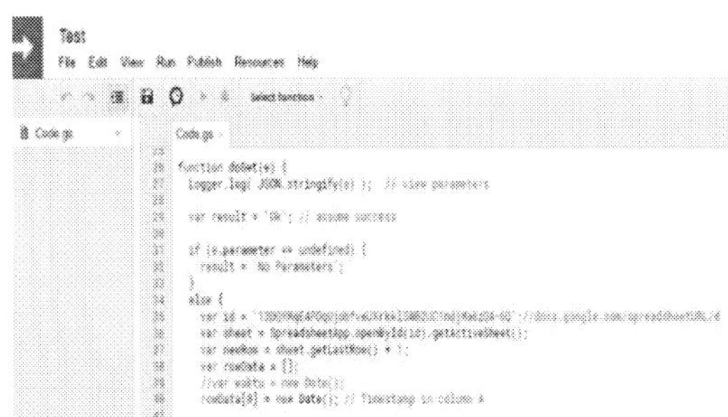

Fig. 6. Google script.

Next, the Google spreadsheet is saved in other to receive the web app URL to be used in the next step. Then the Web app URL need to be saved.

The current web app URL that is shown in Figure 7 will be used for PushingBox so that the PushingBox able to communicate with the Google app scripts. For the last step of the IoT is configuring PushingBox. The need for using the PushingBox API is to convert the HTTP transmitted data into Google compliant HTTPS encrypted data. The web app URL will be used now for PushingBox to send the data to a right Google sheet. First, a service must be created in PushingBox to let the user interface with any API of any product. Figure 8 shows how to create a service in PushingBox.

DOI: 10.1051/matecconf/201714001026

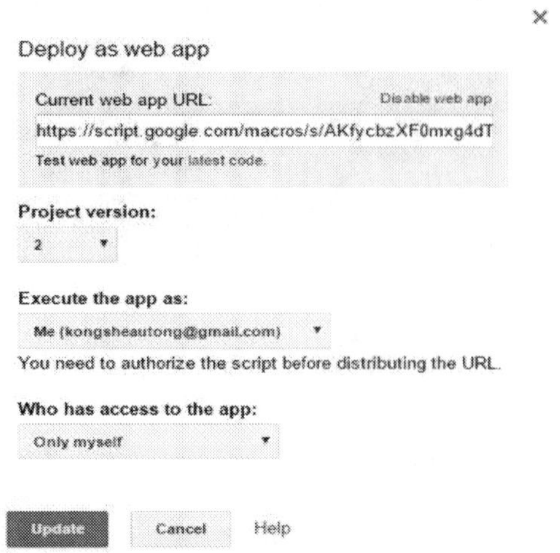

Fig. 7. Save the web app URL.

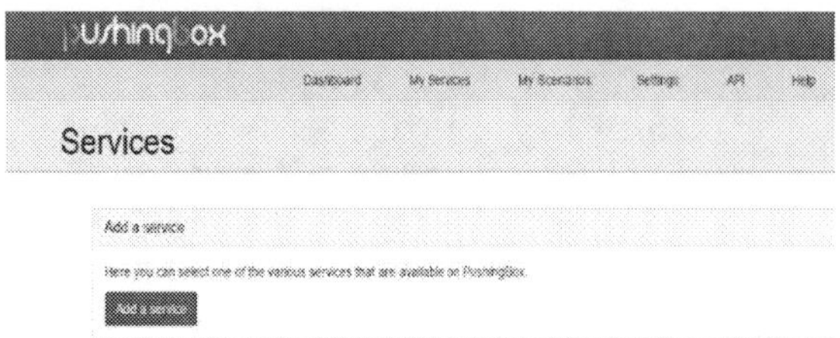

Fig. 8. PushingBox services.

Next, the web app URL that is shown in Figure 7 will now be used in the PushingBox while creating the services. Figure 9 shows the web app URL of Google scripts will be filled in the column of Root URL. There are two options able to choose in the column of method, which is Get or Post. For this project, the PushingBox need to fetch the data from the ESP8266, therefore, the option that chosen need to be Get.

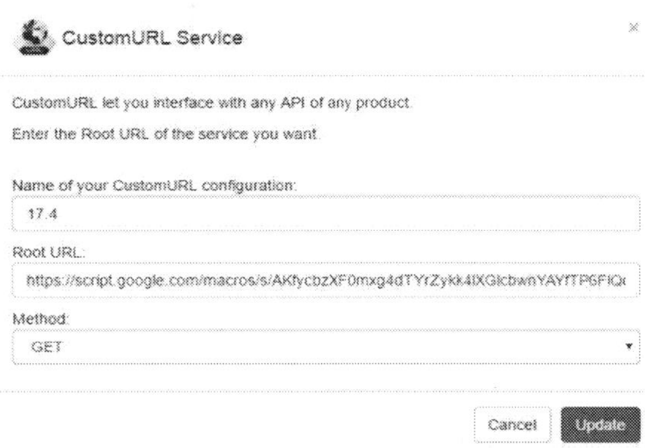

Fig. 9. Key in the web app URL in the Root URL.

Then a scenario is created and user able to receive a Device ID as shown in Figure 11. Device ID will be used in the program code of Arduino IDE as shown as Figure 10. The device ID is used to locate the right service of PushingBox device. The yellow highlight section in Figure 10 is the place that placing the Device ID. The Device ID can be changed for other different services.

```
#include <ESP8266WiFi.h>
#define D1 5
const char* ssid = "Tong";
const char* password = "58648851";

const char WEBSITE[] = "api.pushingbox.com"; //pushingbox API server
const String devid = "v262EAE71022AF16"; //device ID from Pushingbox
```

Fig. 10. Device ID.

Fig. 11. Device ID with command.

Figure 11 is the place that placing the command string to tell the PushingBox which data it should fetch and what data should be posted. The data that with the symbol of $Distance$ is means that the data is getting from the microcontroller. User also able to set a constant data that desire to be posted to the Google spreadsheet, the data must be wrote without the symbol of $. Lastly, through those processes, the data is able to transfer wirelessly to the Google spreadsheet successfully and accurately.

3 Results and discussion

Figure 12 shows the result of the prototype that is made for this project. The prototype contains a Sharp GP2Y0A21YK0F distance sensor, ESP8266 NodeMCU, camera, resistor and transistor.

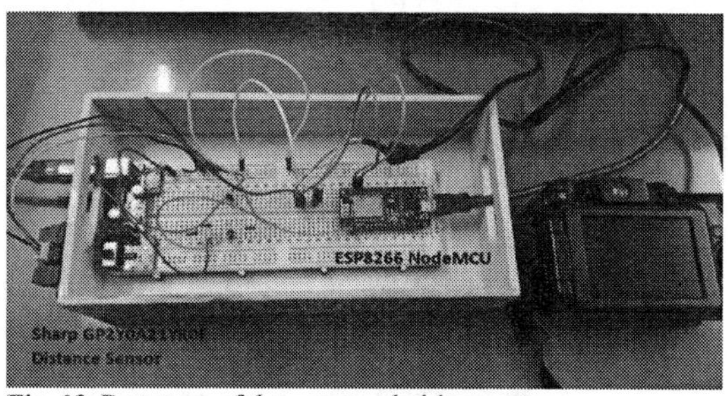

Fig. 12. Prototype of the smart rash driver system.

Figure 13 shows the result that the ESP 8266 NodeMCU has connected to the Internet. It will show "Credentials accepted! Connected to wifi" at the serial communication of Arduino IDE. After the successful connection, the data will be able to be transferred wirelessly.

Figure 14 shows the condition for the serial communication of Arduino IDE when the ESP 8266 NodeMCU unable to connect to a Wi-Fi Server. If the ESP 8266 NodeMCU did not connect to an Internet, the serial communication will keep showing "..........." until it connected to an Internet.

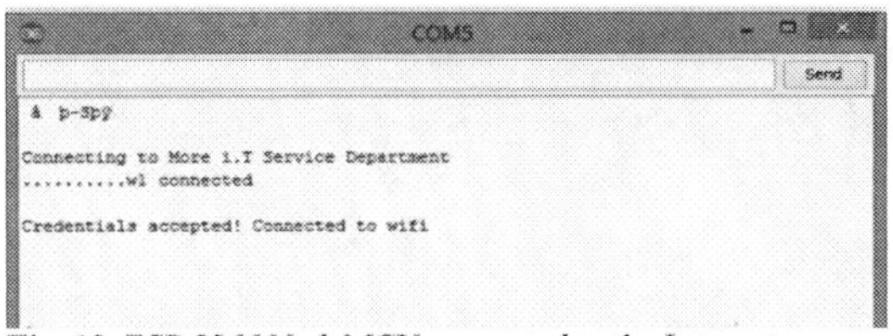

Fig. 13. ESP 8266 NodeMCU connected to the Internet.

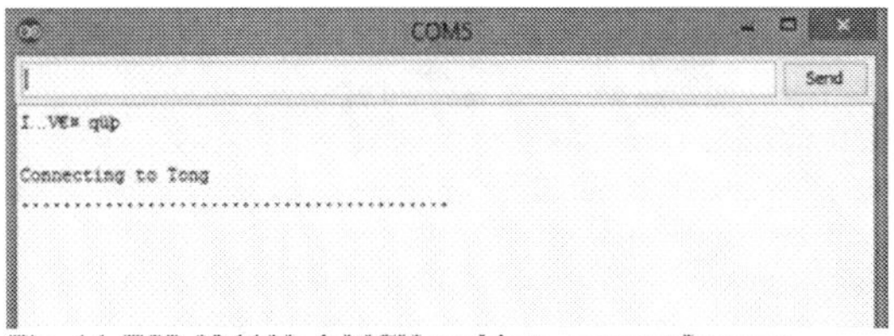

Fig. 14. ESP 8266 NodeMCU unable to connect Internet.

Figure 15 shows the result of the data that are uploaded to the Google spreadsheet. The data uploaded to the Google spreadsheet consist of vehicle speed, date and time, lane number and the location of speeding. As for testing purpose, the system has been set to capture the vehicle with speed over 0.5 km/h.

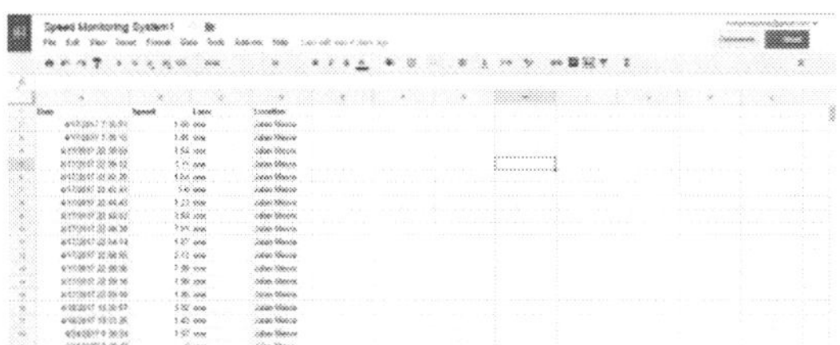
Fig. 15. Result of speeding upload to Google spreadsheet.

4 Conclusions

This work demonstrated a smart rash driver system that manage to capture of the speeding car, calculating the speed and transfer the data to online database via IoT through PushingBox service. Compared to the existing AES, this system does not need LAN cable to transfer the data besides of it is a portable system (easy to install and operate). Based on the evaluation, the data and photo collected were in real time condition and successfully transferred to the database when speeding car triggered the sensors. Therefore, this system is expected to help the authorities to reduce the road accidents in the future with an efficient monitoring system.

References

1. Bernama, Malay Mail Online (2017). URL: http://www.themalaymailonline.com/malaysia/article/average-of-18-people-killed-daily-in-road-accidents-in-2015-says-liow [accessed: 2017-01-13]
2. T. Yi Liang, The Star Online (2016). URL: http://www.thestar.com.my/news/nation/2016/04/18/transport-ministry-almost-half-a-million-road-accidents-in-2015/ [accessed: 2017-01-15]
3. N. H. Rodzi, The Star Online (2015). URL: http://www.thestar.com.my/news/nation/2015/05/02/accident-family-death/ [accessed: 2017-01-15]
4. R. Krishnan and R. U. R. Sohadi, J. Heal. Transl. Med. **2**, 39 (1997).
5. A. Kareem, Malaysian J. Med. Sci. **10**, 31 (2003).
6. L. Abdullah N. Zamri, J. Emerg. Trends Comput. Inf. Sci. **3**, 225 (2012).
7. M. G. Masuri, K. A. M. Isa, M. P. M. Tahir, Procedia - Soc. Behav. Sci. **38**, 213 (2012).
8. WHO, HEALTH PROFILE : MALAYSIA (2014). URL: http://www.worldlifeexpectancy.com/country-health-profile/Malaysia [accessed: 2017-01-17]
9. WHO, Global Status Report on Road Safety 2015 (2015). URL: http://www.who.int/violence_injury_prevention/road_safety_status/2015/en/ [accessed: 2017-01-17]
10. E. Zarulazam, H. Evdorides, IATSS Res. (2017)
11. H. Sivanandam, The Star Online (2015). URL: http://www.thestar.com.my/news/nation/2015/11/16/parliament-contract-aes/ [accessed: 2017-01-21]
12. M. Carvalho, Y. Meikeng, H. Sivanandam, The Star Online (2014). URL: http://www.thestar.com.my/news/nation/2014/10/30/aes-helps-reduce-accidents-says-liow-cases-at-14-locations-dropped-30-since-implementation/ [accessed: 2017-01-21]
13. Bernama, TheStar Online (2016). URL: http://www.thestar.com.my/news/nation/2016/04/29/more-aes-cameras-necessary-to-nab-speedsters/ [accessed: 2017-01-21]
14. J. M. Harrington, Occup. Environ. Med. **58**, 68 (2001).
15. A.G. Foord, W.G. Gulland, Process Saf. Environ. Prot. **84**, 3, 171-173 (2006).
16. J. M. Haight and R. G. Caringi, Int. J. Risk Assess. Manag. **7**, 708 (2007).
17. J. M. Haight, V. Kecojevic, Process Saf. Prog. **24**, 45 (2005).
18. S. Borsay, Transmit ESP8266 Data to Google Sheets (2016). URL: https://www.hackster.io/detox/transmit-esp8266-data-to-google-sheets-8fc617 [accessed: 2017-01-25]
19. M. Mehta, Int. J. Electron. Commun. Eng. Technol. **6**, 7 (2015).
20. R. L. Tirupatamma, A. Pulagam, Int. J. Eng. Sci. Res. Technol. **5**, 634 (2016).
21. K. Chooruang, P. Mangkalakeeree, Procedia Comput. Sci. **86**, 160 (2016).
22. T. A. Abdulrahman, O. H. Isiwekpeni, N. T. Surajudeen-Bakinde, A. O. Otuoze, Procedia Comput. Sci. **94**, 473 (2016).
23. B. Yong, Z. Xu, X. Wang, L. Cheng, X. Li, X. Wu, Q. Zhou, J. Parallel Distrib. Comput. (2017).

 DOI: 10.1051/matecconf/201714001027

Brain Computer Interface: Assessment of Spinal Cord Injury Patient towards Motor Movement through EEG application

Syahrull Hi-Fi Syam[1], Heba Lakany[2], R.B. Ahmad[3] and Bernard A. Conway[2]
[1]Polytechnic of Tuanku Syed Sirajuddin, 02600, Perlis, Malaysia
[2] Biomedical Engineering, University of Strathclyde G4 0NW, Glasgow, United Kingdom.
[3]Faculty of Informatics and Computing, Universiti Sultan Zainal Abidin (UniSZA), Kuala Terengganu, Malaysia

Abstract. Electroencephalography (EEG) associated with motor task have been comprehensively investigated and it can also describe the brain activities while spinal cord injury (SCI) patient with para/tetraplegia performing movement with their limbs. This paper reviews on conducted research regarding application of brain computer interface (BCI) that offer alternative for neural impairments community such as spinal cord injury patient (SCI) which include the experimental design, signal analysis of EEG band signal and data processing methods. The findings claim that the EEG signals of SCI patients associated with movement tasks can be stimulated through mental and motor task. Other than that EEG signal component such as alpha and beta frequency bands indicate significance for analysing the brain activity of subjects with SCI during movements.

1 Introduction

Spinal cord injury (SCI) is an incapacitating disorder that affects physical and psychological health which lead to loss of motor and sensory function [1], increased risk of depression, on-going pain syndromes, profound lifestyle change, drug dependence as well as increased hospital admissions and decreased general health [2,3]. SCI may be caused either by trauma or disease and consequences of the SCI may drive the patient to be discriminated, isolated and restricted from normal daily life activity due to the motor and sensory disabilities.

Despite the limitation of independency, there are number of modalities which can offer the SCI population an assessment tool and a significance level of recovery through invasive or non-invasive method. For instance stem cells method, engineering approach and brain computer interface (BCI) approach. Among these 3 approaches BCI offer more applicable choice due to its reliability and promising results. This is because the stem cells method restrains by financial issue [4] and lack of independent validation's results of the various stem cell treatments [5]. Apart from that engineering approach induced modest behavioral which need further investigation.

BCI applies electroencephalography (EEG) signal to communicate with augmentative and assistive devices which offer an alternative to those who suffered from motor and sensory disabilities due to the SCI. BCI cover various type of application and one of it is derivation information from EEG signal regarding performing motor task and imaginary performing motor task [6]. In order to derive this information, it is essential to improve the EEG signal processing techniques and selection of features. By achieving this, the speed and accuracy of communication of the interface devices can be improved [7] and the training time of using it can be reduced.

2 Experimental Design

In recent years, there have been advances in a numbers of research investigating and developing successful BCI endeavors to improve the speed and accuracy of communication of these interfaces. In order to achieve this goal, multidimensional approaches and modalities have been implemented involving mental task [6], [8], [9] and various motor tasks utilising finger [10]-[14], hand [15]-[17],[19], foot [16], [18], [19], and tongue movement [19].

Craig and Nguyen [8] design four different mental tasks for patients with chronic SCI. The mental tasks which were chosen to evoke changes in brain activity were intended to generally give results which tended to favor one hemisphere over the other, as the intention of the research is to use the differences in EEG over the two hemispheres to help with the classification. The results established that the 'stop-start' classifier has achieved a total accuracy of 94%, while the 'left-right' classifier has achieved approximately 77.5% accuracy.

Studies carried by Erbil & Ungan [14], investigate whether the suppression that is known to occur during the onset of a single movement would persist or otherwise habituate when the movement is continuously repeated for a long period of time. They designed experimental protocol which required subjects to perform repetitive simultaneous extension-flexion of the fingers II-V in one hand, continuously for a period of at least 30 s with interval of rest of 30 s. The finding demonstrated that the sustained and continuous movements are conducted through quite different processes and modulation of the beta amplitude, rather than the changes in alpha amplitude, appeared to be more closely correlated with the execution of a continuous movement.

Apart from that Boi et al [17], conduct a research exploring the effectiveness combinations of computational methods for prediction of movement intention preceding the production of self-paced right and left hand movements. They designed a protocol

syahrull30@gmail.com

which require subjects to perform self-paced movements consisting of three key strokes with either hand. The results indicated that the discrimination accuracy was as high as 75% by utilizing one of the better combinations of independent component analysis (ICA), power spectral density estimation (PSD) and support vector machine (SVM). Results also claimed that beta band EEG activity of the channels over right sensorimotor cortex was most appropriate for discrimination of right and left hand movement intention.

On the other hand, Morash et al [18], investigating on neural signals preceding 4 movement (right hand, left hand, tongue, and right foot movements) and motor imagery in order to predict the occurrence of the four movements/motor imageries. They adapt an experimental protocol that require subjects to perform or kinaesthetically imagine four movements which instructed with a specific stimulus and performed at a second stimulus. The obtained results indicate that accuracy of movement/imagery preparation classification varied between subjects. It also include that high accuracies were associated with classifications relied heavily on the ERD/ERS in order to discriminate movement/imagery planning.

3 Signal Analysis of EEG Band Signal

EEG or brainwaves are electrical signals produced by excitatory and inhibitory potentials of neurons (electrical activity of an alternating type) recorded from the scalp surface after being picked up by metal electrodes and conductive media [20].

Normally EEG activity can be characterised as being within five distinct frequency bands, for instance delta/δ (0.1 – 3 Hz), theta/θ (4 – 7 Hz), alpha/α (8 – 13 Hz), beta/β (14 – 29 Hz) and Gamma/σ (30 – 40 Hz) [21]. Numbers of study on BCI have been using one or a combination of these frequency bands in order to access variety of brain signal component such as visual evoked potentials [22], [23], slow cortical potentials [24], [25], P300 evoked potentials [26], [27], mu and beta rhythms [28],[29] and cortical neuronal action potentials [30].

Antonio Castro et al [31], extract the movement-related brain potentials (MRBPs) consisting of the readiness potential, pre-motion positivity, motor potential and reafferent potential in order to reflect brain electrical activity associated with processes of movement preparation and movement execution.

On the hand, Boord et al [32], compared brain wave activity (frontal, central, parietal, occipital and temporal regions) between the able-bodied and SCI subjects utilizing the 8–13Hz magnitude, peak amplitude and frequency of 8–13Hz levels. Reporting in his other work [33] regarding comparing the peak frequency (θ-α band) and power (δ, θ, α and β bands) between able-bodied, no-pain, and pain groups; peak θ-α band frequency (PTAF) was calculated as the frequency of peak power between 4 and 13Hz.

Apart from that Yong woong et al [34], examining mu and beta event-related desynchronization/ synchronization (ERD/ERS) during motor imagery tasks with varying movement duration and the potential impacts of movement duration on ERD/ERS patterns. This study measured ERD/ERS through mu (8-13 Hz) and beta bands (14-30 Hz). The ERD/ERS was quantified following standard calculation of Pfurtscheller and Aranibar [35].

According to Jean-Michel Guerit's article [36], on the main principles and indications of EEG and evoked potential (EP) neuromonitoring in the operating room; which claim that neuromonitoring in the operation room can assess the depth of anesthesia and prevent ischemic brain or spinal cord damage, mechanical injuries of the brain. At the same time it also may help anesthesiologist to control the level of sedation and aid surgeons to prevent neurological complication when they are performing an operation.

In Heba and Conway [37] studies, motion related potentials (MRP's) were recorded from the motor cortex area in the left hemisphere of brain through channel C3. MRP's were analyzed from α (8 – 12 Hz), β (18 – 24 Hz) and σ (28 – 37 Hz) bands and the results from this study indicate that the different direction of recorded EEG signals of wrist movement can be discriminated using Euclidean distance classifier.

4 Data Processing Methods

Recorded EEG data are always being contaminated by artifacts, interfering noises due to the electronic amplifier, power line interference and external interference plus the skull which acts as a spatial low-pass band filter that eliminates much of the information generated intracortically [38]. Thus in order to build BCI that apply real time EEG signal as medium to communicate/ control augmentative , it is essential to improve the EEG signal processing techniques (pre-processing, features extracting and features classifying).

Pre-processing procedure is carried out as for removing the unwanted signal and noise by using spatio-spectro-temporal filters. These filters can be simple frequency filters (low pass filter, band pass filter), independent component analysis (ICA) and common spatial pattern. For instance Heba and Conway [37] used low pass filter at 50 Hz for removing eye blink and movement artifacts. As for same purpose for pre-processing process, Yongwoong et al [34] also have implemented band pass filtered at 8-30Hz and Boord et al [32] adopt ICA for removing eye blink and saccade artifacts.

Features extraction involves simplifying the amount of resources required to describe a large set of data accurately. In BCI endeavors, feature extraction taxonomy can be grouped broadly as time domain, frequency domain and time-frequency domain. Heba and Conway [37] use continuous wavelet transform (CWT) for extracting spatio-temporal characteristic of EEG during movement transition and movement planning. Navarro `et al [38] extract features mainly on time based feature which include 50ms window averages of rectified EEG amplitude, dominant frequency, time-domain variance over 2s response and total power over 2s response (using Welch's method) in order to classify wrist movement (flexion, extension, pronation, supination). In the other study by Antonio Castro [40] which focus in analysing the relationships between movement preparation and movement execution, the amplitude of each one of the MRBPs were measured as the dependent variable.

Features classifying is a process to assign automatically a class (mental task performed by BCI user) to the feature vector which previously extracted using algorithm known as classifiers. In BCI field most of the classifiers belongs to linear classifier, neural network, nearest neighbour and classifier combination. Craig and Nguyen [8] have implemented neural network classifier and suggested the inclusion of adaptive training of the classifier. In their work the classifier was able to generalize the result from subjects (5) and learn to recognize them with an overall success rate of 82.4%. In the other study by Heba and Conway [38], classification of the extracted features vector was done by measuring Euclidean distance between the test sample and the representation vector. The classification's results are promising and prove that the suitability of the technique for the application.

5 Conclusion

This paper specifies the major role played by EEG as essential medium in BCI endeavors which offer alternative way of communicating and interacting with outside environment among the impairments community population. The promising and convincing results from the recent research demonstrate that, in the future BCI will be the main option for non-invasive method for controlling neuroprosthesis in neural impairments community such as SCI. Future works will focus more on investigating and discovering the effective modalities which can stimulate the organization of brain functional networks activated in presence of motor impairment and enhance the processing techniques on manipulating neural activity which can be volitionally modulated to sub serve training control strategy.

References

1. Krishnan R. V., Mythuswamy R., & Sankar V., "Spinal cord injury repair research: A new combination treatment strategy," International Journal of Neuroscience, 108: 201–207, 2003.
2. Craig A., Hancock K., Dickson H., & Chang E., "Immunizing against depression and anxiety following spinal cord injury," Archives of Physical Medicine and Rehabilitation, 79: 375–377, 1998.
3. Widerstrom-Noga E., Felipe-Cuervo E., & Yezierski R., "Relationships among clinical characteristics of chronic pain after spinal cord injury," Archives of Physical & Mental Rehabilitation, 82: 1191-1197, 2001.
4. Simon Frantz., "Embryonic stem cell pioneer Geron exits field, cut losses," Nature Biotechnology, 30: 12-13, 2012.
5. Michael G., & Reaz Vawda., "Cellular treatments for spinal cord injury: The time is right for clinical trials," Neurotherapuetics, 8: 704-720, 2011.
6. Lakany H., & Conway B.A., "Comparing EEG patterns of actual and imaginary wrist movements - a machine learning approach," Proceedings of the first ICGST International Conference on Artificial Intelligence and Machine Learning AIML, ICGST, Cairo, Egypt, 05(5): 124-127, 2005.
7. Vaughan, T.M., "Brain-computer interface technology: a review of the second international meeting," IEEE Transactions on Neural Systems and Rehabilitation Engineering, 11:94–109, 2003.
8. Craig. D.A., Nguyen. H.T., "Adaptive EEG Thought Pattern Classifier for Advanced Wheelchair Control," Engineering in Medicine and Biology Society, 29th Annual International Conference of the IEEE: 2544 – 2547, 2007.
9. J.A. Pineda., B.Z Allison., and A. Vankov., "The Effects of Self-Movement, Observation and Imagination on μ Rhythms and Readiness Potentials (RP's): Towards a Brain-Computer Interface (BCI)," IEEE Transactions on Rehabilitation Engineering, 8:219-222, 2000.
10. Stancák, A., Pfurtscheller, G., "Desynchronization and recovery of b rhythms during brisk and slow self-paced finger movements in man," Neuroscience Letters, 196: 21-24, 1995.
11. Cassim, F., Szurhaj, W., Sediri, H., Devos, D., Bourriez, J.L., Poirot, I., Derambure, P., Defebvre, L., Guieu, J.D., "Brief and sustained movements: differences in event-related (de)synchronization (ERD/ERS) patterns," Clinical Neurophysiology, 111: 2032-2039, 2000.
12. Haeger-Ross, C., Schieber, M.H., "Quantifying the independence of human finger movements: comparisons of digits, hands and movement frequencies," Journal of Neuroscience, 20(22):8542-8550, 2000.
13. Li, Y., Gao, X., Liu, H., Gao, S., "Classification of single-trial electroencephalogram during finger movement," IEEE Transactions on Biomedical Engineering, 51(6): 1019-1025, 2004.
14. Erbil, N., Ungan, P., "Changes in the alpha and beta amplitudes of the central EEG during the onset, continuation, and offset of long-duration repetitive hand movements," Brain Research, 1169: 44-56, 2007.
15. Pfurtscheller, G., Neuper, C., Pichler-Zalaudek, K., Edlinger, G., Lopes da Silva, F., "Do brain oscillations of different frequencies indicate interaction between cortical areas in humans?," Neuroscience Letters, 286:66-68, 2000.
16. Neuper, C., Pfurtscheller, G., "Evidence for distinct beta resonance frequencies in human EEG related to specific sensorimotor cortical areas," Clinical Neurophysiology, 112:2084-2097, 2001.
17. Bai, O., Lin, P., Vorbach, S., Li, J., Furlani, S., Hallett, M., "Exploration of computational methods for classification of movement intention during human voluntary movement from single trial

EEG," Clinical Neurophysiology, 118:2637-2655, 2007.

18. Müller-Putz, G.R., Zimmermann, D., Graimann, B., Nestinger, K., Korisek, G., Pfurtscheller, G., "Event-related beta EEG-changes during passive and attempted foot movements in paraplegic patients," Brain Research, 1137:84-91, 2007.
19. Morash, V., Bai, O., Furlani, S., Lin, P., Hallett, M., "Classifying EEG signals preceding right hand, left hand, tongue, and right foot movements and motor imageries," Clinical Neurophysiology, 119:2570-2578, 2008.
20. Niedermeyer E, F. H. Lopes da Silva., "Electroencephalography: Basic principles, clinical applications and related fields,". 3rd edition, Lippincott, Williams & Wilkins, Philadelphia, 1993.
21. Guyton. A.C. Textbook of Medical Physiology, Sixth Edition, W.B.Saunders Co., Philadelphia, Pa.6: 675-680, 1981.
22. Sutter, E.E., Tran, D., "The brain response interface: communication through visually induced electrical brain responses," Journal of Microcomputer Applications 15:31-45, 1992.
23. Middendorf, M., McMillan, G., Calhoun, G., Jones, K.S., "Braine computer interfaces based on the steady-state visual-evoked response," IEEE Transaction on Rehabilitation Engineering 8 (2):211-214, 2000.
24. Rockstroh, B., Elbert, T., Canavan, A., Lutzenberger, W., Birbaumer, N., "Slow Cortical Potentials and Behavior," second ed.. Urban and Schwarzenberg, Baltimore, MD, 1989.
25. Birbaumer, N., "Slow cortical potentials: their origin, meaning, and clinical use," In: van Boxtel, G.J.M., Bo€cker, K.B.E. (Eds.), Brain and Behavior Past, Present, and Future, Tilburg University Press, Tilburg, pp. 25-39, 1997.
26. Donchin, E., Spencer, K.M., Wijesinghe, R., "The mental prosthesis: assessing the speed of a P300-based brain computer interface," IEEE Transaction on Rehabilitation Engineering 8:174-179, 2000.
27. Li, Y., Nam, C.S., Shadden, B., Johnson, S., "A P300-based Brain-Computer Interface (BCI): Effects of Interface Type and Screen Size," International Journal of Human-Computer Interaction 27 (1):52-68, 2011.
28. Neuper, C., Müller-Putz, G.R., Scherer, R., Pfurtscheller, G., "Motor imagery and EEG-based control of spelling devices and neuroprostheses," Progress in Brain Research 159:393-409, 2006.
29. Neuper, C., Scherer, R., Wriessnegger, S., Pfurtscheller, G., "Motor imagery and action observation: modulation of sensorimotor brain rhythms during mental control of a brain-computer interface," Clinical Neurophysiology 120:239-247, 2009.
30. Kennedy, P.R., Bakey, R.A.E., Moore, M.M., Adams, K., "Direct control of a computer from the human central nervous system," IEEE Transaction on Rehabilitation Engineering 8:198-202, 2000.
31. Antonio Castro, Fernando Diaz and Geert J. M. van Boxtel, "How does a short history of spinal cord injury affect movement-related brain potentials?," European Journal of Neuroscience, 25:2927-2934, 2007.
32. P Boord, Y Tran, J Middleton, A Craig, "Levels of brain wave activity (8-13 Hz) in persons with spinal cord injury," Spinal Cord. Houndsmills, 42(2): 73, 2004.
33. P Boord, P J Siddall, Y Tran, D Herbert, et al, "Electroencephalographic slowing and reduced reactivity in neuropathic pain following spinal cord injury," Spinal Cord. Houndsmills, 46(2): 118-124, 2008.
34. Yongwoong Jeon, Chang S. Nam, Young-Joo Kim, Min Cheol Whang, "Event-related (De) synchronization (ERD/ERS) during motor imagery tasks: Implications for brainecomputer interfaces," International Journal of Industrial Ergonomics, 41:428-436, 2011.
35. Pfurtscheller, G., Aranibar, A., "Evaluation of event-related desynchronization (ERD) preceding and following voluntary self-paced movement," Electroencephalography and Clinical Neurophysiology 46:138-146, 1979.
36. Jean-Michel Guérit, "Neuromonitoring in the operating room: why, when, and how to monitor?," Electroencephalography and Clinical Neurophysiology, 106(1): 1-21, 1998.
37. Heba Lakany and B.A. Conway, "Classification of Wrist Movements using EEG-based Wavelets Features," Proceedings of IEEE Engineering in Medicine and Biology, 27:5405-5407, 2005.
38. I. Navarro, F. Sepulveda, B. Hubais, "A Comparison of Time, Frequency and ICA Based Features and Five Classifiers for Wrist Movement Classification in EEG Signals,"Proceedings of the 2005 IEEE Engineering in Medicine and Biology,2118-2121, 2005.
39. Guger,W.H., Ramoser, H. & Pfurtscheller,G., "Real-time EEG analysis with subject-spesific spatial pattern for brain computer interface," IEEE Transaction on Rehabilitation Engineering, 8(4), 2000.
40. Antonio Castro, Fernando Diaz and Geert J. M. van Boxtel, "How does a short history of spinal cord injury affect movement-related brain potentials?," European Journal of Neuroscience, 25: 2927-2934, 2007.

DOI: 10.1051/matecconf/201714001028

Comparing Common Average Referencing to Laplacian Referencing in Detecting Imagination and Intention of Movement for Brain Computer Interface

Syahrull Hi-Fi Syam[1], Heba Lakany[2], R.B. Ahmad[3] and Bernard A. Conway[2]
[1]Polytechnic of Tuanku Syed Sirajuddin, 02600, Perlis, Malaysia
[2] Biomedical Engineering, University of Strathclyde G4 0NW, Glasgow, United Kingdom.
[3]Faculty of Informatics and Computing, Universiti Sultan Zainal Abidin (UniSZA), Kuala Terengganu, Malaysia

Abstract. Brain-computer interface (BCI) is a paradigm that offers an alternative communication channel between neural activity generated in the brain and the user's external environment. This paper investigates detection of intention of movement from surface EEG during actual and imagination of movement which is essential for developing non-invasive BCI system for neuro-impaired patients. EEG signal was recorded from 11 subjects while imagining and performing right wrist movement in multiple directions using 28 electrodes based on international 10-20 standard electrode placement locations. The recorded EEG signal later was filtered and pre-processed by spatial filter namely; Common average reference (CAR) and Laplacian (LAP) filter. Features were extracted from the filtered signal using ERSP and power spectrum and classified by *k*-nearest neighbour (*k*-NN) and quadratic discriminant analysis (QDA) classifiers. The classification results show that LAP filter has outperformed CAR with respect to classification. Classification accuracy ranged from 63.33% to 100% for detection of imagination of movement and 60% to 96.67% for detection of intention of actual movement. In both of detection of imagination and intention of movement k-NN classifier gave better result compared to QDA classifier.

1 Introduction

Neurological disorders such as amyotrophic lateral sclerosis (ALS), brainstem stroke, brain or spinal cord injury, cerebral palsy, muscular dystrophies and multiple sclerosis impair the neural pathway that control the muscle and /or the muscle themselves [1]. Patients loose voluntary control over their body due to these diseases and are driven to live an isolated, discriminated and restricted life in the long run. Their motor and sensory disabilities unable them to live a normal independent daily life.

Despite of losing their voluntary control over their body due to such neurological disorders, which is often a permanent effect, neuro-impaired patients can still communicate with outside world through brain-computer interface (BCI). BCI decodes the brain activities from electroencephalogram (EEG) signal and translates the user's intentions into commands [2] to control and/or communicate with the augmentative and assistive devices without activating any muscle or peripheral nerve.

In the past decades, number of BCI studies has been conducted in order to build successful interface that applies real-time EEG signal as command to control and /or communicate with the outside world [1]. Most of the studies aim to improve the speed and the accuracy of these interfaces by using improved EEG signal processing techniques and feature selection. One of the approaches that can be considered in signal processing techniques is implementing spatial and temporal filtering methods. The most often used spatial filtering methods in BCI studies to enhance the signal-to-noise ratio of EEG signal are Common average reference (CAR) [3, 4, 5] and Laplacian filter (LAP) [6, 7, 8].

In this paper we present the comparison of detecting motor imagery and intention of movement using 2 different spatial filters namely CAR and LAP. Our aim was to investigate which spatial filter would produce better results in detecting motor imagery and intention of movement.

The rest of the paper is structured as follow: Section 2 defines the implementation of experiment protocol and data analysis procedure. Section 3 presents the results of the experiment and the paper ends with a conclusion of the findings in section 4.

2 Method

2.1 Signal Enhancement by Spatial Filtering Methods

EEG recordings are usually contaminated by several sources of artefacts produced by the subject (for example any minor body movement, electromyogram (EMG), electrocardiogram (ECG), eye movements, sweating). Also, technical glitches (for instance, power line, impedance fluctuation, cable movement, broken wire contact or excessive electrode paste/ dried pieces) add to the noise levels of the EEG signal [9]. Thus, it is important to remove any existing noise and artefact from the recorded EEG signal before implementing any further signal processing analysis. In this paper the EEG signal was filtered by spatial filter namely CAR and LAP.

CAR is one of the reference-free techniques that is not affected by problem associated with an actual physical reference [9]. In CAR, the potential at each electrode is measured with respect to the average of all

syahrull30@gmail.com

electrodes. The CAR was computed using (1) given below [10]:

$$V_i^{CAR} = V_i^{ER} - \frac{1}{n}\sum_{j=1}^{n} V_j^{ER} \quad (1)$$

Where V_i^{ER} is the potential between i th electrode and the reference and n is the number electrodes in the montage.

LAP serves as a high pass filter that enhances localized activity while suppressing the diffusion activity [8]. LAP is obtained by subtracting the sum of weighted potential of the surrounding electrodes from the current electrode potential where the weight is electrode distance dependent. The LAP was computed using (2) and (3) given below [10]:

$$V_i^{LAP} = V_i^{ER} - \sum_{j \in Si}^{n} V_j^{ER} g_{ij} \quad (2)$$

Where

$$g_{ij} = \frac{1}{d_{ij}} / \sum_{j \in Si}^{n} \frac{1}{d_{ij}} \quad (3)$$

Si is the set of surrounding electrodes of the *i* th electrode and *dij* is the distance between electrodes *i* and *j* (where *j* is a member of *Si*).

2.2 Experimental Recording Setup

EEG signals were recorded from 11 subjects (9 males and 2 female) and the experimental procedure had been approved by the Departmental Ethics Committee of the Biomedical Engineering Department, University of Strathclyde.

Each subject faced a LCD monitor screen while being seated on a wheelchair at a distance of 1 meter from the screen. Subjects were attached to a manipulandum on to their right side. EEG and EMG recordings were acquired using Curry Neuroimaging Suite 7.0.8 XSB Software and NeuroScan™ Synamps[2] amplifiers from electrodes at 28 scalp locations and 4 bipolar electrodes (Figure 1). During the data recording process, subjects were required to hold the manipulandum and attempt, perform and imagine (kinesthetic imagery) performing right wrist movement (burst, point to point center out movement) towards multiple direction (3,6,9,12) triggered by a visual cue.

The subject had to move the manipulandum rapidly in order to correspond to the cue direction shown in the monitor. On reaching the cue position, subjects had to hold the manipulandum at the cue position for as long as the cue remained visible on the screen and later reposition the manipulandum to the neutral position (0) according to the cue.

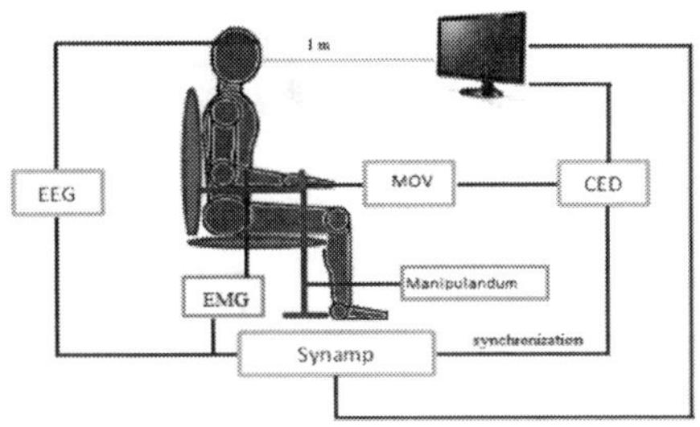

Figure 1: Experimental Recording Set Up.

The visual cue was displayed on the LCD monitor screen which consisted of 5 squares initially placed at the center of the screen whilst the subject held the manipulandum in the neutral position (0). The cue then randomly switched to different directions (3, 6, 9, 12) with a time interval of 10 seconds in between any 2 consecutive cues [11] (Figure 2).

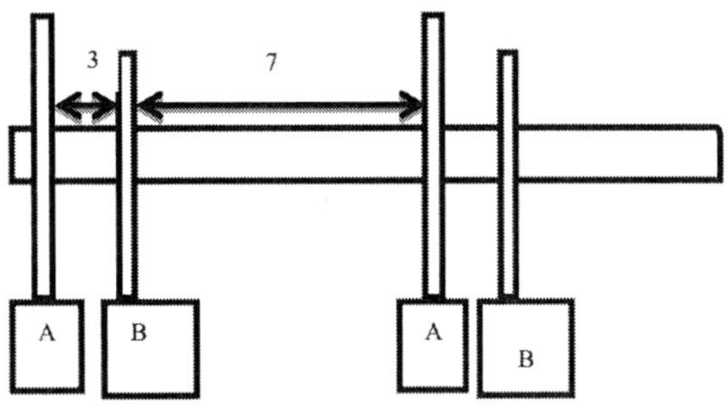

Figure 2: Timeline of Visual Cue Presentation.

In Fig. 2, during phase A, subjects were asked to move or imagine (kinesthetic imagery) moving the manipulandum from neutral position (0) to one of four directions (3, 6, 9, 12) and hold it in that position until the cue was visible on the screen; in phase B, subjects then repositioned the manipulandum back to the neutral position (0) as per the cue. While in the neutral position, subjects were instructed to stay calm and relaxed.

All the participants completed all the trials by imagining the movement and performing the actual movement based on external cue provided to them. Each experiment comprised of a total of 200 trials of both movement and imagery of movement towards 4 different directions, thus, establishing 50 repetitions per direction.

2.3 Data Recording Set Up

EEG, EMG and movement signals were recorded simultaneously during the experiments. EEG signal was recorded from 28 scalp locations located according to international 10-20 system (earlobe reference) using high density montage (Figure 3). EMG signals were recorded from flexor carpi radialis, extensor carpi ulnaris, extensor carpi radialis brevis and extensor carpi radialis longus muscles. The EMG signal was recorded in order to make sure that there is no movement during experiment of imagery of movement. Both EEG and EMG were recorded by Curry Neuroimaging Suite 7.0.8 XSB software using NeuroScan™Synamps² at sampling frequency of 2 KHz.

The movement signal was recorded from 2 precision servo potentiometers that are attached to the manipulandum in order to detect the onset of a movement. The signal recorded using Spike2 software through CED I401 (Cambridge Electronic Design, United Kingdom) at a sampling frequency of 100Hz.

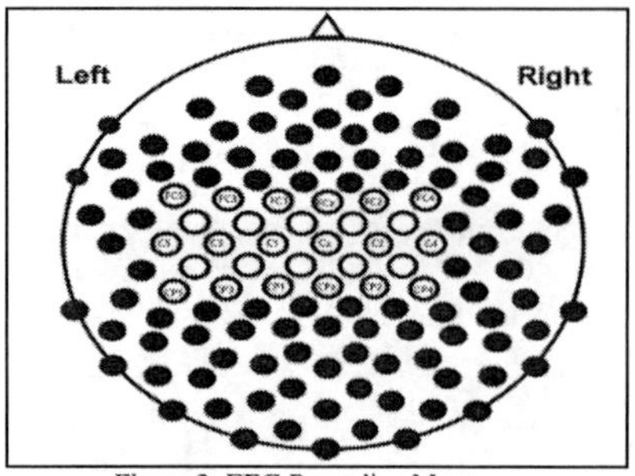

Figure 3: EEG Recording Montage.

2.4 Data Pre-processing

The recorded data from both of the experiments were processed offline. EEG was epoched using EEGLAB software version 12 [12] based on the type of experiments and categorized according to the type of the direction (3, 6, 9, 12). For instance, in experiment of external cue movement, the EEG signal was epoched 3 seconds before and 3 seconds after the onset of the movement (the movement initiation time) whereas, for the experiment where the subjects imagined the movement, the EEG signal was epoched 3 seconds before and 3 seconds after the initiation time (visual cue presentation time).

A Notch filter was applied to the epoched EEG to remove 50Hz power line interference [13] and then a Chebyshev type 2 bandpass filter was used [14]. CAR and LAP methods were applied before the features were extracted.

2.4 Features Extraction and Classification

The features of the interest were Event Related Spectral Perturbation (ERSP) and average of Power Spectrum in alpha (α, 8-12 Hz), beta (β, 13-30 Hz) and gamma (γ, 31-50 Hz) frequency bands. ERSP is a generalization of Event Related Desynchronization (ERD)/Event Related Synchronization (ERS) which helps visualize the entire spectrum in form of baseline-normalized spectrogram. ERSP is computed where each epoch is divided into a number of overlapping windows and spectral power is calculated for each window. The calculated spectral power is then normalized (divided with the baseline spectra calculated from the EEG immediately before each event) and averaged over all the trials. The whole process was performed by EEGLAB [12]. Power Spectrum indicates the distribution level of the signal power in frequency. Average of Power Spectrum in α, β and γ frequency bands were computed using code adapted from the EEGLAB.

Features were extracted from neutral condition and motor imagery/ intention of movement condition for direction of 3, 6, 9 and 12 respectively. For neutral condition, features were extracted 500ms before the subject imagined/performed the wrist movement. On the other hand, for motor imagery, features were extracted 500ms after the initiation time (t=0) whereas for intention of movement features were extracted 500ms prior to the onset of the movement (t=0) [15, 16]. The extracted features of the epoched EEG were spilt into training data set (70%) and testing data set (30%) [19,21]. The training data set and testing data set are randomly [19,23] selected by using Matlab function 'crossvalind' and fed to *k*-Nearest Neighbors (*k*-NN) [17,18] and quadratic discriminant analysis (QDA) [19, 20] classifier for pattern recognition classification. For *k*-NN classifier, '*k*' was set to 8 [22].

3 Results

3.1 Results of Event Related Spectral Perturbation (ERSP)

Figure 4,5,6 and 7 show the ERSP results for detection of motor imagery and intention of movement for direction 3, 6, 9 and 12 at channel C3 using CAR and LAP filters respectively. The results are obtained from subject 1 for both of experiments.

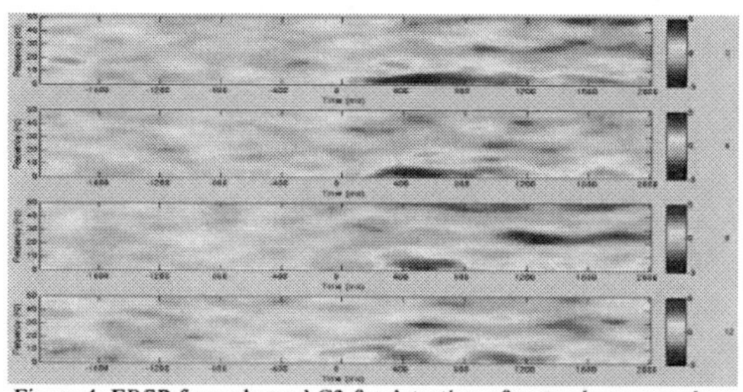
Figure 4: ERSP from channel C3 for detection of motor imagery using CAR Method.

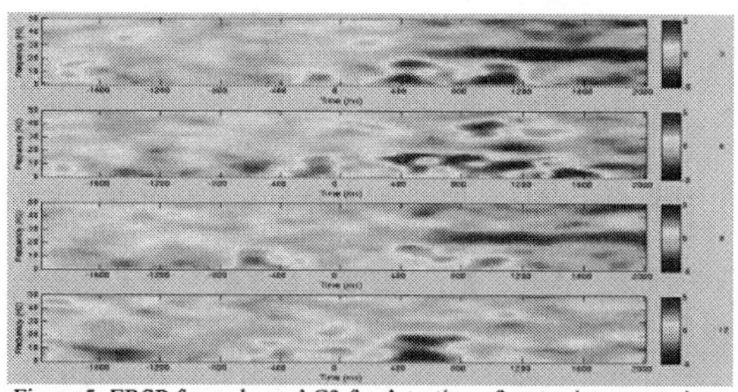
Figure 5: ERSP from channel C3 for detection of motor imagery using LAP Method

From Figures 4 and 5, ERSP using LAP method show more significant result when compared to CAR method in detection of motor imagery towards direction 3, 6, 9 and 12. The decrease in power spectrum as shown by the ERD (blue colour) 400ms after the initiation time (t=0) for all directions. ERD was evidently detected in β

(direction 3, 6, 9 and 12) and γ band (direction 3, 6 and 9).

Also from Figure 6 and 7, ERSP using LAP method again shows more significant result compared to CAR method in detection of intention of movement towards direction 3, 6, 9 and 12. In this paper, intention of movement refers to 500ms prior to the onset of the movement (t=0). ERD was clearly detected in α (6, 9 and 12) and γ band (12, 9 and 3).

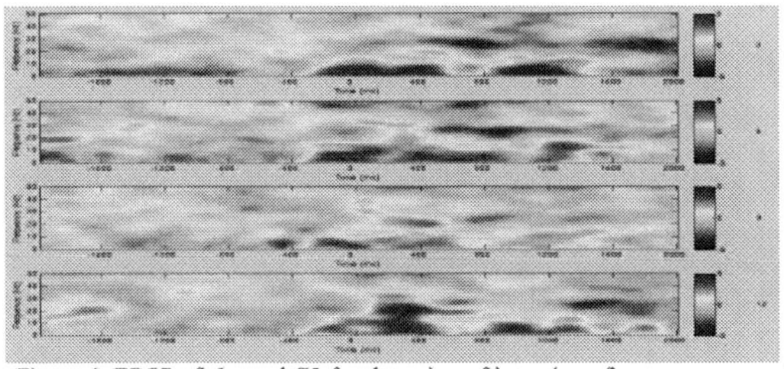

Figure 6: ERSP of channel C3 for detection of intention of movement using CAR Method.

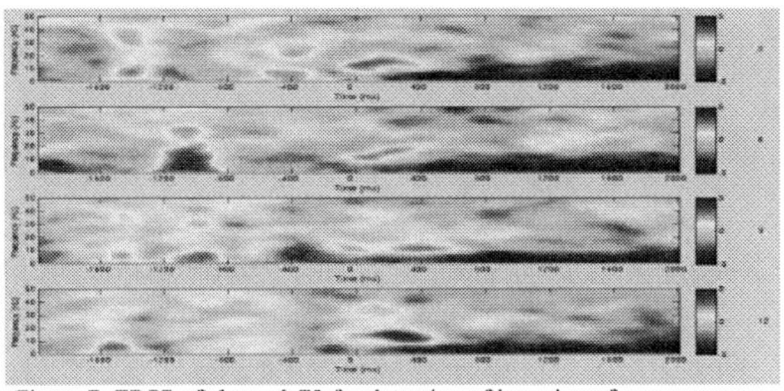

Figure 7: ERSP of channel C3 for detection of intention of movement using LAP Method.

3.2 Results of Power Spectrum

The results for detection of motor imagery are presented in Table 1 and Table 2, while results for detection intention of movement are tabulated in Table 3 and Table 4. The extracted features using CAR method and LAP method for direction toward 3, 6, 9 and 12 in detection of motor imagery and intention of movement are classified using *k*-NN and QDA classifier. In each of the table presented the maximum classification accuracy (Acc), channel of the maximum classification accuracy (Ch), true positive rate (TPR) and false positive rate (FPR).

TABLE 1: Classification results of detection of motor imagery for direction toward 3, 6, 9 and 12 using *k*-NN classifier on average of Power Spectrum in alpha (α), beta (β) and gamma (γ) band feature.

Subject	Parameter	Maximum Classification Accuracy, True Positive Rate (TPR) and False Positive Rate (FPR) for direction 3, 6, 9 and 12 (%)							
		Direction 3		Direction 6		Direction 9		Direction 12	
		CAR	LAP	CAR	LAP	CAR	LAP	CAR	LAP
S1	Acc	73.33	**83.33**	76.67	**80.00**	76.67	**86.67**	76.67	**80.00**
	Ch	CCP3	FC5	FCZ	C3	CCP4	CFC5	CP2	CFC1
	TPR	66.67	**80.00**	93.33	93.33	80.00	80.00	93.33	66.67
	FPR	20.00	13.33	40.00	33.33	26.67	6.67	40.00	6.67
S2	Acc	83.33	**90.00**	86.67	**93.33**	83.33	**93.33**	83.33	**96.67**
	Ch	CFC1	CFC1	CFC1	CFC1	CFC1	CCP4	CFC1	CFC1
	TPR	73.33	86.67	80.00	93.33	66.67	86.67	73.33	93.33
	FPR	6.67	6.67	6.67	6.67	0	0	6.67	0
S3	Acc	83.33	**86.67**	73.33	**80.00**	**83.33**	76.67	73.33	**76.67**
	Ch	C4	FC5	CP4	CFC2	C4	CFC4	FC1	CCP3
	TPR	86.67	93.33	53.33	80.00	86.67	73.33	73.33	73.33
	FPR	20.00	20.00	6.67	20.00	20.00	20.00	26.67	20.00
S4	Acc	73.33	**80.00**	76.67	**80.00**	80.00	**83.33**	76.67	**93.33**
	Ch	CZ	C3	CP4	C3	CZ	FC5	FC4	FC4
	TPR	80.00	73.33	66.67	66.67	60.00	80.00	66.67	93.33
	FPR	33.33	13.33	13.33	6.67	0	13.33	13.33	6.67
S5	Acc	73.33	**96.67**	73.33	**80.00**	76.67	**80.00**	83.33	**86.67**
	Ch	FC4	CFC4	FC3	CCP1	CP1	CCP1	FC3	CFC2
	TPR	60.00	93.33	66.67	73.33	66.67	100	86.67	80.00
	FPR	13.33	0	20.00	13.33	13.33	40.00	20.00	6.67
S6	Acc	70.00	**80.00**	73.33	**86.67**	73.33	**80.00**	73.33	**90.00**
	Ch	CP4	C4	FCZ	CCP3	CCP2	CCP3	C3	CP3
	TPR	66.67	86.67	66.67	80.00	66.67	80.00	80.00	80.00
	FPR	26.67	26.67	20.00	6.67	20.00	20.00	33.33	6.67
S7	Acc	70.00	**83.33**	70.00	**93.33**	73.33	**96.67**	73.33	**93.33**
	Ch	CP3	CFC3	CP3	FC2	CCP4	CP3	CCP3	FC3
	TPR	80.00	93.33	80.00	100	66.67	93.33	66.67	93.33
	FPR	40.00	26.67	40.00	13.33	20.00	0	20.00	6.67
S8	Acc	73.33	**93.33**	70.00	**100**	73.33	**90.00**	70.00	**90.00**
	Ch	CFC5	CP2	C3	FC1	C4	CP3	CPC3	FC5
	TPR	73.33	93.33	73.33	100	86.67	100	80.00	80.00
	FPR	26.67	6.63	33.33	0	40.00	20.00	40.00	0
S9	Acc	73.33	**76.67**	73.33	**76.67**	73.33	**83.33**	73.33	**76.67**
	Ch	CP5	CFC5	FC1	CP5	FC1	CCP5	CFC3	CFC5
	TPR	60.00	80.00	53.33	80.00	86.67	86.67	53.33	80.00
	FPR	13.33	26.67	7.14	26.67	31.58	20.00	6.67	26.67
S10	Acc	70.00	**90.00**	83.33	**86.67**	70.00	**86.67**	73.33	**90.00**
	Ch	CP1	C3	C4	CCP5	CCP5	FC3	CCP5	C3
	TPR	73.33	86.67	100	73.33	60.00	100	73.33	86.67
	FPR	33.33	6.67	33.33	0	20.00	26.67	26.67	6.67
S11	Acc	70.00	**90.00**	76.67	**83.33**	70.00	**93.33**	76.67	**83.33**
	Ch	C2	CP5	CPZ	FC5	C2	CP4	CFC3	CZ
	TPR	80.00	93.33	66.67	86.67	66.67	86.67	100	86.67
	FPR	40.00	13.33	13.33	20.00	26.67	0	46.67	20.00

Tabulated data in Table 1 shows the classification results of direction toward 3, 6, 9 and 12 for detection of motor imagery, dwell within the range of 70.00% to 100.00%. Maximum classification accuracy for these four directions are obtained from different electrodes, namely CFC4 with classification accuracy of 96.67% (direction toward 3), FC1 with classification accuracy of 100.00% (direction toward 6), CP3 with classification accuracy of 96.67% (direction toward 9) and CFC1 with classification accuracy of 96.67% (direction toward 12). In all four directions, features extracted using spatial filter LAP contributed for the maximum classification results.

TABLE 2: Classification results for detection of motor imagery for direction toward 3, 6, 9 and 12 using QDA classifier on average of Power Spectrum in alpha (α), beta (β) and gamma (γ) band feature.

ubject	Parameter	Maximum Classification Accuracy, True Positive Rate (TPR) and False Positive Rate (FPR) for direction 3, 6, 9 and 12 (%)							
		Direction 3		Direction 6		Direction 9		Direction 12	
		CAR	LAP	CAR	LAP	CAR	LAP	CAR	LAP
S1	Acc	73.33	73.33	66.67	**70.00**	66.67	**83.33**	73.33	**80.00**
	Ch	CP1	C3	FC1	CCP1	C4	CFC5	CFC4	CFC1
	TPR	93.33	86.67	86.67	53.33	86.67	80.00	53.33	80.00
	FPR	46.67	40.00	53.33	13.33	53.33	13.33	6.67	20.00
S2	Acc	83.33	**93.33**	86.67	**96.67**	83.33	**93.33**	83.33	**93.33**
	Ch	CFC1	CFC1	CFC1	CFC1	FCZ	CFC1	FC2	CFC1
	TPR	73.33	100	86.67	100	73.33	100	80.00	86.67
	FPR	6.67	13.33	13.33	6.67	6.67	13.33	13.33	0
S3	Acc	76.67	**90.00**	**76.67**	73.33	73.33	**76.67**	73.33	73.33
	Ch	CFC5	CFC5	CP4	CFC4	FC2	CFC4	CFC2	CFC4
	TPR	86.67	80.00	73.33	53.33	80.00	60.00	66.67	60.00
	FPR	33.33	0	20.00	6.67	33.33	6.67	20.00	13.33

S4	Acc	70	**76.67**	73.33	73.33	76.67	**80.00**	80.00	**86.67**
	Ch	CP5	CCP5	CP4	C3	CP2	FC1	FC4	C2
	TPR	53.33	80.00	60.00	53.33	66.67	80.00	80.00	80.00
	FPR	13.33	26.67	13.33	6.67	13.33	20.00	20.00	6.67
S5	Acc	76.67	**90.00**	73.33	73.33	70.00	**86.67**	76.67	**80.00**
	Ch	C4	CFC3	CFC1	CCP1	FC4	CCP1	C3	FC3
	TPR	60.00	86.67	53.33	66.67	66.67	93.33	66.67	80.00
	FPR	6.67	6.67	6.67	20.00	20.00	20.00	13.33	20.00
S6	Acc	63.33	**80.00**	70.00	**80.00**	70.00	**76.67**	73.33	**83.33**
	Ch	FC1	C2	CPZ	FC5	CFC4	CFC1	CCP1	CFC1
	TPR	66.67	73.33	53.33	80.00	60.00	66.67	86.67	86.67
	FPR	40.00	13.33	13.33	20.00	20.00	13.33	40.00	20.00
S7	Acc	63.33	**80.00**	70.00	**80.00**	70.00	**76.67**	73.33	**83.33**
	Ch	FC1	C2	CPZ	FC5	CFC4	CFC1	CCP1	CFC1
	TPR	66.67	73.33	53.33	80.00	60.00	66.67	86.67	86.67
	FPR	40.00	13.33	13.33	20.00	20.00	13.33	40.00	20.00
S8	Acc	70.00	**90.00**	70.00	**96.67**	76.67	**90.00**	70.00	**83.33**
	Ch	C1	FC1	CCP1	CP3	C4	CP2	C2	CP3
	TPR	86.67	86.67	80.00	93.33	86.67	86.67	86.67	73.33
	FPR	46.67	6.67	40.00	0	33.33	6.67	46.67	6.67
S9	Acc	70.00	**76.67**	70.00	**76.67**	70.00	**80.00**	73.33	**73.33**
	Ch	CFC3	FC1	CP3	CP5	CFC3	CFC1	FC3	CCP1
	TPR	80.00	73.33	73.33	80.00	53.33	66.67	60.00	80.00
	FPR	40.00	20.00	33.33	26.67	13.33	6.67	13.33	33.33
S10	Acc	70.00	**93.33**	86.67	**93.33**	70.00	**90.00**	70.00	**96.67**
	Ch	CCP5	C2	CP3	CCP5	CFC5	C2	CCP5	CCP5
	TPR	66.67	93.33	86.67	93.33	93.33	86.67	66.67	100
	FPR	26.67	6.67	13.33	6.67	13.33	6.67	26.67	6.67
S11	Acc	73.33	**96.67**	70.00	**83.33**	70.00	**86.67**	70.00	**83.33**
	Ch	FC5	CPZ	CPZ	CFC3	CZ	CP2	CFC1	CFC3
	TPR	100	100	60.00	80.00	73.33	80.00	80.00	93.33
	FPR	33.33	6.67	20.00	13.33	33.33	6.67	40.00	26.67

Referring to Table 2, the classification results toward four directions for detection of motor imagery lie within the range of 63.33% to 96.67%. Maximum classification accuracies for direction toward 3, 9 and 12 were obtained from electrode CFC1 with classification accuracy of 93.33%. On the other hand, maximum classification accuracy for direction toward 6 was obtained from electrodes CP3 and CFC1 with classification accuracy of 96.67%. The features extracted using spatial filter LAP contributed for the maximum classification results in all direction.

TABLE 3: Classification results of detection intention of movement for direction toward 3, 6, 9 and 12 using *k*-NN classifier on average of Power Spectrum in alpha (α), beta (β) and gamma (γ) band feature.

Subject	Parameter	Maximum Classification Accuracy, True Positive Rate (TPR) and False Positive Rate (FPR) for direction 3, 6, 9 and 12 (%)							
		Direction 3		Direction 6		Direction 9		Direction 12	
		CAR	LAP	CAR	LAP	CAR	LAP	CAR	LAP
S1	Acc	76.67	**83.33**	76.67	**80.00**	80.00	80.00	76.67	**80.00**
	Ch	CCP4	FC1	CCP4	FC5	FC4	FC2	C3	C1
	TPR	80.00	66.67	86.67	73.33	86.67	80.00	60.00	73.33
	FPR	26.67	0	23.33	13.33	26.67	20.00	13.33	13.33
S2	Acc	90.00	**93.33**	**96.67**	90.00	93.33	93.33	90.00	**93.33**
	Ch	CP5	CFC5	CP5	CCP2	CP5	CCP5	CP5	CCP2
	TPR	80.00	93.33	93.33	93.33	86.67	86.67	80.00	100
	FPR	0	6.67	0	13.33	0	0	0	13.33
S3	Acc	80.00	**83.33**	70.00	**80.00**	73.33	**83.33**	73.33	**76.67**
	Ch	CZ	CP4	CP1	CCP5	CCP4	CFC4	CCP3	CFC1
	TPR	80.00	80.00	60.00	86.67	66.67	93.33	60.00	93.33
	FPR	20.00	13.33	20.00	73.33	20.00	26.67	13.33	40.00
S4	Acc	**90.00**	80.00	**80.00**	76.67	80.00	80.00	76.67	**86.67**
	Ch	FC5	CCP1	CFC3	C4	FC5	CFC5	FC5	CP2
	TPR	93.33	93.33	93.33	93.33	60.00	66.67	60.00	73.33
	FPR	13.33	33.33	33.33	40.00	0	0	6.67	0
S5	Acc	83.33	**93.33**	76.67	**90.00**	80.00	**96.67**	76.67	**96.67**
	Ch	CCP1	CCP1	C1	C3	CFC1	C1	CZ	CFC3
	TPR	73.33	86.67	60.00	80.00	73.33	93.33	66.67	100
	FPR	6.67	0	6.67	0	13.33	0	13.33	6.67
S6	Acc	76.67	**90.00**	73.33	**93.33**	70.00	**83.33**	73.33	**96.67**
	Ch	C3	CP3	C4	CP1	FC3	CCP5	CP1	CP1
	TPR	66.67	80.00	60.00	100	40.00	73.33	60.00	100
	FPR	13.33	0	13.33	13.33	0	6.67	20.00	6.67
S7	Acc	73.33	**93.33**	76.67	**96.67**	73.33	**90.00**	76.67	**93.33**
	Ch	C3	CFC5	C5	C4	FC1	CCP5	C5	CFC5
	TPR	66.67	86.67	66.67	93.33	60.00	86.67	60.00	93.33
	FPR	20.00	0	13.33	0	13.33	6.67	6.67	6.67
S8	Acc	70.00	**83.33**	73.33	**93.33**	76.67	**96.67**	70.00	**100**
	Ch	FCZ	FC3	CP5	C5	CP2	CP2	CP3	FC3
	TPR	80.00	73.33	80.00	86.67	93.33	100	80.00	100
	FPR	40.00	6.67	33.33	0	40.00	6.67	40.00	0
S9	Acc	73.33	**90.00**	73.33	**86.67**	73.33	**93.33**	73.33	**90.00**
	Ch	FC3	FC1	CCP2	CFC5	CP3	FCZ	CP3	CP2
	TPR	86.67	86.67	66.67	80.00	53.33	86.67	60.00	86.67
	FPR	40.00	6.67	20.00	6.67	6.67	0	13.33	6.67
S10	Acc	70.00	**73.33**	73.33	**76.67**	73.33	73.33	70.00	**76.67**
	Ch	FCZ	FC5	CP5	CFC2	FCZ	FCZ	CZ	FCZ
	TPR	66.67	66.67	66.67	80.00	80.00	73.33	66.67	66.67
	FPR	26.67	20.00	20.00	26.67	33.33	26.67	26.67	13.33
S11	Acc	70.00	**80.00**	73.33	**83.33**	73.33	**80.00**	90.00	**93.33**
	Ch	CCP3	FC5	CCP5	CP2	FCZ	FC5	FC5	C5
	TPR	73.33	66.67	80.00	73.33	86.67	73.33	100	100
	FPR	33.33	6.67	33.33	6.67	40.00	13.33	20.00	13.33

From Table 3, the classification results of four directions for detecting intention of movement lie within the range of 70.00% to 100.00%. Maximum classification accuracy for these four directions are obtained from different electrodes, namely CFC5 and CCP1 with classification accuracy of 93.33% (direction toward 3), CP5 and C4 with classification accuracy of 96.67% (direction toward 6), C1 and CP2 with classification accuracy of 96.67% (direction toward 9) and FC3 with classification accuracy of 100.00% (direction toward 12). In all four directions, features extracted using spatial filter LAP contributed for the maximum classification results.

TABLE 4: Classification results of detection intention of movement for direction toward 3, 6, 9 and 12 using QDA classifier on average of Power Spectrum in alpha (α), beta (β) and gamma (γ) band feature.

Subject	Parameter	Maximum Classification Accuracy, True Positive Rate (TPR) and False Positive Rate (FPR) for direction 3, 6, 9 and 12 (%)							
		Direction 3		Direction 6		Direction 9		Direction 12	
		CAR	LAP	CAR	LAP	CAR	LAP	CAR	LAP
S1	Acc	70.00	**76.67**	**76.67**	73.33	**86.67**	83.33	73.33	**76.67**
	Ch	CCP4	FC2	CP5	CFC4	FC4	CFC4	C4	CCP1
	TPR	66.67	66.67	80.00	66.67	86.67	80.00	93.33	53.33
	FPR	26.67	13.33	26.67	20.00	13.33	13.33	46.67	0
S2	Acc	80.00	**96.67**	90.00	**93.33**	**96.67**	93.33	90.00	90.00
	Ch	CPZ	CCP3	CP5	CCP2	CP5	CCP2	CP5	CCP5
	TPR	100	100	93.33	86.67	93.33	86.67	80.00	80.00
	FPR	40.00	6.67	13.33	0	0	0	0	0
S3	Acc	**73.33**	70.00	66.67	**70.00**	60.00	**73.33**	66.67	**70.00**
	Ch	C2	CCP3	FC1	CCP5	CCP2	CZ	C4	CCP5
	TPR	66.67	66.67	60.00	73.33	40.00	86.67	40.00	66.67
	FPR	20.00	26.67	26.67	33.33	20.00	40.00	6.67	26.67
S4	Acc	76.67	**80.00**	73.33	73.33	**83.33**	80.00	86.67	**90.00**
	Ch	FC5	CFC5	FC3	C3	CP1	CFC5	C1	FC2
	TPR	80.00	86.67	73.33	86.67	73.33	73.33	80.00	86.67
	FPR	26.67	26.67	26.67	40.00	6.67	13.33	6.67	6.67
S5	Acc	86.67	**90.00**	76.67	**96.67**	83.33	**93.33**	76.67	**86.67**
	Ch	FC3	CCP1	CP1	CFC1	C1	CCP3	CZ	CFC3
	TPR	86.67	93.33	73.33	93.33	73.33	86.67	73.33	93.33
	FPR	13.33	13.33	20.00	0	6.67	0	20.00	20.00
S6	Acc	70.00	**90.00**	70.00	**96.67**	73.33	**83.33**	66.67	**96.67**
	Ch	CCP4	CP4	C4	FC4	FC3	CFC5	CFC3	C5
	TPR	53.33	100	73.33	93.33	66.67	86.67	73.33	93.33
	FPR	13.33	20.00	33.33	0	20.00	20.00	40.00	0

 DOI: 10.1051/matecconf/201714001028

S7	Acc	80.00	**96.67**	73.33	**96.67**	73.33	**93.33**	73.33	**96.67**
	Ch	CP5	CCP5	C3	CCP5	CCP4	CFC5	CCP3	CCP5
	TPR	86.67	100	60.00	100	66.67	86.67	66.67	100
	FPR	26.67	6.67	13.33	6.67	20.00	0	20.00	6.67
S8	Acc	**73.33**	70.00	70.00	**93.33**	73.33	**93.33**	73.33	**93.33**
	Ch	CCP2	CP2	FCZ	CP1	CP4	FC2	CP1	CP3
	TPR	80.00	86.67	66.67	93.33	86.67	93.33	73.33	86.67
	FPR	33.33	46.67	26.67	6.67	40.00	6.67	26.67	0
S9	Acc	73.33	**86.67**	73.33	**86.67**	63.33	**90.00**	70.00	**80.00**
	Ch	CFC1	CP1	CCP1	CP2	C3	FCZ	C3	CP1
	TPR	53.33	100	73.33	80.00	73.33	80.00	60.00	80.00
	FPR	6.67	26.67	26.67	6.67	46.67	0	13.33	20.00
S10	Acc	70.00	70.00	70.00	70.00	70.00	**70.00**	66.67	**73.33**
	Ch	CFC2	CP4	FC1	CCP1	C4	CCP4	CZ	FC2
	TPR	60.00	60.00	60.00	53.33	66.67	73.33	80.00	60.00
	FPR	20.00	20.00	20.00	13.33	26.67	33.33	46.67	13.33
S11	Acc	70.00	80.00	70.00	76.67	63.33	66.67	93.33	96.67
	Ch	CCP5	FC3	CFC2	FC2	C2	C5	FC5	CFC5
	TPR	66.67	73.33	66.67	60.00	66.67	66.67	93.33	93.33
	FPR	26.67	13.33	26.67	6.67	40.00	33.33	6.67	0

Table 4 indicates the classification results of four directions for detecting intention of movement lie within the range of 60.00% to 96.67%. Maximum classification accuracy for these four directions are obtained from different electrodes, namely CCP3 and CCP5 with classification accuracy of 96.67% (direction toward 3),CFC1, CCP5 and FC4 with classification accuracy of 96.67% (direction toward 6), CCP3,CFC5 and FC2 with classification accuracy of 93.33% (direction toward 9) and CCP5 and C5 with classification accuracy of 96.67% (direction toward 12). In all four directions, features extracted using spatial filter LAP contributed for the maximum classification results.

4 Conclusion

In this paper we have demonstrated the feasibility of detecting motor imagery and intention of movement using right wrist movement in multiple directions namely, direction towards 3, 6, 9 and 12. This is significantly supported by ERSP results that evidently detected ERD and ERS in both intention and imagination of movement in all directions.

The classification results from this paper also highlights the comparison of using two type of spatial filters namely CAR and LAP, wherein the LAP filter outperforms CAR. This finding is supported by Ng and Raveendran (2007). In their study which based on motor imagery paradigm, LAP outperformed other spatial filter using ERD/ERS from different hemispheres. On the other hand our findings contradict the finding by Alhaddad (2012). In his work (based on P300 paradigm) CAR outperformed LAP. Nonetheless, both CAR and LAP referencing are superior to the ear reference [10]. Apart from that the classification results, the importance of implementation of high density electrodes is also highlighted.

References

1 J. R. Wolpaw, N. Birbaumer, D. J. McFarland, G. Pfurtscheller, and T. M. Vaughan, "Brain–computer interfaces for communication and control," Clin. Neurophysiol., vol. 113, no. 6, pp. 767–791, Jun. 2002.

2 J. D. R. Millán, R. Rupp, G. R. Müller-Putz, R. Murray-Smith, C. Giugliemma, M. Tangermann, C. Vidaurre, F. Cincotti, a Kübler, R. Leeb, C. Neuper, K.-R. Müller, and D. Mattia, "Combining Brain-Computer Interfaces and Assistive Technologies: State-of-the-Art and Challenges," Front. Neurosci., vol. 4, no. September, pp. 1–15, Jan. 2010.

3 K. K. Ang, J. Yu, and C. Guan, "Extracting effective features from high density nirs-based BCI for assessing numerical cognition," 2012 IEEE Int. Conf. Acoust. Speech Signal Process, pp. 2233–2236, Mar. 2012.

4 M. J. Alhaddad, "Common Average Reference (CAR) Improves P300 Speller,"International Journal of Engineering and Technology, vol. 2, no. 3, 2012.

5 X. Lei, P. Yang, P. Xu, T. Liu, and D. Yao, "Common Spatial Pattern Ensemble Classifier and Its Application in Brain-Computer Interface," Journal of Electronic Science and Technology of China, vol. 7, no. 1, pp. 17–21, 2009.

6 C. Sannelli, C. Vidaurre, K. Müller, and B. Blankertz, "Common Spatial Pattern Patches - an Optimized Filter Ensemble for Adaptive BCIs," International Journal of Bioelectromagnetism, vol. 13, no. 3, pp. 161–162, 2011.

7 T. Solis-Escalante, G. Müller-Putz, and G. Pfurtscheller, "Overt foot movement detection in one single Laplacian EEG derivation," J. Neurosci. Methods, vol. 175, no. 1, pp. 148–53, Oct. 2008.

8 J. Lu, D. J. McFarland, and J. R. Wolpaw, "Adaptive Laplacian filtering for sensorimotor rhythm-based brain-computer interfaces.," J. Neural Eng., vol. 10, no. 1, p. 016002, Feb. 2013.

9 Teplan, M, "Fundamentals of EEG Measurement," Measurement Science Review, 2(2),1-11,2002.

10 D. J. McFarland, L. M. McCane, S. V. David, and J. R. Wolpaw, "Spatial filter selection for EEG-based communication," Electroencephalogr. Clin. Neurophysiol., vol. 103, no. 3, pp. 386–394, Sep. 1997.

11 G. Pfurtscheller and F. H. Lopes da Silva "Event-related EEG/MEG synchronization and desynchronization: basic principles," Clin. Neurophysiol., vol. 110, no. 11, pp. 1842–57,1999.

12 Delorme, A. &Makeig, S"EEGlab: An Open Source Toolbox For Analysis Of Single-Trial EEG Dynamics Including Independent Component Analysis," Journal Of Neuroscience Methods, 134,9-21,2004.

13 M. Ferdjallah and R. E. Barr. "On the Unit Circle for the Removal of Power line Noise from Biomedical Signals," vol. 41, no. 6, pp. 529–536, 1994.

14 V. B. Deepa& P. Thangaraj. "A study on classification of EEG Data using the Filters," vol. 2, no. 4, pp. 94–96, 2011

15 H. Lakany and B. a Conway, "Classification of Wrist Movements using EEG-based Wavelets Features.," Conf. Proc. IEEE Eng. Med. Biol. Soc., vol. 5, pp. 5404–5407, 2005.

16 Lakany, H. and Conway, B.A. "Comparing EEG patterns of actual and imaginary wrist movements – A machine learning approach," Proceedings of the first ICGST International Conference on Artificial Intelligence and Machine Learning AIML 05, 5 . ICGST, Cairo, Egypt, pp. 124-127, 2005.
17 I. Dokare and N. Kant, "Performance Analysis of SVM, k-NN and BPNN Classifiers for Motor Imagery," vol. 10, no. 1, pp. 19–23, 2014.
18 C. Liu, H. Wang, and Z. Lu, "EEG Classification for Multiclass Motor Imagery BCI," Chinese Control and Decision Conference, pp. 4450–4453, 2013.
19 S. Bhattacharyya, A. Khasnobish, S. Chatterjee, A. Konar, and D. N. Tibarewala, "Performance analysis of LDA, QDA and KNN algorithms in left-right limb movement classification from EEG data," Int. Conf. Syst. Med. Biol. ICSMB 2010 - Proc., no. December, pp. 126–131, 2010.
20 O. Carrera-león, J. M. Ramirez, V. Alarcon-aquino, M. Baker, D. D. Croz-baron, and P. Gomez-gil, "A Motor Imagery BCI Experiment using Wavelet Analysis and Spatial Patterns Feature Extraction," pp. 18–20.
21 C. Elkan, "Evaluating Classifiers", University of San Diego, California, retrieved [01-05-2013] from http://cseweb.ucsd.edu/~elkan/250B/classifiereval.pdf
22 R. Palaniappan, "Biological Signal Analysis," Ventus Publishing, Denmark, 2010.
23 H. Hwang, K. Kwon, and C. Im, "Neurofeedback-based motor imagery training for brain – computer interface (BCI)," vol. 179, pp. 150–156, 2009.
24 S. C. Ng and P. Raveendran, "Comparison of different Montages on to {EEG} classification," 3rd Kuala Lumpur Int. Conf. Biomed. Eng. 2006, vol. 15, pp. 365–368, 2007.

Broadband Circular Polarization Antenna Based on Inverted Suspended Circular Patch with Square Slot

Hamizan Abu Bakar, *Mohamad Zoinol Abidin* Bin Abd Aziz, *Badrul Hisham* Ahmad, *Nornikman* Hassan

Centre for Telecommunication Research and Innovation (CeTRI), Faculty of Electronics and Computer Engineering Universiti Teknikal Malaysia Melaka (UTeM), 76100 Durian Tunggal, Melaka, Malaysia

Abstract. In this paper, an inverted suspended circular patch with square slot design is proposed for broadband circular polarized antenna. The antenna was designed based on L-probe technique and air gap of 10 mm distance between substrate and copper ground. L-probe technique is used in this antenna design to feed the 50 Ohm SMA connector signal through the feedline to the antenna structure. The proposed antenna had been designed and simulated by using computer simulation technology (CST) software. This antenna design is targeted for wireless local area network (WLAN) application which operating frequency at 2.4 GHz. Antenna performances in term of return loss, resonant frequency, bandwidth, gain, axial ratio, directivity, total efficiency and radiation pattern at the design frequency are analyzed. Result of simulation show that the proposed antenna achieved axial ratio below 3 dB which is 0.62 dB with return loss -35.60 dB at targeted frequency. Comparison result of simulation and measurement show that the return loss bandwidth (RLBW) and realized gain of the antenna at 2.4 GHz frequency decreases from 556 MHz to 354 MHz with gain 7.76 dB to 4.54 dB. However, the proposed antenna covers a wider bandwidth of 2.4 GHz where both simulation and measurement bandwidth is over 350 MHz.

1 Introduction

Microstrip patch antennas have been widely used due to their advantages of low profile, ease of fabrication, in-expensive, light weight and ease of integration with active Radio Frequency (RF) devices [1]. The basic and most commonly shape used to design microstrip antenna is the rectangular patch antenna. As in [2], the antenna is construct based on a rectangular patch design with two cutting slots to get an E-shaped antenna design which yields circular polarization. Instead of rectangular is the basic and most commonly used shape, microstrip patch antenna could be in lot of various shapes such as circular, square, elliptical and triangular [3].

However, microstrip patch antenna is restricted with narrow bandwidth and poor gain. As a way to solve this issue, several modifications need to be done are included adding slot [4,5], fed by L-probe [6-10] and others. The major advantage of L-probe fed is the ability of the design to improve bandwidth no less than 5% from its centre frequency [7] while keeping reasonable radiation properties. In [7], the simulated impedance bandwidth is 11.04% covering from 2.444 GHz to 2.729 GHz is obtained at -10 dB return loss.

The polarization of microstrip patch antenna may be categorized into linear polarization (LP) and circular polarization (CP). In a circular polarized antenna, the plane of polarization rotates in a circle forming one complete turn during one period of the wave. A circular polarized wave radiates signal in both the horizontal and vertical planes and all planes in between. Circular polarization is regarded as the well-known polarization schemes mainly because it can offer better mobility and weather penetration than linear polarization [11,12]. Other than that, CP also has the advantages of the reduction of "Faraday rotation", mitigation of multipath fading and immunity of polarization mismatching between transmitting and receiving antennas [13]. Microstrip patch antenna in their basic form normally provides linear polarization, but by certain modification it can provide circular polarization. The circular polarized antenna can be obtained by using single or double feed [1,14], truncated corner [2,15], slot [15], double layer.

In this work, a wideband L-probe inverted suspended circular patch with square slot antenna is presented. The wideband was obtained by the separation of air gap between the substrate and copper ground plane. The proposed circular patch antenna was designed for circular polarization at 2.4 GHz frequency.

2 Antenna Design

In this study, the antenna was designed using an FR4 board with dielectric constant, ε_r of the substrate = 4.4, tangent loss, tan δ = 0.019, substrate thickness, h = 1.6 mm and copper thickness, t = 0.035 mm. To represent as L-probe feed, the antenna was feed with the SMA coaxial probe connector from ground layer through the feedline of the antenna structure. The width (W_f) and length (L_f)

* Corresponding author: mohamadzoinol@utem.edu.my

DOI: 10.1051/matecconf/201714001029

of feedline is 3.1 and 13 mm respectively. The feedline and the circular patch with square slot antenna were designed at the bottom substrate of the antenna which is thus defined as inverted suspended. Where at the upper substrate is just a plain substrate without any patch layer. The antenna design and its ground plane were separated by an air gap with 10 mm distance.

2.1 Configuration of inverted suspended circular patch antenna with square slot design

Configuration of the inverted suspended circular patch antenna design is demonstrated as in Fig. 1. As shown in Fig. 1 (a) and (b) is the front view (plain substrate) and back view (with feedline and circle patch) of the antenna respectively. While, in Fig. 1 (c) and (d) shows the enlargement view and vertical view of the antenna where the square slot position can clearly be seen in Fig. 1 (c) before being subtract with the circular patch design.

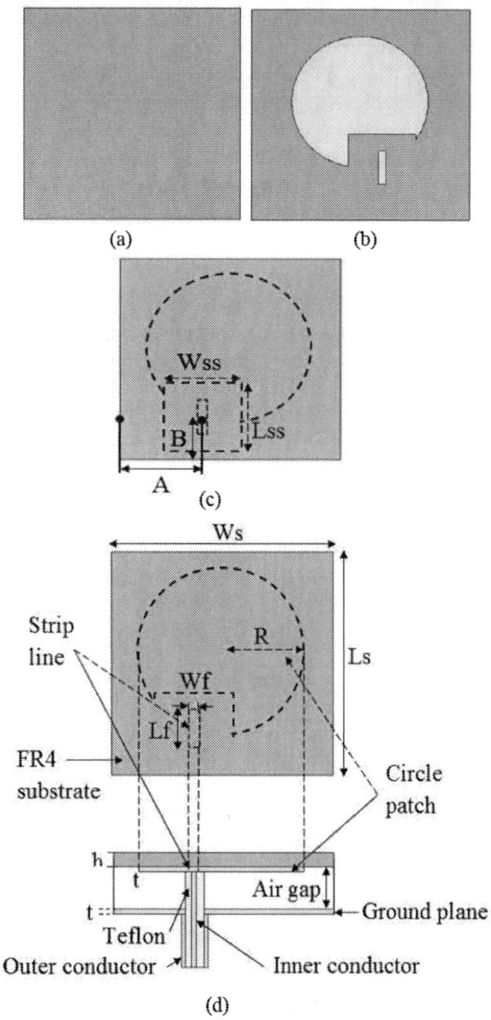

Fig. 1. Configuration of L-probe inverted suspended circular patch with square slot antenna (a) front view (b) back view (c) enlargement view and (d) vertical view.

2.2 Prototype view of fabricated inverted suspended circular patch antenna with square slot design

As demonstrated in Fig. 2 (a) and (b) is the fabricated prototype front view and side view of the antenna. As shown in Fig. 2 (a), the antenna width and length of substrate and copper ground layer is the same which is 80 x 80 mm.

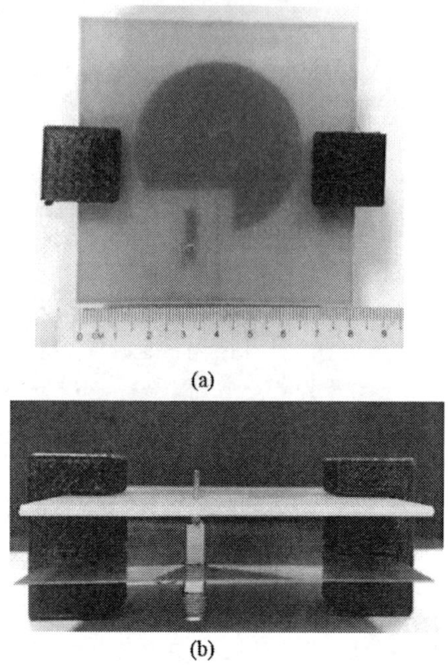

Fig. 2. Fabricated prototype of L-probe inverted suspended circular patch with square slot antenna (a) front view and (b) side view.

The length and width of the substrate are indicated as L_s and W_s, whereas L_{ss} and W_{ss} are used for the square slot. The size of square slot was decided according to the best result performance of parametric study. The ideal dimension for the antenna is tabulated as in Table 1.

Table 1. Dimension of L-probe inverted suspended circular patch with square slot antenna design.

Design Parameter	Dimension (mm)	Description
Air gap	10	Air gap
h	1.6	Thickness of substrate
t	0.035	Thickness of copper
A	32	Middle of feed to substrate
B	20	Middle of feed to substrate
R	25	Radius of circle patch
L_f	13	Length of feed
L_{ss}	25	Length of square slot
L_s	80	Length substrate
W_f	3.1	Width of feed
W_{ss}	25	Width of square slot
W_s	80	Width of substrate

3 Result and discussion

This section discusses the antenna parameters which include the return loss (RL), resonant frequency (f_r), return loss bandwidth (RLBW), axial ratio (AR), directivity, realized gain, total efficiency, axial ratio bandwidth (ARBW) and radiation pattern of the antenna. The antenna performance at 2.4 GHz frequency were analysed and investigated. Comparison of simulation and measurement result of return loss bandwidth, resonant frequency and realized gain at 2.4 GHz frequency is demonstrated and tabulated as in Fig. 3 and Table 3.

3.1 Return loss (RL), resonant frequency (f_r) and bandwidth (BW)

Return loss of antenna need to be more than -10 dB in order to get more than 90% antenna efficiency. According to the simulation and measurement result as demonstrated in Figure 3, the return loss of the designs at 2.4 GHz frequency achieved below than -10 dB which is -35.6 dB and -20.3 dB with bandwidth (BW) greater than 350 MHz. The resonant frequency of the simulation result is at 2.398 GHz with -35.8 dB return loss, while the measurement result shifts a little bit to higher frequency which is at 2.42 GHz with return loss dropped slightly to -21.5 dB.

The simulation bandwidth of the antenna is 556 MHz which ranging from the frequency of 2.20 GHz until 2.76 GHz. While the bandwidth of measurement is 354 MHz, it is less compared than the simulation bandwidth. However, both simulation and measurement bandwidth still cover a wider operation bandwidth at 2.4 GHz which is over 350 MHz.

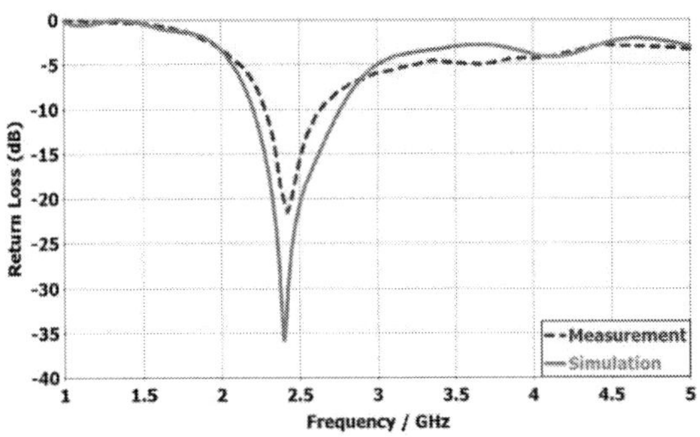

Fig. 3. Comparison of simulated and measured return loss for the inverted suspended circular patch with square slot antenna design.

3.2 Directivity (Dir.) and total efficiency (Tot. Effc.)

Fig. 4 illustrates the simulation view for the graph of directivity (dBi) over frequency. Based on the simulation results, the directivity for the antenna design at 2.4 GHz is 7.94 dBi. The highest directivity of the antenna is at 2.5 GHz with 8.18 dBi respectively. The directivity above 8 dBi covered from 2.42 GHz until 2.61 GHz frequency range.

Based on Fig. 5, the highest total efficiency of the antenna is at 2.4 GHz which is -0.18 dB. Total efficiency must be more than -3 dB to get more than 50% antenna efficiency at 2.4 GHz frequency. Total efficiency of the antenna also can be calculated based on result of gain minus directivity.

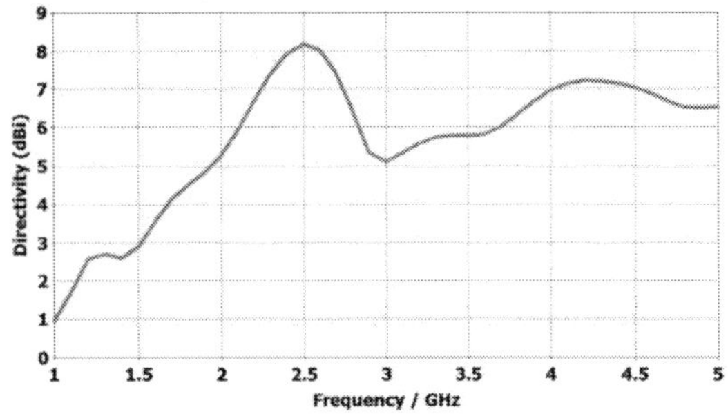

Fig. 4. Simulated directivity for inverted suspended circular patch with square slot antenna design.

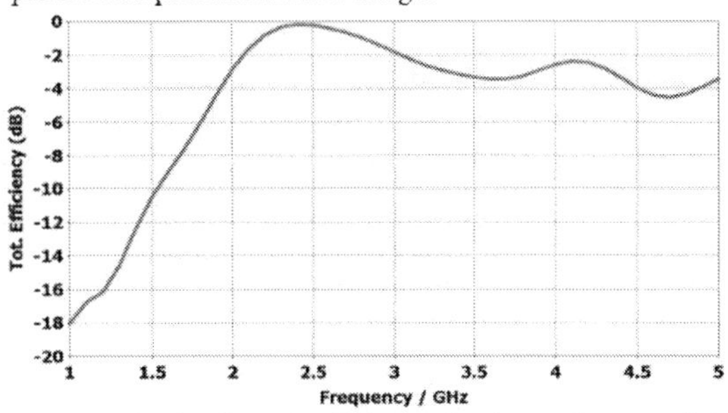

Fig. 5. Simulated total efficiency for inverted suspended circular patch with square slot antenna design.

3.3 Axial ratio (AR)

In Fig. 6, the polarization for the antenna may be identified according to the axial ratio analysed by the CST software. The axial ratio of circular polarized (CP) should be below than 3 dB. Based on Fig. 6, the minimum axial ratio is at 2.1 GHz frequency which is 0.07 dB. While at 2.4 GHz, the axial ratio for the antenna design is 0.62 dB. The axial ratio bandwidth (ARBW) below than 3 dB is 1120 MHz which is covered from frequency range of 1.37 – 2.49 GHz. It was indicated that for this inverted suspended circular patch with square slot antenna design, it can operate and presented a good circular polarization at 2.4 GHz.

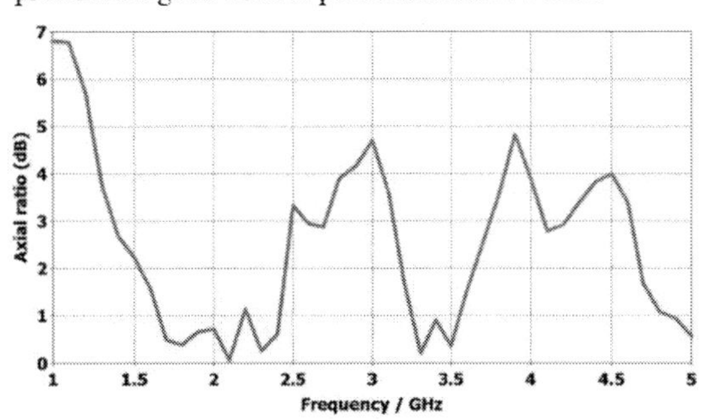

Fig. 6. Simulated axial ratio for inverted suspended circular patch with square slot antenna design.

DOI: 10.1051/matecconf/201714001029

3.4 Radiation pattern

Radiation pattern of antenna is measured in far field region. Fig. 7 demonstrate the comparison of simulated and measured radiation pattern of the antenna at 2.4 GHz when phi = 0°, phi = 90° and theta = 90°. The radiation pattern gained from measurement result is quite similar with the simulation result.

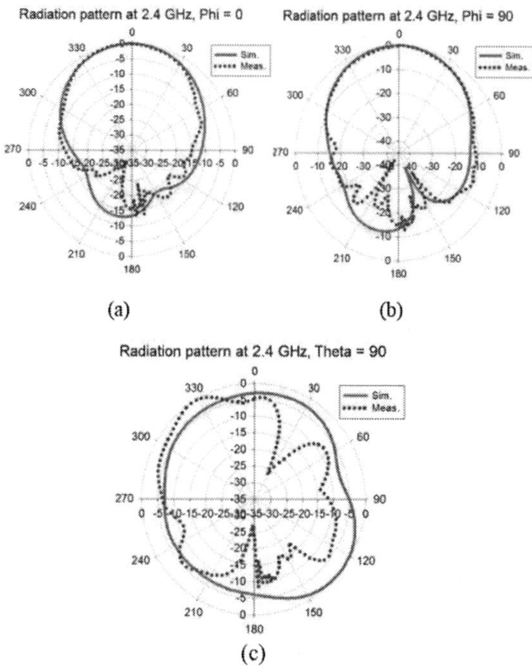

Fig. 7. Comparison of simulated and measured radiation pattern for inverted suspended circular patch with square slot antenna design at 2.4 GHz frequency (a) phi = 0° (b) phi = 90° (c) theta = 90°.

Table 2 shows the result of axial ratio bandwidth (ARBW), axial ratio (AR), directivity and total efficiency at 2.4 GHz. While, Table 3 shows the result of return loss bandwidth (RLBW), resonant frequency (f_r), return loss and realized gain at 2.4 GHz. According to Table 3, the simulation result of gain is 7.76 dB and based on calculation of gain for the data taken from measurement result, the gain is 5.54 dB respectively. The comparisons of simulation and measurement results for the gain show that the gain for measurement is reduced around 2.22 dB. This might be due to the reduction of return loss for measurement result at 2.4 GHz from -35.6 dB to -20.3 dB which is around -15.3 dB less compared to the simulation result.

Table 2. Performance result of ARBW, AR, directivity and total efficiency of the inverted suspended circular patch with square slot antenna design.

Ant. Design	Axial Ratio Bandwidth (ARBW)		2.4 GHz	2.4 GHz	2.4 GHz
	Freq. Range *(GHz)*	***BW (GHz)***	***AR (dB)***	***Dir. (dBi)***	***Tot. Effc. (dB)***
Sim.	1.37-2.49	1.12	0.62	7.94	-0.18

Table 3. Comparison of simulation and measurement result of return loss bandwidth (RLBW), f_r, RL and realized gain at 2.4 GHz.

Ant. Design	Return Loss Bandwidth		Resonant Freq. (f_r)		2.4 GHz	2.4 GHz
	Freq. range *(GHz)*	***BW (MHz)***	**Freq *(GHz)***	***RL (dB)***	***RL (dB)***	***Gain (dB)***
Sim.	2.20-2.76	556	2.398	-35.8	-35.6	7.76
Meas	2.28-2.63	354	2.42	-21.5	-20.3	4.54

4 Conclusions

In this paper, antenna with the operating frequency at 2.4 GHz which is for WLAN application is proposed. The antenna was designed based on the L-probe technique of inverted suspended circular patch with a square slot and air gap substrate between FR4 substrate and ground plane. Result performance of the proposed antenna such as return loss, resonant frequency, bandwidth, RLBW, axial ratio, ARBW, realized gain, directivity, total efficiency and radiation pattern at 2.4 GHz are analyzed. From the observation on the simulation result at 2.4 GHz, axial ratio of the antenna is circular polarized where it is 0.62 dB with return loss -35.60 dB and total efficiency above -3 dB which is -0.18 dB. The bandwidths of both simulation and measurement result for the antenna is more than 350 MHz thus meet the requirement of the broadband antenna operation. The results of measurement for the gain were reduced compared to the simulation result which is from 7.76 dB to 4.54 dB. Overall, the simulation and measurement result for the proposed design meet requirement for circular polarized antenna and broadband operation. The utilization of L-probe fed with a thick air-filled substrate leads to bandwidth enhancement. In future, in order to assemble as more compact or smaller sized antenna, size reduction can be achieved by incorporating more slots on the circular patch antenna design or with the integration of metasurface (MS) structure. Antenna with large bandwidth, compact in size and can operate as circular polarization is a major consideration in the practical application for wireless communication system.

Acknowledgements

The researchers wanted to be thankful to Universiti Teknikal Malaysia Melaka (UTeM) for helping in acquired the data and material in the development for our tasks. We also would like to thank to the Government of Malaysia which offer MyBrain15 program for financing this tasks under the research grant TRGS/1/2014/FKEKK/02/1/D00001 from Ministry of Higher Education (MOHE). Furthermore, we equally show appreciation to the anonymous referees whose comments resulted into a much better presentation of our tasks.

References

1. Balanis, C. A., *Antenna Theory: Analysis and Design 4th Edition*, John Wiley & Sons, Inc., New York, (2016).
2. D. Prajakta, B. Chandan and S. Tazeen. *Design of modified E-shape microstrip patch antenna for circular polarization*. International Journal of Emerging Technology and Advanced Engineering, vol. 3, Issue 10, pp. 247-250 (2013).
3. C. Hannachi and S. O. Tatu, *Performance Comparison of 60 GHz Printed Patch Antennas with different Geometrical Shapes using Miniature Hybrid Microwave Integrated Circuits Technology*. IET Microwaves, Antennas and Propagation, Volume 11, Issue 1, Pages 106-112 (2017).
4. R. K. Pandey and T. Shanmuganantham, *Miniaturization and Bandwidth Enhancement of a Wideband Slotted Microstrip Patch antenna*. IEEE International Conference on Circuit, Power and computing Technologies (ICCPCT), Article number 7530263, (2016).
5. R. Mishra, R. G. Mishra, P. Kuchhal and R. K. Chaurasia. *Optimization of the Dimension of Circular slotted Rectangular Microstrip Antenna for Wide Band wireless applications*. Interbational Conference on Recent Advances and Innovations in Engineering (ICRAIE), Article number 7939468, (2016).
6. K. S. Phoo, M. Z. A. Abd. Aziz, B. H. Ahmad, M. A. Othman, M. K. Suaidi and F. Abd. Malek. *Investigation of Broadband Inverted Suspended Rectangular Patch Linear Polarized Antenna*. 2014 IEEE International Conference on Control System, Computing and Engineering (ICCSCE), IEEE, pp. 236-240 (2014).
7. M.T Ali, N. Nordin, N. Ya'acob and M. N. Md Tan. *Design of Wideband Microstrip Patch Antenna Using L-Probe Fed at 2.6 GHz*. International Conference on Computer and Communication Engineering (ICCCE), pp. 961-965 (2012).
8. L. Wang, Y. X. Guo and W. Sheng, *A 60-GHz Wideband L-probe Patch Antenna Array with Gain Enhanced Structure based on LTCC Technology*. Asia-Pacific Microwave Conference Proceedings (APMC), Pages 151-153 (2012).
9. J. Jia, C. Lu, Z. Li and Y. Jie, *Design of Broadband Circularly Polarized High Gain Microstrip Antenna with L-shaped Probe Feed*. International Conference on Microwave and Milimeter Wave Technology (ICMMT), Volume 3, Pages 1007-1010 (2012).
10. M. T. Ali, N. Nordin, I. Pasya and M. N. M. Tan, *H-Shaped Microstrip Patch Antenna using L-probe Fed for Wideband Applications*. Proceedings of 6th European Conference on Antennas and Propagation (EuCAP), Article number 6206166, Pages 2827-2831 (2012).
11. Cai, Y.M., Li, K., Yin, Y. Z. and Hu, W., *Broadband Circularly Polarized Printed Antenna with Branched Microstrip Feed*. IEEE Antennas and Wireless Propagation Letters, Vol. 13. 674-677 (2014).
12. Wu, J. J., Ren, X. S., Li, Z. X. and Yin, Y. Z., *Modified Square Slot Antennas for Broadband Circular Polarization*. Progress In Electromagnetics Research C, Vol. 38, 1-14 (2013).
13. S. Gao, Q. Luo, and F. Zhu, circularly Polarized Antennas. Hoboken, NJ, USA: Wiley (2013).
14. Xu, Y.X., Gong, S.X. and Guan, Y. *Design of a Double Feed circularly Polarized Rectangular Microstrip Patch Antenna on a Circular Asymmetrical Carrier*. International Conference on Microwave and Milimeter Wave Technology (ICMMT) , Volume 3, Pages 965-968 (2012).
15. Deshmukh, A. A., Singh, D. and Verma, P., *Analysis and Design of Circular Slot Cut Corner Truncated Rectangular Microstrip Antenna*. 2nd International Conference on Advances in Electrical, Electronics, Information, Communication and Bio-Informatics, IEEE – AEEIC, Pages 627-631 (2016).

Smart Bin: Internet-of-Things Garbage Monitoring System

Mustafa M.R[1], and *Ku Azir* K.N.F[1]*

[1]Universiti Malaysia Perlis, ENAC Research Cluster, 02600, Arau, Perlis, Malaysia

Abstract. This work introduces the design and development of smart green environment of garbage monitoring system by measuring the garbage level in real time and to alert the municipality where never the bin is full based on the types of garbage. The proposed system consisted the ultrasonic sensors which measure the garbage level, an ARM microcontroller which controls system operation whereas everything will be connected to ThingSpeak. This work demonstrates a system that allows the waste management to monitor based on the level of the garbage depth inside the dustbin. The system shows the status of different four types of garbage; domestic waste, paper, glass and plastic through LCD and ThingSpeak in a real time to store the data for future use and analysis, such as prediction of peak level of garbage bin fullness. It is expected that this system can create greener environment by monitoring and controlling the collection of garbage smartly through Internet-of-Things.

1 Introduction

From 1950 until 2017, Malaysia's population has rapidly increase 74% which makeup into Malaysia current population which is 31, 164, 177 where 77% of the population is staying at urban. According to United Nations Population Fund (see www.unfpa.org), by 2030 five billion of people will be lived in urban areas, therefore, there is no surprise where Malaysian produce an average of 30,000 tons of waste every day and only 5% percent of it is recycled. According to Ministry of Urban Wellbeing, Housing and Local Government shows that these waste are resulting in tremendous land and air pollution for the environment, health problems for communities and bottlenecks to the economic growth. Taken together, the problem of poor waste management in Malaysia is one of the nation's biggest issues to date.

Waste can be divided into two categories, liquid or solid waste, both can be hazardous. Both of these waste can be group into organic, re-usable and recyclable waste. Mainly, liquid waste came from a point source or non-point source discharges such as wash water from homes, liquids used from cleaning in industries and waste detergents. Meanwhile, solid waste is any garbage, refuse or rubbish that make from home. These include old car tires, old newspapers, broken furniture and even food waste.

Generally, in Malaysia waste management is monitor by the state council. Locally, in Perlis the waste management managed by Idaman company which was given the responsibility by the state council to collect the waste. The domestic wastes are collected twice a week while the recycled wastes are collected once a week. Though, the waste collection is consistent however the current collection does not allow the local municipal to know the status of the garbage bin either full or empty. This practice of garbage collection become irregular and not relevant, once the increasing of state's population. Whilst, it does not have a systematic schedule to collect every type of garbage, the overloaded garbage will attract animals and insects. So, it will create unhygienic condition for surrounding environment and creates bad small which can lead in spreading some deadly disease and human illness. Current garbage collection is inefficiencies, time waste and required a huge amount of human energy. This is because the garbage collectors need to check whether the garbage is full or not according to the fix schedule. The objectives of the project are to design a prototype of Internet-of-Thing (IoT) garbage monitoring system and alert the garbage collectors the fullness of the bin by identify the level of garbage based on the depth of the bin.

2 Garbage Monitoring System

Waste management is the activities and actions required to manage waste from its origin to its final disposal. This incorporates in addition to other things, collection, transport, treatment and disposal of waste together with monitoring and regulation. It also includes the legal and regulatory framework that relates to waste management encompassing guidance on recycling.

The term normally relates to all kinds of waste, whether generated during the extraction of raw materials, the processing of raw materials into intermediate and final products, the consumption of final products, or

* Corresponding author: fazira@unimap.edu.my

other human activities, including municipal, agricultural, and social. Waste management is intended to reduce unfriendly effects of waste on health, the environment or aesthetics. Waste management practices are not uniform among countries developed and developing nations, regions urban, rural area, sectors residential and industrial. Thus, many researchers and industry are trying their best to creates a resolution of finding smarter way to manage waste pollution.

Prakash and Prabu [1] have design IoT Base Waste Management for Smart City for solving the trash over flow which create unhygienic condition and bad smell around the surrounding. This project has two part that are transmitter section and receiver. The 8051microcontroller, RF Transmitter and sensors in the transmitter section that attached to the dustbin. The receiver section using RF Receiver, Intel Galileo, and Web Browser is used. This system can detect waste level in dustbin and avoid the overflow of dustbin. Meanwhile, Tey Ja Sin and et al., [2] design a sustainability waste management. This concept was design based on the interview to six industry practitioner in both private and government sector in Malaysia. S.S.Navghane et al., [3] use the dustbin that interfaced with microcontroller based system having IR wireless systems along with central system showing current status of garbage, on mobile web browser with html page by Wi-Fi. Hence the status will be updated on to the html page.

Smart Garbage System (SGS) [4] is proposed to reduce the amount of food waste. The system exchange information with each other using wireless mesh networks, and a router and server collect and analyse the information for service provisioning and been operated as pilot project in Gangnam district. The results showed that the average amount of food waste could be reduced by 33%. "Smart Bin" by [5] was designed to manage the waste collection system based on smart city. The network of sensors enabled smart bins connected through the cellular network generates a large amount of data, which is further analysed and visualized at real time to gain insights about the status of waste around the city. Meanwhile, Catania and Ventura [6] proposed smart waste collection by using smart-M3 platform to improve and optimize the handling of solid urban waste by introduce 'green points' for encouraging citizens to recycle. On the other hand, Medvedev et., al [7] IoT components such as RFIDs, sensors, cameras and actuators for efficient waste collection by proposed Decision Support Syste (DSS) for efficient waste collection in Smart Cities. The system handles ineffective waste collection by incorporates a model for data staring between truck drivers on real time and dynamic route optimization.

3 Methodology

Fig 1 shows the application of this project that demonstrates a system based on Internet-of-Thing (IoT) that allows the waste management to monitor based on the level of the garbage depth inside the dustbin. The system let users being alert the level of garbage on four types of garbage; domestic waste, paper, glass and plastic.

Fig. 1. Implementation of proposed waste management system

Meanwhile, Fig 2 shows the proposed system through a block diagram. The proposed system is using ultrasonic sensor as input and placed at the maximum level of the garbage bin. The system consists the ultrasonic sensor which measure the garbage level and an ARM microcontroller which controls system operation whereas everything will be connected to ThingSpeak. At the same time, the level of garbage also will display on LCD to allow user to know the level of garbage in the dustbin without open it. The four ultrasonic sensors connect to ARM microcontroller to detect the level of garbage of each bin based on the depth of the bin. At the same, these four ultrasonic sensors connect to ESP8266 wifi module to make sure the data transfer and display on ThingSpeak. The LCD are interfaces with ARM microcontroller will display the percentage of the garbage for each bins. In this work, the system will try to monitor the depth of the garbage based on garbage type. The domestic waste does not to wait the bin to be 100% full as the longer it will be in the bin; the longer the domestic waste will be rotten and create unpleasant environment.

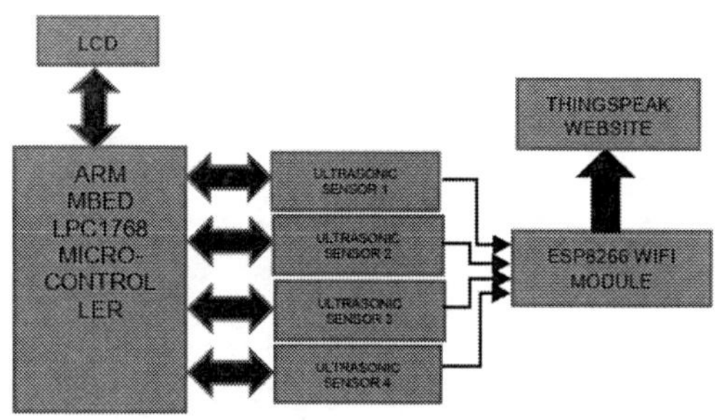

Fig. 2. Block diagram of the proposed system

Fig 3 shows the circuit diagram and each component connect each other. Here, four ultrasonic sensors are connected to ARM microcontroller and ESP8266 wifi module using logic level converter. The function of logic level converter is to reduce the voltage from ultrasonic sensor 5V to 3.3V. This is because every PIN in ESP8266 wifi module can accept 3.3V only. If the wifi module accept more than 3.3V it will make the wifi module burn. In this system, the ultrasonic sensor need at least 5V to generate the data and display the data on LCD. To make connect ultrasonic sensor to wifi module, its need logic level converter to reduce the voltage.

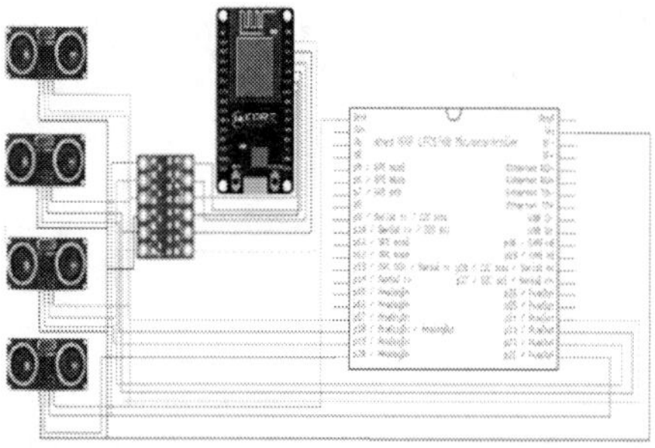

Fig. 3. Circuit Diagram

These four sensors then connected to ESP8266, a low-cost Wi-Fi chip with full TCP/IP stack which give any microcontroller access own network Wi-FI. It requires a minimal external circuitry and integrates a integrates a 32-bit Tensilica MCU, standard digital peripheral interfaces, antenna switches, RF balun, power amplifier, low noise receive amplifier, filters and power management modules all in one small package. The, mbed NXP LPC1769 is used as the microcontroller as the controller of the system. The date than been collected then send to ThinkSpeak to analyse and visualise uploaded data.

4 Result and Discussion

Fig 4 shows a complete circuit of IoT garbage monitoring system. This system is using ARM microcontroller and ARM application board that attach with the garbage bins to show the real application. This circuit are made from several components which comprises of ARM board, ultrasonic sensor, ESP8266 wifi module, LCD that interface with ARM microcontroller, rechargeable battery, breadboard and some wires as connection.

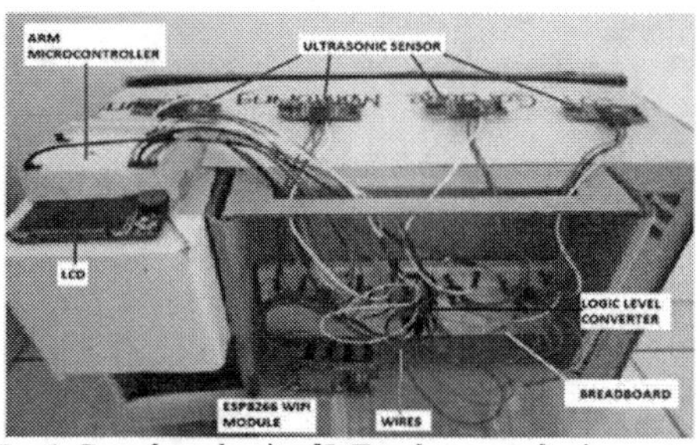

Fig. 4. Complete circuit of IoT garbage monitoring system

The system is evaluated by testing the emptiness and fullness of the garbage bin. Fig 5 (a) shows that the garbage bin is empty thus level of garbage is empty. Data will then display the percentage of the fullness of the bin on LCD that attach at the bin to alert the users percentage of the fullness of the bin.

Fig 5(a) Fig 5(b)

Fig 5(a) show emptiness of the garbage bin and the level of garbage as shown on LCD in Fig 5(b).

At the same time, data from sensor will send to ThingSpeak via ESP8266 wifi module. The ThingSpeak will shown the data in the real time as shown in Fig 6. The data on the ThingSpeak show the zero value because no garbage in the bin. So, the waste management can monitor based on the level of garbage depth inside the dustbin in ThingSpeak.

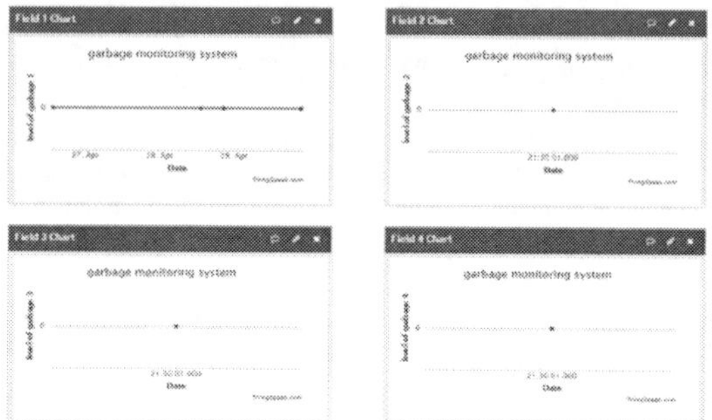

Fig. 6. The data on ThingSpeak for each bin

Fig 7(a) shows the dustbins are full with garbage. The sensor at the bin will detect the level of garbage and will convert it to the percent. The sensor will connect to microcontroller that control system operation. Moreover, data for each bin will display on LCD (Fig 7(b)) that attach at the bin to alert the user how many percent of the garbage in the bins.

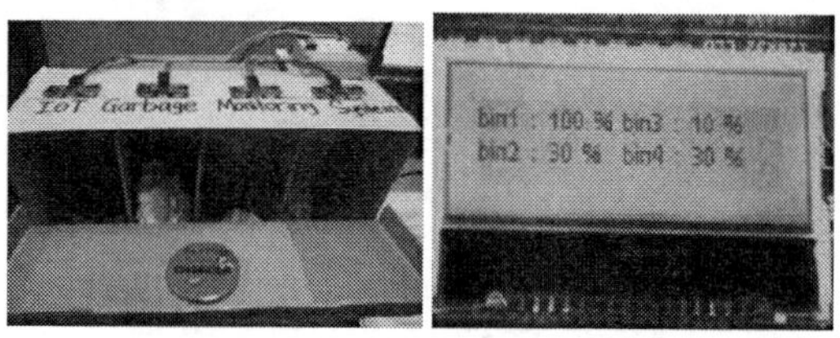

Fig 7(a) Fig 7(b)

Fig 7(a) show the garbage bin fill in with garbage and the level of garbage as shown on LCD in Fig 7(b).

The data on the ThingSpeak will shows the percentage for each bins to make sure the waste management can monitor it as shown in Fig 8. If the bin is full, the waste management can inform to collector to collect the garbage. So, collector can do their duties without any problem.

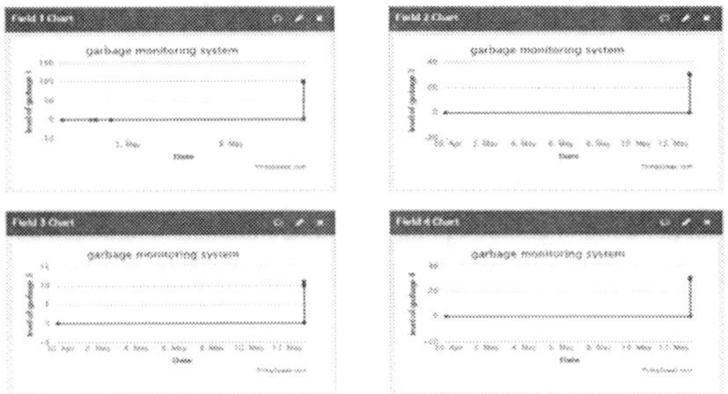

Fig. 8. The data on ThingSpeak once the bin been full with garbage

5 Conclusions

A practical system for monitor the level of garbage is being presented in this paper. This project implementing real time waste management system by using sensors to check the level of garbage in the dustbin. In this system, the information of the dustbin can have accessed from anywhere and anytime. This system will help inform the status of each dustbins in real time. So, waste management can send the garbage collector to pick up the garbage when the dustbin is full. The range of ultrasonic sensor can detect distance is between 2cm until 400cm.This sensor will compare the depth of the dustbin to show the level of garbage in the bin. This sensor will collect the data and sent to microcontroller to display on LCD. At the same time, this sensor will send data to ThingSpeak via ESP8266 wifi module. The data in ThingSpeak will show the data in real time. Therefore, waste management can be monitor.

However, if there is a limited coverage or availability of network, the wifi module will not be working thus will disturb the whole system. A proposed future development of the communication hardware need to be done in the future. Additionally, the development of the system can improve by designing a system that can identify different type of garbage by using image recognition in avoiding mistakenly put the garbage at the wrong bin.

By implementing this proposed system, it will reduce cost, man power and indirectly reducing traffic in that place.

References

[1] Parkash and V. Prabu, "IOT based waste management for smart city," *International Journal Research Compututer and Communciation Engineering.*, vol. 4, no. 2, 2016.

[2] A. Omran, A. Mahmood, and H. a Aziz, "Current practice of solid waste management in Malaysia and its disposal," *Environmental Engineering Management Journal*, vol. 6, no. 4, pp. 295–300, 2007.

[3] S. S. Navghane, M. S. Killedar, and V. M. Rohokale, "IoT Based Smart Garbage and Waste Collection Bin," *Internatonal Journal of Advanced Research in Electronics and Communication Engineering,* vol. 5, no. 5, pp. 1576–1578, 2016.

[4] I. Hong, S. Park, B. Lee, J. Lee, D. Jeong, and S. Park, "IoT-based smart garbage system for efficient food waste management.," *Scientific World Journal.*, vol. 2014, p. 646953, Aug. 2014.

[5] N. Sharma, N. Singha, and T. Dutta, "Smart Bin Implementation for Smart Cities," *International. Journal. Science Engineering. Research.*, vol. 6, no. 9, pp. 787–791, 2015.

[6] V. Catania and D. Ventura, "An approch for monitoring and smart planning of urban solid waste management using smart-M3 platform," in *Proceedings of 15th Conference of Open Innovations Association FRUCT*, 2014, pp. 24–31.

[7] A. Medvedev, P. Fedchenkov, A. Zaslavsky, T. Anagnostopoulos, and S. Khoruzhnikov, "Waste management as an IoT-enabled service in smart cities," in *Lecture Notes in Computer Science (including subseries Lecture Notes in Artificial Intelligence and Lecture Notes in Bioinformatics)*, 2015, vol. 9247, pp. 104–115.

MATEC Web of Conferences 140, 01031 (2017)
ICEESI 2017
DOI: 10.1051/matecconf/201714001031

Impedance Analysis of Mobile Phone Antenna in the Presence of User's Hand

Nur Basyirah A Rahman*, *A.A.*Al-Hadi, and *Saidatul Norlyana* Azemi

School of Computer and Communication Engineering, UniMAP, 02600 Arau, Perlis, Malaysia

Abstract. User interaction (UI) has been identified to degrade the performance of the antenna. A system that measures impedance characteristics of a planar inverted-F antenna (PIFA) when there are user interactions is designed using Keysight VEE Pro. The operational frequencies of the antenna are 900 MHz and 1800 MHz at low band and high band, respectively. The obtained results are compared and then a new impedance matching circuit is proposed and designed by using ADS (Advanced Design System). The proposed matching circuit at each operating frequency is a basic lumped circuit using the L network. Both simulation and experimental work have been done.

1 Introduction

Mobile phones are now a necessity need for many people throughout the world. Advanced mobile phones are able to receive and place phone calls, store data, take picture and also can be used as walkie-talkie. As the technology advanced, the impact of mobile phones increased accordingly. Therefore, the contact of mobile phones and human body has increased, too. The interaction between users and mobile antenna is considered as very crucial.

It is an immense challenge to ensure that the antenna in mobile terminal meets the crucial criteria. The fundamentals that the antenna should meet are radiation characteristic to ensure the consistency of communication, a wide bandwidth to discourage distortions of transmitted signals, minimal impact on users and minute sensitivity of the antenna parameters to the imminence of users and of course, it should be in small, compact size [1]. Specific Absorption Rate (SAR) is another significant criterion to mull over. SAR measures the amount of power captivated by body tissue when exposed to electromagnetic radiation [2].

Various studies have been done previously that showed the performance of the antenna in mobile terminal are significantly affected by the presence of users, especially when the mobile antenna is being held by the users, close to their heads [1], [3]–[7]. It is produced feeble performance of mobile terminal; the output power and radiation efficiency are worsened by huge mismatch. This mismatch is likely to happen because of the dependency of frequency on the antenna impedance.

Planar inverted-F antenna (PIFA) is a basic antenna that is being used in a wide range of application, especially mobile devices. Besides having simple structure and uncomplicated fabrication, it is also inexpensive to manufacture and inconspicuous, as it is in small volume [8]. The structures of PIFA make it pertinent to be used as a multi-band antenna, hence relevant in mobile devices [2], [9], [10]. The antenna involves is a dual-band PIFA.

This paper is to analyse antenna impedance in free space and also in the presence of human hands and heads, consequently to propose impedance matching circuit to improve the antenna impedance mismatch.

2 Impedance analysis

The first stage of the analysis was designing a system to capture the impedance value of the antenna in mobile terminal, and the experiment was conducted when the antenna was in free space and with user interaction (UI). This system was built within a simulation tool, VEE Pro software. The antenna used in this stage is PIFA and the specification will be elaborated more in the next section. The second stage was proposing impedance matching circuit when mismatch occurred in the measurement done in the first phase. This circuit utilized the usage of Advanced Design System (ADS).

2.1. Software development for impedance measurement

This stage involved the usage of simulation tool and an antenna prototype, namely Keysight VEE and planar inverted-F antenna (PIFA). These two components worked closely with the network analyzer in this stage of the project. VEE or Visual Engineering Environment is a graphical programming language for building test and measurement application [11].

Keysight VEE was used to implement a system to capture the value of impedance when the antenna was in

* Corresponding author: basyirahrahman@gmail.com

free space and when human was present. Keysight VEE allows users to alter parameters without reconfiguring programs, thus making it easy to use to measure and record the impedance value at different frequency range, 900 MHz and 1800 MHz, respectively.

When the program was executed, it captured the impedance and S_{11} (dB) at the specified operating frequencies. The program could only be simulated when the computer was connected to a network analyzer using a GPIB cable or LAN cable.

The PIFA prototype used was a dual-band antenna, fabricated for mobile terminal with 100(L) x 40(W) x 0.035(H) mm3 dimension. The antenna prototype was supplied with 50 ohm coaxial feeding structure. Once the program was successfully simulated without error, the antenna was connected to the network analyzer and the antenna was calibrated to the desired frequency range of 700 MHz – 2000 MHz for easy impedance measurement.

Once the setup was all fixed, the program code was simulated. The first simulation of the program, the antenna was left in free space, as shown in Fig. 1, and without user interaction (UI), as shown if Fig. 2. The next simulation was then carried out with UI and this process was done 20 times. The data were automatically saved in the computer and they could be retrieved to be compared.

Fig. 1. The antenna in free space

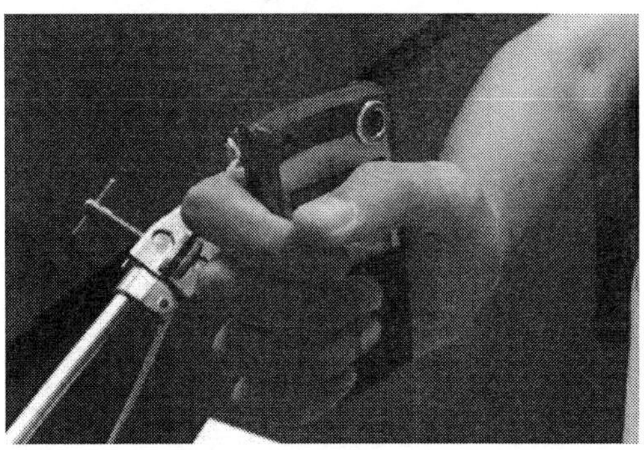

Fig. 2. User gripping the antenna

2.2. Impedance matching circuit

In this stage, an impedance matching circuit network was built by using Keysight ADS (Advanced Design System). It provides a built-in design environment [12] and this stage utilizes the workspace to design schematics to display data obtained from network analyzer.

The measurement data were used to compare and contrast the impedance mismatch when user interactions were present. To correct the mismatch, an impedance matching circuit was proposed and designed based on the measurement obtained in the first stage.

The proposal of the impedance matching circuit was made easy by using the feature provided in ADS, the Smith Chart Utility. When the impedance was matched to the desired value, a circuit was automatically designed and the values of the inductor and capacitor, if any, were given.

3 Results

The data from the simulation using VEE Pro software for impedance characteristics of antenna with and without user interactions were recorded and analysed, and matching circuits were proposed accordingly.

3.1. Impedance measurement

The simulation of the programming in VEE Pro software and the experiments done were focused on the variation of the antenna impedance with user interaction (UI). The data recorded using VEE Pro software is shown in Table 1 and Table 2.

Table 1. Data recorded from VEE Pro software at 900 MHz

Experiment	Impedance (Ohm)	S11 (db)
Free Space	56.20030, -25.70119	-12.3
1	7.18044, -9.41670	-2.4
2	73.42599, -30.76016	-10.3
3	114.01299, -47.13789	-6.7
4	107.93769, -82.02252	-4.9
5	10.02922, -17.07871	-3.1
6	9.02490, -17.28456	-2.8
7	10.83449, -15.96367	-3.4
8	10.83825, -18.32352	-3.3
9	9.04438, -15.85823	-2.9
10	108.00266, -45.95788	-6.9
11	69.37858, -40.83315	-8.9
12	7.76002, -8.78291	-2.6

13	157.55675, -48.81924	-5.1
14	51.25545, -12.28277	-9.1
15	94.54969, -65.05639	-6.1
16	75.99115, -52.71152	-7.3
17	97.31188, -24.59574	-8.9
18	6.85889, -7.77914	-2.3
19	8.47236, -5.16774	-2.9
20	7.87053, -6.77857	-2.7

Table 1 and Table 2 show the values of impedance when the antenna was in free space and with user interactions at 900 MHz and 1800 MHz, and the values of S_{11} in dB. The simulation was first done on the antenna without user interaction. Then the experiments were done with 20 users, recording the impedance and S11 for each user at both frequencies. The measurements were executed using freestyle grips. The antenna was placed in a casing to make the gripping easier. Each user gripped the antenna differently, providing a variation of impedance (Ω) and S11 values.

Table 2. Data recorded from VEE Pro software at 1800 MHz

Experiment	Impedance (Ohm)	S11 (db)
Free Space	26.79966, 33.13157	-12.3
1	14.82247, 53.02929	-2.4
2	28.80263, 24.62985	-8.1
3	24.65929, 13.18249	-8.5
4	22.76062, 23.49544	-6.5
5	24.09021, 49.19400	-4.1
6	26.35174, 45.86049	-4.7
7	23.59299, 46.91540	-4.2
8	20.62890, 46.00855	-3.8
9	22.34369, 52.98932	-3.5
10	18.69014, 22.75773	-5.4
11	30.35454, 24.68291	-8.5
12	13.14519, 49.63192	-2.3
13	23.93606, 13.63711	-8.2
14	27.97118, 26.95319	-7.4
15	28.57141, 25.01479	-7.9
16	29.50213, 21.57662	-8.9
17	36.77092, 26.56201	-9.7
18	14.69248, 58.97473	-2.1
19	18.86003, 62.04056	-2.5
20	18.62644, 63.33371	-2.4

Mobile antenna is usually designed to operate ideally in free space. These experiments proved that having user interactions cause impedance mismatch that varies with frequencies at high band and low band. Human body is made of tissues that are highly likely of having high permittivity that causes the detuning of antenna. Each user also has different hand size and different strength in gripping the antenna that absorbs different amount of power, leading to the decrease of efficiency.

3.1 S-Parameters and smith charts

The variation of S-parameters and the smith chart for impedance variation at low band and high band are shown in Fig. 3, Fig. 4, Fig. 5 and Fig. 6. The single black line iss the measurement made in free space, and the coloured lines are the measurement made on 20 different users. Based on the smith chart, it can be seen that user interactions have caused several severe impedance mismatch.

It is duly noted that the impedance at 900 MHz and 1800 MHz measured in free space were not properly matched. There are several circumstances that can be considered as the result of the impedance mismatch in free space. One of the factors is the fabrication tolerance. There might have been a minor miscalculation in the antenna size or a slight error during fabrication. A loose antenna port and the material used as the antenna casing should be taken into consideration too. The size of the antenna must be accurate and the port must be properly intact for the antenna to operate at high performance in free space.

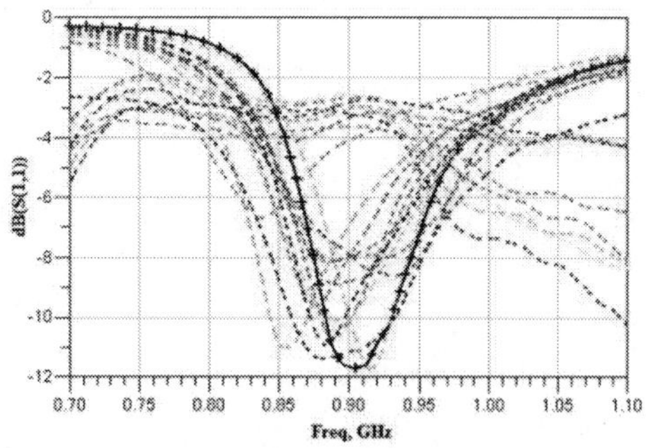

Fig. 3. Frequency (GHz) against S11 (dB) of antenna at low band

DOI: 10.1051/matecconf/201714001031

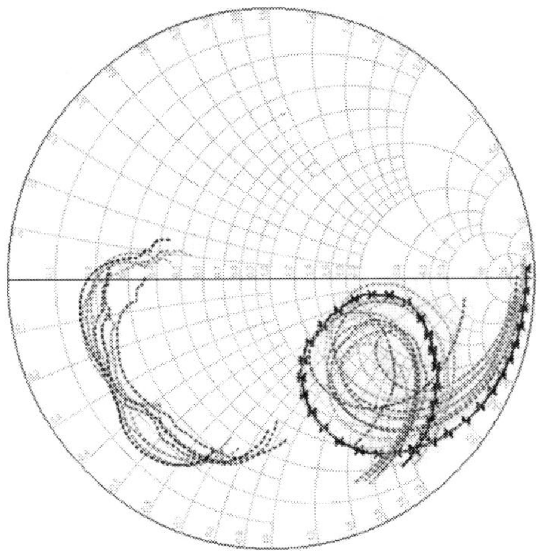

Fig. 4. Smith chart for impedance variation at low band

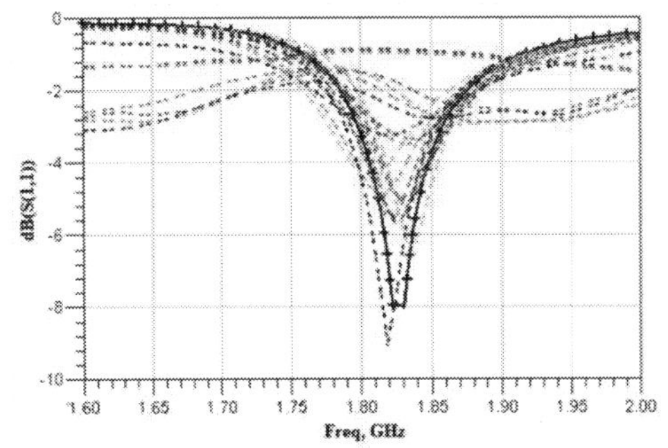

Fig. 5. Frequency (GHz) against S11 (dB) of antenna at high band

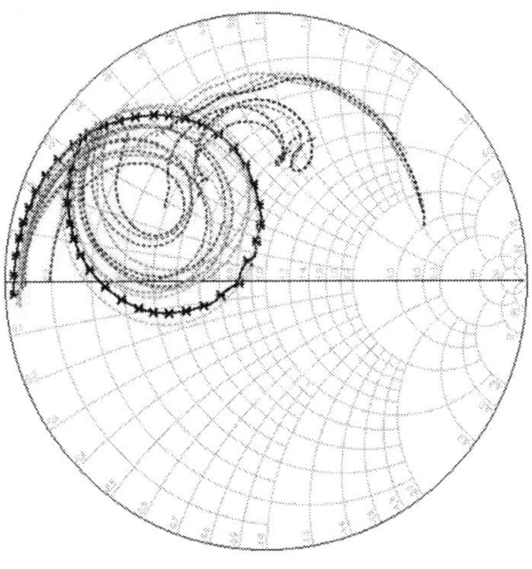

Fig. 6. Smith chart for impedance variation at high band

3.3 Impedance matching circuit

Fig. 7 shows that the impedance mismatch at 900 MHz is more likely to happen in the 51.25Ω – 157.55Ω impedance range. An average impedance value of 91.42Ω is selected.

The average Z value of 91.4203 – j43.2616Ω is inserted in the space provided and then a matching circuit consists of an inductor (L) and a capacitor (C) is designed, as shown in Fig. 8. The values of the inductor and capacitor used in the circuit are 10.0299nH and 1.0003pF respectively, as shown in Fig. 9.

Fig. 10 shows that the impedance mismatch at 1800 MHz is more likely to happen in the 13.15Ω – 36.78Ω impedance range. An average impedance value of 23.62Ω is selected.

The average Z value of 23.6178 + j37.3144Ω is inserted in the space provided. Another matching circuit consists of two capacitors (C) is designed, as shown in Fig. 11. The value of C1 is 2.2502pF and the value of C2 is 0.8317pF, as shown in Fig. 12.

The results from the experiments done on the effect of user interactions (UI) on mobile antenna and the possible causes of antenna when the antenna is being held by user have been analyzed and discussed. Matching circuits for mismatched impedance at low band and high band have also been designed. This study has successfully achieved its objectives to design a system that measure the impedance of antenna in mobile antenna, to study the behavior of impedance in free space and in the presence of user interaction and to propose impedance matching circuit for each operational frequency.

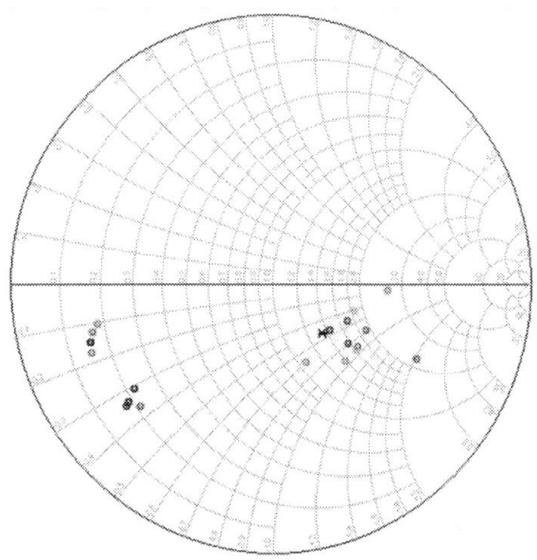

Fig. 7. Impedance variation at 900 MHz

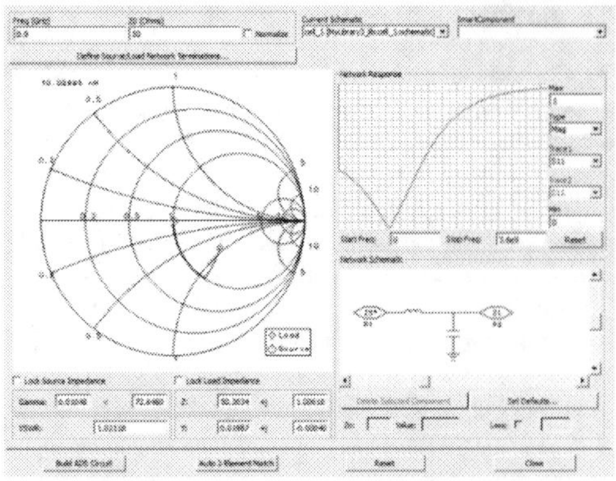

Fig. 8. Designing matching circuit at 900 MHz

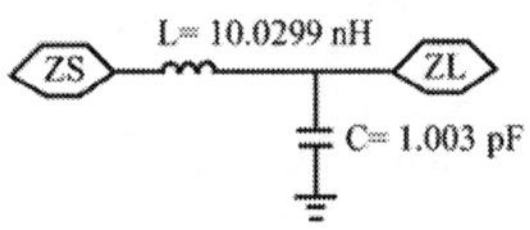

Fig. 9. Impedance matching circuit 900 MHz

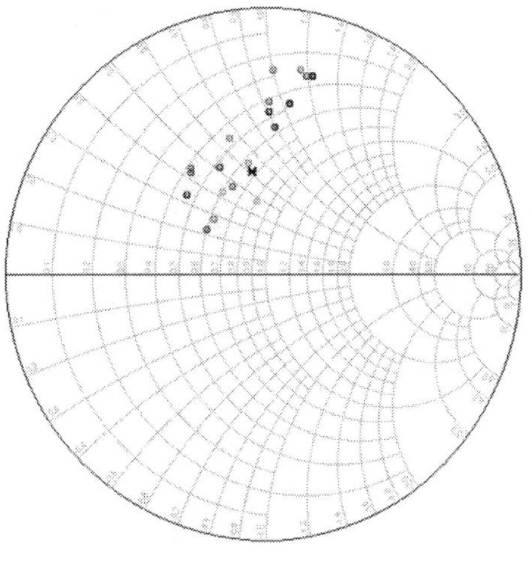

Fig. 10. Impedance variation at 1800 MHz

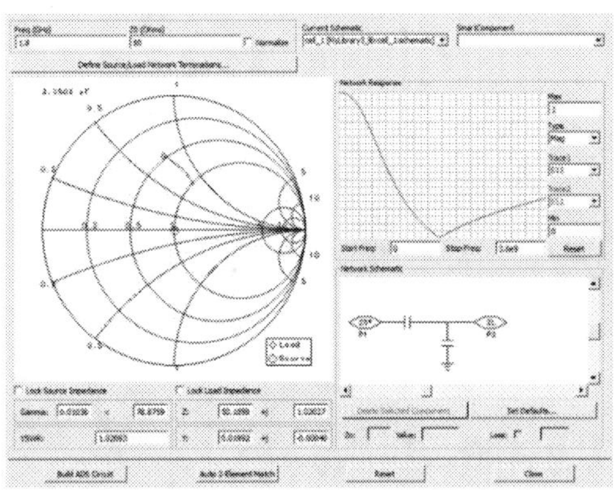

Fig. 11. Designing matching circuit at 1800 MHz

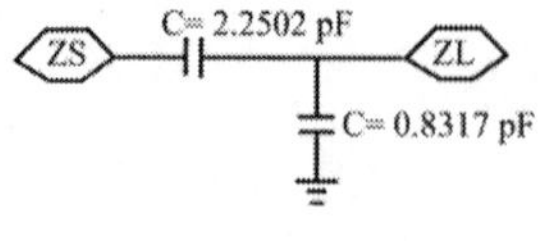

Fig. 12. Impedance matching circuit at 1800 MHz

Conclusion

The variation of impedance and S_{11} measurement caused by user interactions were because each user had different method of holding or gripping the antenna. The size of user's hand and the strength of their grips also affected the performance of antenna, in some cases, very severely. The absorbed power by each user's hand was different too.

The measurements of impedance obtained from the experiments were then used to propose matching circuits for impedance mismatch at each band; low band and high band. These matching circuits are highly important in tuning mismatched impedance at mobile antenna terminal. They help to keep the antenna to be in good performance even when there is user interaction.

The proposed matching circuit in this study, a simple lumped circuit using the L network, fully utilized the Smith Chart Utility feature on ADS. The circuit designed at low band is made up of an inductor and a capacitor, while at high band, the circuit designed is made up of two capacitors of different values.

The authors would like to thank the Ministry of Science, Technology and Innovation for the financial support under eSciencefund (Reference Code: 01-01-015-SE0258)

References

1. R. Przesmycki, M. Bugaj, L. Nowosielski, M. Wnuk, Dual-band mobile phone antenna for RF protection of human head, *2011 SBMO/IEEEMTT-S Int. Microw. Optoelectron. Conf. (IMOC 2011)*, 235-238 (2011)
2. Z. Ying, Antennas in cellular phones for mobile communications, *Proc. IEEE,* **100**, 2286-2296 (2012)
3. K. Boyle, M. Leitner, Mobile phone antenna impedance variations with real users and phantoms, *Final Progr. B. Abstr. - iWAT 2011 2011 IEEE Int. Work. Antenna Technol. Small Antennas, Nov. Struct. Innov. Metamaterials*, **1**, 420-423 (2011)
4. K. Zhao, S. Zhang, Z. Ying, T. Bolin, S. He, SAR Study of different MIMO antenna designs for LTE application in smart mobile Phones, 2-3 (2012)
5. M. Lin, C. Chiu, Human-body effects on the design of card-type UHF RFID tag antennas (2011)
6. Q. Guo, R. Mittra, F. Lei, Z. Li, J. Ju, J. Byun, Interaction between internal antenna and external antenna of mobile phone and hand effect, *Antennas Propagation, IEEE Trans.*, **61**, 862-870 (2013)
7. K. R. Boyle, Y. Yuan, L. P. Ligthart, Analysis of mobile phone antenna impedance variations with

user proximity, *IEEE Trans. Antennas Propag.*, **55**, 364-372 (2007)
8. N. A. Saidatul, P. J. Soh, Y. Sun, D. Lauder, and A. A. H. Azremi, Multiband fractal PIFA (Planar Inverted F Antenna) for mobile phones, 671-675 (2010)
9. P. Tsai, D. Lin, H. Lin, and P. Chen, Printed monopole slot antenna for internal multiband mobile phone antenna, **55,** 3690-3697 (2007)
10. C. Rowell and E. Y. Lam, Mobile-phone antenna design, *IEEE Antennas Propag. Mag.*, **54**, 14-34 (2012)
11. Agilent Technologies, *Agilent VEE Pro VEE Pro Users Guide* (2014)
12. Keysight Technologies, *Keysight Technologies EDA Advanced Design System* (2014)

DOI: 10.1051/matecconf/201714001032

The Effects of Metallic Loop-Like Accessory Worn on the Human Body on SAR at 2.45 GHz

H. H. Zainal[1], H. A. Rahim[1], Soh Ping Jack[2], K.N.Abdul Rani[3], H. Lago[2], M.S.Zulkefli[3], and N.F.N.Yah[1]*

[1]Bioelectromagnetics Research Group (BioEM),
[2]Advanced Communication Engineering (ACE),
School of Computer and Communication Engineering, Universiti Malaysia Perlis (UniMAP), Kampus Pauh Putra, Perlis, Malaysia
[3]Dept of Electronics Engineering Technology, Faculty of Engineering Technology,Universiti Malaysia Perlis,Malaysia
[4]Centre of Diploma Studies (CDS),Universiti Malaysia Perlis (UniMAP), Campus UniCITI Alam Sg. Chuchuh,Perlis,Malaysia

Abstract. This paper presents the Specific Absorption Rates (SAR) in the human body with a monopole antenna.The distance between the antenna and the body were varied at different distances. The parameters (εr, σ) used in the human body set according to the standard tissue equivalent liquids recommended by the IEEE and FCC. The simulations were made by means of CST Microwave Studio software at frequencies of 2.45GHz .The effect of the body on the SAR calculation in the body were measured. The SAR values were recorded in term of SAR for 10g of tissue. The TM is positioned against the metallic loop-like accessory, place on the left wrist of the generic arm at a varied distance from the cylindrical phantom.Numerical analysis conducted using a broadband textile monopole antenna (TM) with variations of orientation and distance showed that SAR values increased when the TM is horizontally polarized.

1 Introduction

Recently, wearable textile antenna is important in providing comfort to users in wireless body area network(WBAN).The main requirement for wearable textile antennas is the use of flexible materials for ease integration into clothing.

A wearable textile antenna requires such as it must be comfortable, lightweight and inexpensive for their common use to become actuality. Because the wearable antenna will be placed in the closeness of human body, it is necessary to consider the interaction between the human body and antenna.

In additional, when the antenna is mounted on the body, the reflection coefficient , radiation pattern and efficiencies of antenna are afflicted due to the lossy nature of human skin. Therefore, the antenna's performance will be degraded when it is located on the body compared to its distance away from body.

To protect the human being body from radio wave exposure, antennas used in WBAN's must have low SAR based on the standard SAR limit which is 2W/kg averaged over 10g of tissue following IEC standard for wireless device. Thus, this project is performed in order to evaluate the realistic position and distances of metallic loop like accessory worn on the body against the wearable textile antenna.

Specific absorption rate is a measure of heat syndication in a body medium by an antenna. It describes the_absorption of the electromagnetic energy by human tissues and is defined as follows[1]:

$$\text{SAR} = \frac{\sigma\,|E|^2}{\rho} \quad (W/kg) \qquad (1)$$

where E is the electric field induced in the tissue, V/m , σ is the conductivity of the tissue, S/m and ρ is the tissue density, kg/m3.

Most of previous research has been focusing on the effect of metallic objects. Metallic objects such as spectacles ,earing and ring is no longer a symbol of wealth, but it is a necessity to every individual. A dental implant positioned near the source had higher effect on local SAR,which has also been observed in previous study[2]. The authors reported that the metal implant positioned in second premolar in the mandible and the first molar in the maxillary at 900MHz affected the current density more than other position.

In[3], it concludes that different value of dielectric properties inside the body give influence on the performance of antenna. The performance of antenna (return loss and radiation pattern) is significantly affected by the presence of the body.

In previous research, the authors have found that part of energy radiated by the mobile phone antenna is absorbed by the human body, head and hand[4]. SAR of mobile phone consumed by humans is very influenced by the mobile network carrier, characteristics of mobile phone and antenna, antenna positioning, and the extended power from the mobile phone[5].

In[6],they described that a metallic object can increase or decrease SAR values. Obviously, the maximum SAR can be increased up to 6 times comparing with no object condition by a ring shape object, 4.7 times by a triangle and 5 times by a square object.

Therefore, to investigate the interaction between the wearable device, i.e., antenna and the human body, a cylindrical phantom model is used in the simulation to human body structure. Numerical simulations are conducted using CST Microwave Studio based on the Finite Integral Technique (FIT). The generic human body consists of three cylindrical bricks, representing the upper human body torso and two arms.

* Corresponding author: *haslizarahim@unimap.edu.my*

DOI: 10.1051/matecconf/201714001032

2 Method

In this paper ,textile monopole antenna is employed as the radiating source, which operates at 2.45 GHz. The antenna consists of 0.17mm thickness of metallic layer (copper) with dielectric constant Er = 1.43,loss tangent of 0.025 and thickness of 1.7 mm[7]. A detail description on the antenna design can be found in[7] . The SAR values resulting from the radiating textile monopole antenna are benchmarked against two scenarios, which are with and without presence of metallic loop-like accessory1Watts input power was applied to the textile monopole antenna for both frequencies.

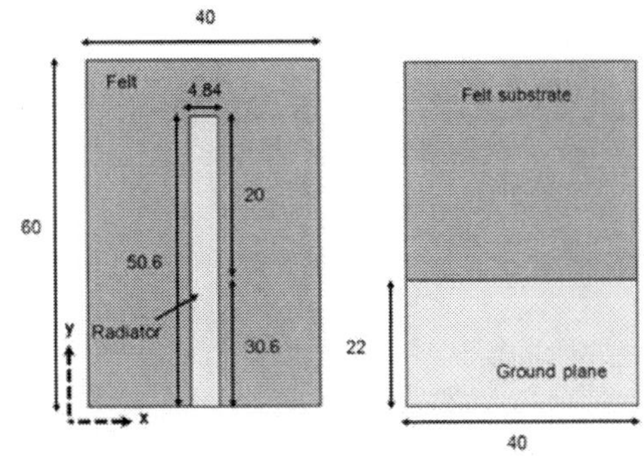

Figure 1 : TM antenna[7]

2.1 Human Torso Model

In this paper, the human body is modeled as cylindrical phantom made available by CST Microwave Studio. The frequency dependent parameter (ε_r, σ) use in human body tissue were the same as recommended by IEEE and FCC[8] , as listed in Table 1

Table 1.Dielectric Properties of Body Tissue at 2.45

Frequencies ,GHz	ε_r	σ(S/m)
2.45	53.57	1.81

2.2 The Metallic Loop like Accessory Items

In order to study the impact of the metallic items on the human body tissues, metallic loop like accessory were modeled as conducting objects (silver) for both frequencies respectively. However in this study, the edge of metallic loop like accessory was closely attached to the cylindrical model (Figure 2).

The distance between the human arm and textile antenna was varied to 4, 8, 12, 16 and 20 while the metallic loop like accessory was fixed at the same position in each case investigated. TM antenna was oriented in two different direction, i.e. ,vertical and horizontal .

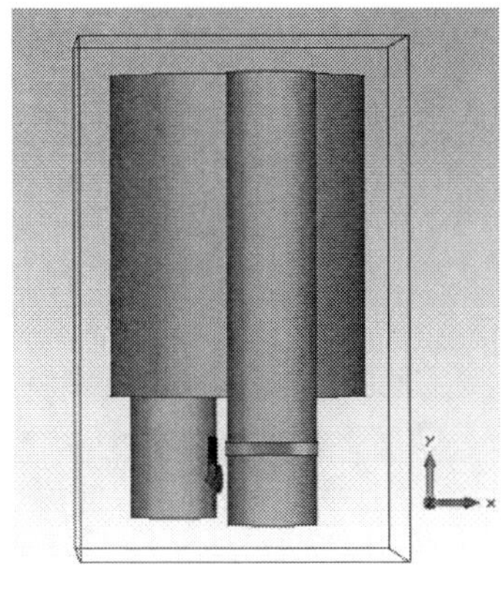

Figure 2 : TM in position horizontally

3 Result

Figure 3 shows the peak SAR values inside the human body with and without the presence of metallic loop like accessory at frequency 2.45GHz in term 10g SAR.The highest SAR value when the antenna horizontally orientation at distances 4 mm which is 23.69 W/Kg. The max SAR value decrease 90% when the antenna vertically orientation.That can be conclude,the amount of energy absorbed inside the human body is noticeably decreased when the distance between the metallic loop like accessory and the human body is increased.

Figure 3 clearly showed that a metallic loop like accessory that was worn in the close proximity to the radiating source noticeably affects the SAR distribution. The maximum SAR generally appears in the area where the metallic loop like accessory is in the closest distances from the body and occurs when antenna is horizontally polarized.

For example, the value of max 10g SAR (4.79 W/Kg) occurs when antenna is horizontally polarized and mounted 16 mm from the metallic item but when the antenna vertically polarized and placed 16 mm from metallic loop like accessory the value of SAR is small which is 0.65 W/Kg only.It shows the antenna orientation could decreased the SAR value by approximately 7 times compared to horizontally case.

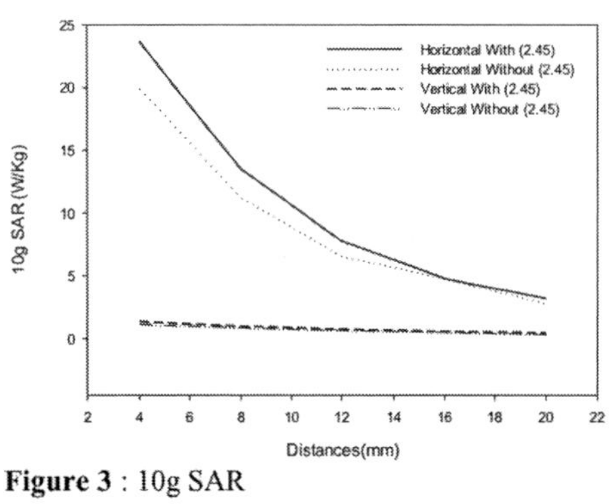

Figure 3 : 10g SAR

DOI: 10.1051/matecconf/201714001032

Figure 4 show the variations on antenna resonant frequency due to the presence of metallic loop like accessory at frequency 2.45GHz. It can be clearly seen that the presence of metallic loop like accessory has detuned the resonant frequency up to 3% at 2.45 GHz .However , the detuning effect may vary depending on the distance between the antenna and metallic loop like accessory. The worst case of S_{11} is observed when TM is horizontal polarization, which is parallel to the human body model with the distance to the metallic item, i.e, 8 mm and 12 mm.

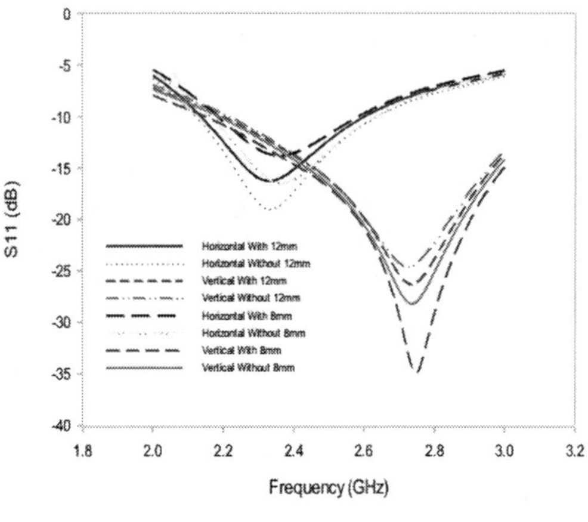

Figure 4 : S_{11} at 2.45GHz

Figure 6 show the changes of the gain for two situation, i.e.,1) with metallic loop like accessory,2) without metallic loop like accessory with two orientation such as horizontally and vertically. The gain value increase for entire vertically polarized antenna case investigated.

In addition, the results have shown that the antenna gain in frequency 2.45 GHz is significantly improved as the separation distance between the antenna and human body, distance increase in all position.

From the plot, the maximum gain of more than 6 dBi was obtained between 4 and 20 mm. From the Figure 5, the results show that the gain of the initial before attached with metallic loop like accessory with horizontally polarized was 3.9 dBi. However, in the presence of metallic loop like accessory ,the gain of the antenna gain was increased by 1.5 times.

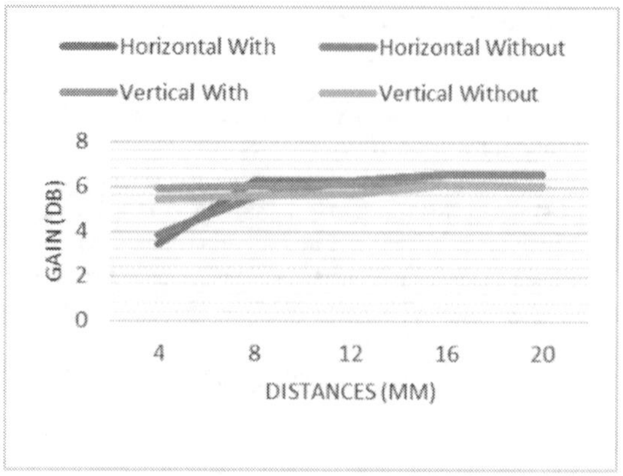

Figure 5 : Gain of antenna in horizontal and vertical orientation at 2.45GHz

4 Conclusion

The result presented in this paper shows that the presence of metallic loop like accessory affects the SAR and the reflection coefficient of the planar wearable textile monopole. Besides, the antenna orientations and locations produce a prominent effect on SAR value. But,the value of gain also increased when the vertically polarization. SAR is increased when the TM antenna is horizontally orientated than when the antenna is vertically polarized in frequency 2.45GHz.

References

1. Islam, M. T., Abidin, H. Z., Faruque, M. R. I. & Misran, N. Analysis of Materials Effects on Radio Fre- Quency Electromagnetic Fields in Human Head. *Prog. Electromagn. Res.* **128,** 121–136 (2012).

2. Safari, M. & Abdolali, A. Dental Implants and Mobile-Phone Use. 43–51

3. Source, A. E. Comparison on the Effect of Homogeneous and Inhomogeneous Body on Antenna Performance and SAR. 62–64 (2012).

4. Zhu, C., Wu, D., Ping, L. & Liang, W. The Effects of Head Parameters on the Specific Absorption Rate in the Human Head. 423–427 (2015).

5. Elias, N. A., Samsuri, N. A., Rahim, M. K. A., Othman, N. & Jalil, M. E. Effects of Human Body and Antenna Orientation on Dipole Textile Antenna Performance and SAR. 132–136

 DOI: 10.1051/matecconf/201714001032

(2012).

6. Sriprasoet, C. Effects of Ring Metallic Objects on Specific Absorption Rate in Human Head Caused by Mobile Phone. 1–6 (2015).

7. Rahim, H. A., Abdulmalek, M., Soh, P. J. & Vandenbosch, G. A. E. Evaluation of a broadband textile monopole antenna performance for subject-specific on-body applications. *Appl. Phys. A* **123,** 97 (2017).

8. Ronald, S. H. *et al.* Designing Asian-Sized Hand Model for SAR Determination at GSM900/1800: Simulation Part. *Prog. Electromagn. Res.* **129,** 439–467 (2012).

MATEC Web of Conferences **140**, 01033 (2017) DOI: 10.1051/matecconf/201714001033
ICEESI 2017

Performance Analysis of Congestion Control Mechanism in Software Defined Network (SDN)

M. Z. A. Rahman[1], N. Yaakob[1*], A. Amir[1], RB Ahmad[2], S.K.Yoon[1], A.H. Abd Halim[1]

[1]School of Computer and Communication Engineering (SCCE), Universiti Malaysia Perlis (UniMAP), 02600 Arau, Perlis, Malaysia
[2] Faculty of Informatics and Computing, Universiti Sultan Zainal Abidin (UniSZA), Kuala Terengganu, Malaysia

Abstract. In the near future, the traditional networks architecture will be difficult to be managed. Hence, Software Defined Network (SDN) will be an alternative in the future of programmable networks to replace the conventional network architecture. The main idea of SDN architecture is to separate the forwarding plane and control plane of network system, where network operators can program packet forwarding behaviour to improve the network performance. Congestion control is important mechanism for network traffic to improve network capability and achieve high end Quality of Service (QoS). In this paper, extensive simulation is conducted to analyse the performance of SDN by implementing Link Layer Discovery Protocol (LLDP) under congested network. The simulation was conducted on Mininet by creating four different fanout and the result was analysed based on differences of matrix performance. As a result, the packet loss and throughput reduction were observed when number of fanout in the topology was increased. By using LLDP protocol, huge reduction in packet loss rate has been achieved while maximizing percentage packet delivery ratio.

1 Introduction

Recently, the dramatic increased of Internet capacity network traffic due to the increment used of mobile devices, Internet protocol television (IPTV), server virtualization and cloud services are becoming more complicated for a traditional architecture network to handle high traffic. In traditional network, device such as router, switch and firewall has their own control plane and data plane. As a result, the network cannot handle higher user capacity because the forward plane and control plane are in the same device [1]. Thus, more dynamic and flexible network is required to easily managed the traffic.

The communication networks are growing rapidly and becoming more complex to be managed and handled. This is especially true with traditional network architecture, network systems and protocol stack which is difficult to provide adequate solutions to the contemporary networking demands. Hence, Software Defined Network (SDN) is a promising approach to computer networking that separates data plane and control plane combined with centralized controller to allow network operator to program the packet forwarding. The SDN architecture is as shown in Fig. 1

One of the major problems of congestion in network traffic is the reduction in the Quality of Service (QoS) that cause tremendous loss of several network aspects

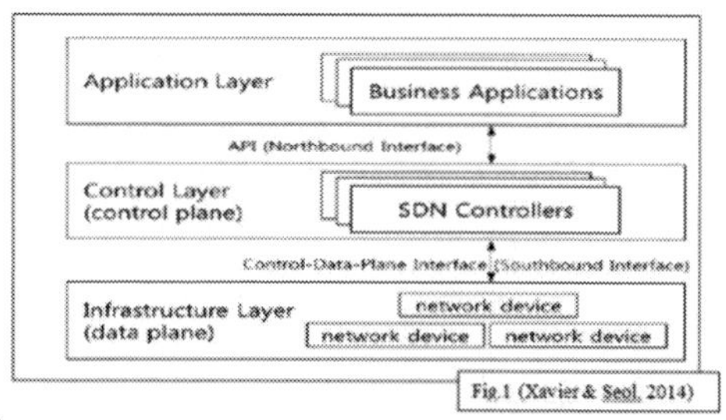

Fig. 1. SDN Architecture [11].

that lead to traffic network congestion which can be detailed out as follows.

* Corresponding author: naimahyaakob@unimap.edu.my

1.1 Congestion cause high number of packet loss and lower network performance.

Congestion in SDN must be avoided because it may cause severe network performance. One of the effects of congestion is the increase in the number of packet loss. High number of packet loss causing the node to retransmit the data and this contribute to low packet delivery ratio (PDR). The time for packet to reach at destination node will also be delayed causing end-to-end delay to increase and the throughput to decrease.

Hence, the data might not be valid once it retransmit to the destination node due to high latency occurs. Therefore, avoiding congestion can reduce the afore-mentioned effects and improve the overall network performance.

1.2 Congestion reduces the QoS in SDN

The high number of users are sending the data simultaneously to the SDN controller, it will not only cause high buffer usage but also cause congestion.

This is because high redundancy in packet transmission will increase the traffic and the possibility of buffer overflow that can lead to congestion. Thus, the congestion can affect to packet loss rate increase, delay increase, low throughput, so cannot provide high QoS to users, user may experience interruption in the communication especially in the real-time applications. In addition, huge number of packets will be wasted in sending the redundant information data.

Therefore, there is need to find a better way to improve packet delay ratio, throughput while reducing packet loss rate and latency.

1.3 Lack of congestion control technique in SDN.

SDN mainly deal with high total data information which cannot be afforded high congestion in network. This is because the congestion can cause high packet loss rates and may cause buffer overflow. Therefore, resolution for congestion control mechanisms need to maintain and control network congestion efficiently in SDN.

1.4 Contribution in SDN

Moreover, in this project contribution to analyse the congestion avoidance in SDN. Part of my project aim at the understanding cause network congestion in SDN. Moreover, this research aims to analyse the avoidance of congestion in SDN's network traffic. Finally, research about cause happened congestion from the sender to the SDN controller is important to the future network community for further growth.

2 Software Define Network

The authors in [1][2][3][4] defined the SDN as decoupled the control plane from data plane, difference from legacy networks which were difficult and complex to be managed since the traditional way implements protocol to allow nodes to work with other nodes to exchange information. Nowadays, the new paradigm network architecture can be performed more effective, flexible and easier to manage by using single centralized controller.

The SDN architecture is important sections to be explained because there are tightly related on each level planes. Moreover, each plane has different functions and mechanisms to work in one network system.

2.1. Application Plane

For the application layer, SDN application are programmed so that it can directly be programmable to communicate with their network demands and the behaviour network to controller via Northbound (API) [5]. This is used to communicate between the controller and services and application running over the network. The SDN application consists of one SDN application logic and one or more Northbound Interface (NBI). In the future, another layer of abstracted network control might be accommodated to provide one or more higher level of NBI. SDN's NBI is allocated between SDN application and SDN controller which fulfil the requirement and control the network behaviour.

2.2 Control Plane

In control plane, suitable and right place SDN controller located where SDN controller as centralized entity in change to translating requirement from first layer to third layer and as agent to SDN application with provide abstract of network view. In SDN controller consist of one or more Northbound Interface (NBI) agent link to NBI driver, SDN control logic and SDN Control-Data-Plane Interface (CDPI).

2.3 Data Plane

The data plane consists of the logical network devices [5]. At this level, the forwarding and data processing occurs. An SDN data path comprises a Control Data Plane Interface (CDPI) and one or more set of traffic forwarding engines and zero or more traffic processing functions. For single (physical) network element, it may consist of one or more SDN data path. Moreover, an

SDN data path can also be knowns as across multiple physical network element because logical definition is not fixed.SDN Control to CDPI is allocated between SDN controller and SDN data path, which function as controller to all forwarding operations, capability advertisement and reporting.

2.3 Open Flow

OpenFlow is one of the protocols used in Mininet which is an open source emulator for SDN [6]. It is an open standard interface that enables researchers to run experimental protocols in real SDN network. To allow researchers to handle the experiment, OpenFlow is provided with standardized features and it based on Ethernet switch but can be used in routers and wireless access point.

In a traditional switch, the data plane and control plane occur in the same device which data plane acts as the fast packet forwarding and the control plane acts as high-level decision maker. Hence, an OpenFlow decouples these two planes where data plane still works on switch, then the control plane is separated to make high level decision, normally a standard server. The communication between the control path and SDN controller via a secure channel use OpenFlow protocol.

3 CONGESTION CONTROL IN SDN

One of the methods to control congestion in SDN is implemented using Link Layer Discovery Protocol (LLDP) as shown in Fig. 2. According to the work in [7], LLDP has been proposed where SDN controller sends packets as Packet-out message during process link discovery. The packets will be sent to all switches in the network and when SDN switch receive LLDP packet, it sends the packet to all other switches directly connected to it. After other SDN switches receive an LLDP packets, they send packet to the SDN controller as Packet-in message. The main purpose for this proposed technique is to allow SDN controller to analyse which connected switches with it and can collect data transmission between SDN controller and switches.

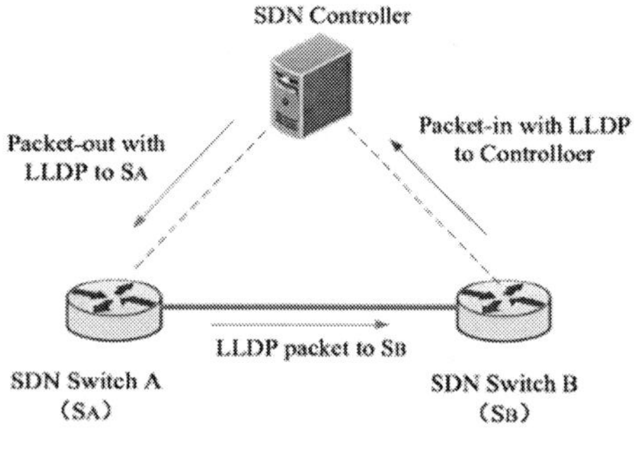

Fig.2. LLDP for SDN topology [7]

As mention before, the first action from controller is sending out the packet as Packet-out message with LLDP packet to the first switch. At the beginning, the second switches will first broadcast its LLDP packet as Packet-out message to find another switch neighbours.

For the first-time operation, the LLDP packet from controller will send at shortest path way to switches. Compared with the OpenFlow (OF) method, OF will send the high number of packets to enable all packets to discovery all switches in the SDN topology.

Hence, the packet will randomly travel to all switches and after that all the packets will identify the switch. Packets will flow through all switches before arriving at the controller because by using OF protocol, there is no forwarding rule in the switches. Let say in the experiment that has a large topology, the OF needs take longer time to arrive at the controller. At the neighbours of switch, LLDP packet will enable FlowTable to send the LLDP packet directly to controller.

3.1 Link Layer Discovery Protocol (LLDP)

In this technique, Packet-out-messages will be sent to all other switches and from the switch, LLDP packets will be sent as Packet-in-message to controller.

This is done because there is no corresponding forwarding rule in the switch's FlowTable. As being discussed in previous section, OpenFlow is an as open source protocol to enable communication between controller and data plane. The OpenFlow protocol takes longer to travel from one switch to another switch before the packet is sent back to the controller. Since it might take longer time to travel, the possibility for the data to be lost is high due to high chances of collision and dropped because of congestion. In this experiment, the method is known as link discovery technique based on LLDP protocol.

The method performs by using LLDP protocol will take the shortest time for packet travel in the link before the packet is retransmitted to the controller and it can analyse which switch are connected directly to each other. This approach is one of the efforts to reduce packet collision and packet drop. The following sections explain about the concept of LLDP.

In the process of implementing LLDP protocol in SDN topology, there are several objectives need to be satisfied. First, LLDP protocol should be able to reduce the number of packet loss during the transmission. This is because each packet loss causing retransmission of the data and every retransmission process will waste the time for a controller to analyse the entire network. Second aims, is to minimize the number of transmissions. This is because by reducing the number of transmission packet will reduce the possibility packet collision with other packet and at the same time percentage of packet loss will be reduce.

 DOI: 10.1051/matecconf/201714001033

3.2 Flowchart of LLDP

From the Fig.2(a), the first packet of LLDP was started at SDN Controller. Then, when first packet LLDP move to first switch, the LLDP packet was tagged as Packet-out message. The Packet-out message will checking another available switch (as S^n). If another switch is available, the packet with LLDP from available switch will sending out Packet-out message to another switch neighbours.

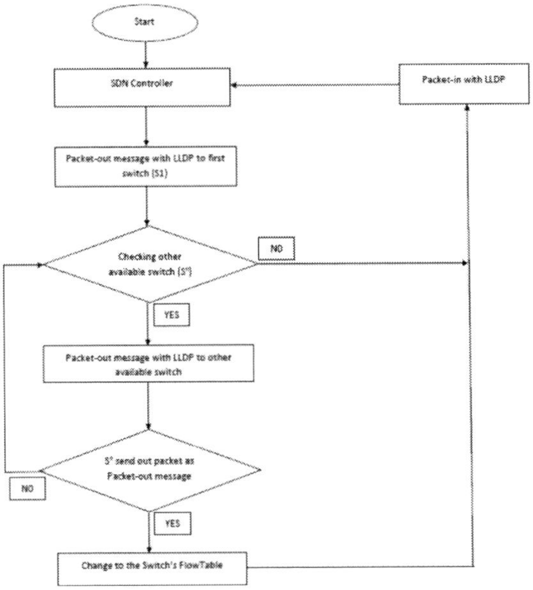

Fig.2(a). Flowchart of LLDP.

If no another switch neighbours receive the LLDP packet, the received switch will sending out the packet as Packet-in with LLDP to the SDN Controller.

4 Methodology

4.1 Simulation Tools

For this research, the emulator tools using for simulating real network devices is Mininet. It was installed in open sources operating system which is Ubuntu 14.04 LTS 64-bit. Mininet is a popular open source network emulator that is capable to combine all components for testbed in SDN. This emulator was created in Phyton language and provide a Phyton application program interface for user customize any level of topology. Moreover, Mininet can virtualize a single operating system. Its hosts can be operated in standard Linux operating system and Linux can support ODL controller.

The installation of Mininet require open source operating system so that Mininet is compatible in Linux. Since it runs a real Linux kernel and network stack, it can be compatible to ready hardware devices with less changes of command line interface (CLI)

4.1.1 OpenDayLight.

OpenDayLight (ODL) of Linux Foundation is open source software platform for SDN and NVF (Network Function Virtualization). Many IT companies such as Cisco, IBM and Dell are participating in ODL project to improve the quality of SDN and NFV. ODL supports OpenFlow, OpFlex and many different network devices control protocol and it also has the advantage with respect to portability since it uses standardized model.

4.1.2 Mininet.

Mininet provides the low-cost testbed for SDN, Authors in [8] using Mininet as SDN testing platform. This is because Mininet allows to develop network of virtual host, SDN controller and switches in very huge scale of topology that up to thousand nodes and it can be perform the test easily. Moreover, Mininet also allows creating the emulator as real-world network scenario and it can emulate SDN controller and OpenFlow devices by using open source operating system such as (Linux Ubuntu server version 14.04. An author in [9], creates the SDN controller using Phyton programming language, hence some C programming language cannot be used to custom the topologies. Authors in [10], was stated that the Mininet is a low cost SDN testbed that can be used for educational and research purpose.

4.2 Simulation Environment

In this experiment, the default command is used for simulation with three different fanout on ODL and virtual network system environment. In table 1 shows the different number of switch for each number of fanout.

Fig. 3. Number of fanout.

Number of fanout	Controller	Switch	Host	Link
1	1	4	1	4
2	1	15	16	30
3	1	40	81	120
4	1	85	256	340

5 Result and Discussion

There are several performance metrics that are being analysed to measure the performance of the proposed method. Those performance metrics are:

1. Packet Loss Rate
2. Packet Delivery Ratio
3. Throughput

For this experiment, only one variable parameter has been tested which is the number of fanout in SDN topology. The first part will present the result of SDN with varying number of packet, while the second part will explain the results types of topology.

The graph shown in the following subsections prove the occurrence of congestion in SDN network before using LLDP protocol. Since we are using Mininet emulator, the topology is already created for basic concept of SDN which consist of controller, switches and many hosts. To implement the first experiment to get variety number of packets, the type of topology was developed in same pattern but difference number of fanout.

The first parameter that we have measured in order to study the effectiveness of LLDP method is packet loss rate. The following section shows the result obtained through the simulation in Mininet.

5.1 Packet Loss Rate.

Referring to Fig. 4, the topology of SDN architecture without LLDP protocol gives highest rate of packet loss. This shows that, congestion is possible to occur in a typical SDN architecture.

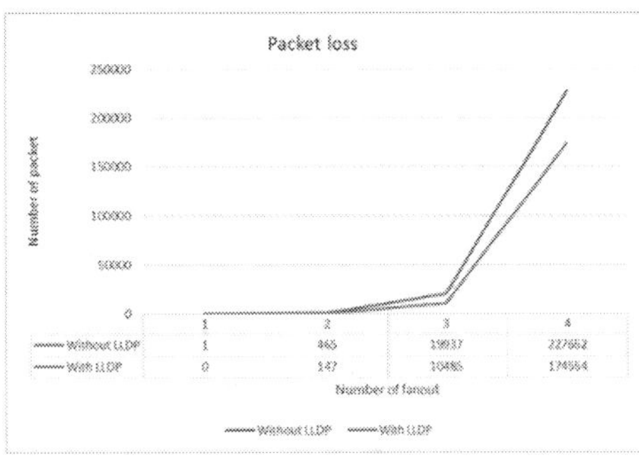

Fig. 4. Graph shows the rate of packet loss for different number of fanout

However, as the LLDP protocol was applied in the topology, the rate of packet loss is reduced. This implies that the congestion has been successfully controlled. From the figure, the both methods show increasing pattern of packet loss rate which reflects that as the number of fanout increases, the number of packet loss also increases. However, when the LLDP protocol is implemented, the number of packet loss is reduced.

This is due to less packet collision occurs when the transmission of packet to other host. To make sure that the number of packet loss is really decreasing when using LLDP, the percentage of the packet loss is calculated and compared. Fig. 5 shows the comparison of packet loss rates in both cases (with and without LLDP protocol) where the pattern for LLDP is decreasing. The congestion is successfully controlled using LLDP since the percentage of packet loss is less than 70% which is the threshold value [6] for the large SDN topology that can be accepted. Basically, the pattern of the packet loss should be decreasing as the number of fanout is lower.

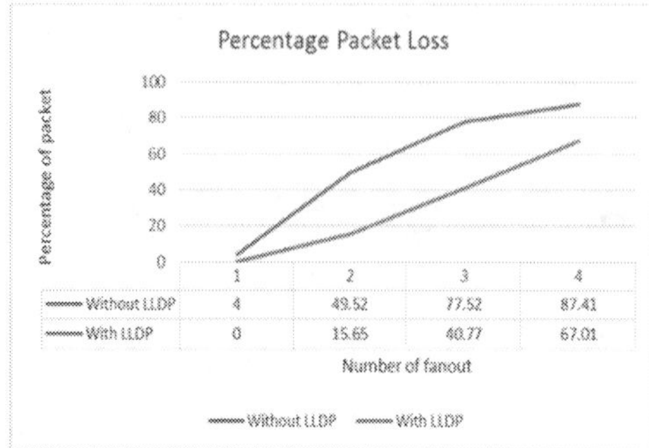

Fig. 5. Graph shows the percentage of packet loss with and without LLDP protocol.

This is because the decreasing of number of nodes in an SDN topology will reduces the number of packets, hence minimize the probability of having collision with other packets while at the same time reduces the congestion. Less congestion means less number of packet are is lost or dropped.

From Fig. 4, the number of packet loss at four number of fanout is very high which reaches 227652 packets especially when there is no LLDP protocol applied. This is because congestion in SDN is highly likely to occur in the increasing number of switches is 85. However, as the number of fanout decreases, the number of packet loss is also reduced. This shows that number of fanout is one, perform 0% number of packet loss when the LLDP protocol was implemented. Moreover, the graph shows decreasing pattern of packet loss value in varying number of fanout with the reduction number when using LLDP protocol. This implies that the LLDP protocol improves the performance of by successfully reduces the collision between packets in the network and thus avoid congestion from happening.

5.2 Packet Delivery Ratio (PDR).

Packet Delivery Ratio is illustrated in Figure 6, which shows the percentage of PDR in both cases (with and without LLDP). As can be seen in the figure, LLDP (which is represented by the red line) achieves 100% percentage of PDR when the number of fanout is zero. This is because no packet loss or packet drop occurs in SDN after LLDP protocol was implemented into the topology. On the other hand, the collected data without the implementation of LLDP protocol shows 92.30% percentage of PDR due to the process of link discovery which takes long period of time because the switch now is not controlled by the controller, thus the packets will travel to each switch before the packet arrive at controller.

However, as the number of fanout increases from 2 to 4, the percentage of PDR is decreasing slightly due to the increasing number of switches in the SDN topology. The implementation of LLDR protocol into SDN topology really increases the percentage of PDR when the number of fanout is two, where the comparison shows that SDN with LLDP is achieve 42.5% PDR.

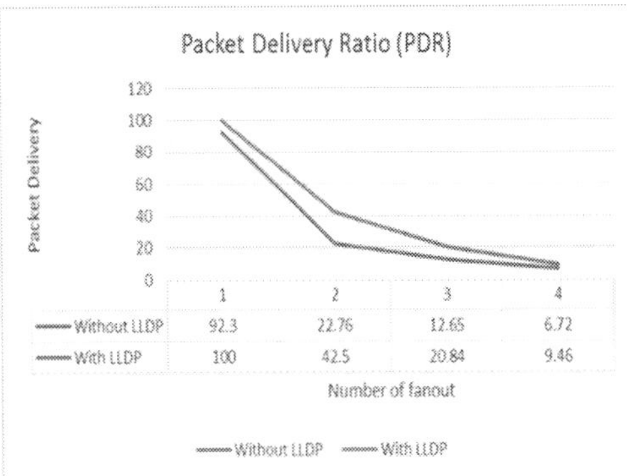

Fig. 6. Red line shows percentage of packet delivery ratio in the presence of LLDP protocol and blue line shows the percentage of packet delivery ratio without LLDP.

This is because LLDP packet will sent back to the controller as Packet-In message, since there is no corresponding forwarding rule in the switch's FlowTable, while without LLDP is 22.76%. Moreover, the pattern is decreasing as the number of node increase when the number of fanout is four. It reflects higher number of switches and the increase in the period link discovery can decrease the PDR.

5.3 Throughput.

Throughput is the average or mean of the packet that successfully send to the destination through a communication link over a period (second). From Figure 7, both graphs show the pattern of decreasing for the throughput. For the original method without the LLDP protocol, the throughput value is in the range of 20 to 100. While for SDN with the LLDP protocol, the throughput is in the range of 50 to 150.

According to the figure 7, the throughput reading shows the decreasing pattern as the number of fanout increase because when the number of nodes increase, more transmission occurs in the topology and it can increase the packet collision. Then number of packets to be dropped or lost will also be higher. Hence, the number of packet successfully sent per time slot will be decreased. The results obtained is then compared with theoretical value and we can conclude that the results comply with the theory. This is due to the more transmission occurs in the network in the sense that the switch that is implemented with LLDP will help sending the Packet-In message to the controller and controller can analyse which switches are connected directly to each other. On the other hand, the switch that is not be implemented with LLDP need takes long period time travel during process of link discovery. As a conclusion, the lower number of nodes to transmit the packet, the higher network throughput

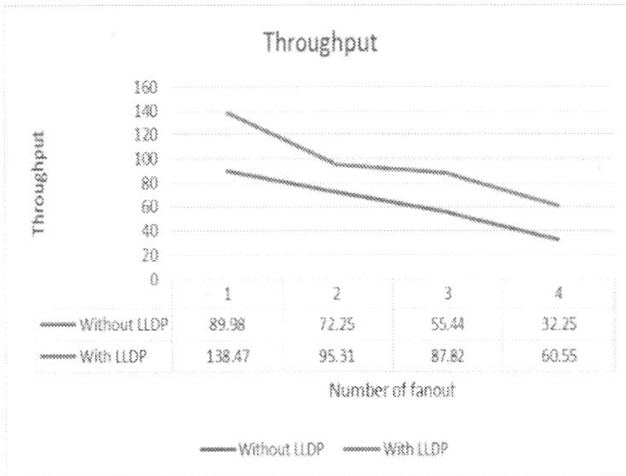

Fig. 7. Red line shows throughput for implement LLDP protocol and blue line shows the throughput without LLDP.

6 Conclusion

Software Defined Network is rapidly growing that has changed the current network topology. It not only handles hundreds switch but it need handle more than thousands of switches to able a billion devices and user in the future. Therefore, to make very big topology, the high possibilities the congestion in the SDN networking can happened. Thus, controller is one of the most important part in develop SDN topology because it used to manage and monitoring entire network from any congestion and traffic load.

SDN controller normally used to collect important information regarding the occurrence of congestion in network link that caused by the larger number of data sends more than the capability and it can make the higher the possibility for congestion to occur. The effect of congestion can be seen by the degradation in the overall quality and the increase in the packet loss rates leading to packet drops, buffer delay and end-to-end delay. As SDN have very limited resource, the congestion network issue is critical to be solved. Concerning on this issue, researchers was introduced Link Layer Discovery Protocol (LLDP) to control congestion in the SDN topology. The experiment already conducted to see the effect of LLDP towards the SDN topology.

DOI: 10.1051/matecconf/201714001033

7 Future Work

Here are some suggestions that can be considered to improve future projects. This research has focused on the method to avoid congestion and the technique used to deal with the subsequent issues that could affect performance of SDN and its underlying platforms. Even though it has achieved desirable performances, possible future studies could be a platform to further improve our proposed solution for using high speed transmission rate in Mininet emulator. Based on experiment, it is more specified only to several topologies and considered as small topology compared if the SDN applied in real network topology that needs huge and capability switches to connect with each other. Therefore, an extension of our work including for high speed transmission rate and buffer limiter should be taken care for the next research investigations.

Acknowledgment

The research reported in this paper is supported by Research Acculturation Grant Scheme (RAGS) [Grant number: 9018-00082]. The author would also like to express gratitude to the Malaysia Ministry of Higher Education (MOHE) and University Malaysia Perlis for the facilities provided

References

1. H. Kim, J. Kim, and Y.-B. Ko, "Developing a cost-effective OpenFlow testbed for small-scale Software Defined Networking," 16th Int. Conf. Adv. Commun. Technol., pp. 758–761, 2014.
2. S. Song, J. Lee, K. Son, H. Jung, and J. Lee, "A congestion avoidance algorithm in SDN environment," Int. Conf. Inf. Netw., vol. 2016-March, pp. 420–423, 2016.
3. Y. Huang, M. Lee, X. Huang, and C. Hsu, "Minimizing Flow Initialization Latency in Software Defined Networks," pp. 303–308, 2015.
4. R. Horvath, D. Nedbal, and M. Stieninger, "A Literature Review on Challenges and Effects of Software Defined Networking," Procedia Comput. Sci., vol. 64, pp. 552–561, 2015.
5. Open Networking Foundation, "SDN Architecture Overview," Onf, pp. 1–5, 2013
6. P. M. Mohan, D. M. Divakaran, and M. Gurusamy, "Performance study of TCP flows with QoS-supported OpenFlow in data center networks," IEEE Int. Conf. Networks, ICON, pp. 1–6, 2013.
7. Z. Shu et al., "Traffic Engineering in Software-Defined Networking: Measurement and Management," vol. 3536, no. c, 2016.
8. D. Kumar and M. Sood, "Software Defined Networks (S.D.N): Experimentation with Mininet Topologies," Indian J. Sci. Technol., vol. 9, no. 32, 2016.
9. C. Decusatis, A. Carranza, and J. Delgado-caceres, "Modeling Software Defined Networks using Mininet," Proc. 2nd Int. Conf. Comput. Inf. Sci. Technol. Ottawa, Canada –, no. 133, pp. 1–6, 2016.
10. J. L. M. N. Weerawardhana, N. J. A. W. Chandimal, and A. Bandaranayake, "SDN Testbed for Undergraduate Education," vol. 2015, 2015.
11. H. F. Xavier and S. Seol, "A Comparative Study on Control Models of Software-Defined Networking (SDN)," vol. 7, no. 32, pp. 1747–1753, 2014.
12. M. Chen, Y. Qian, S. Mao, W. Tang, and X. Yang, " Software-defined mobile networks security," Mobile Netw. Appl., doi : 10.1007/s11036-015-0665-5.
13. F. Pakzad, M. Portmann, W. L. Tan, and J.Indulska, "Efficient topology discovery in software defined networks," in *Proc. 8th Int. Conf. Signal Process. Commun. Syst. (ICSPCS)*, Dec.2014, pp. 1-8.
14. S. Chowdhury, M. F. Bari, R. Ahmed, and R. Boutaba, "PayLess : A low cost network monitoring framework for software defined networks," in *Proc. IEEE Netw. Oper. Manage. Symp. (NOMS)*, May 2014, pp. 1-9.
15. S. A. Mehdi, J. Khalid, and S. A. Khayam, "Revisiting traffic anomaly detection using software defined networking," in *Proc. Int. Symp. Recent Adv.* Intrusion Detection, Menlo Park, CA, USA, Sep. 2011, pp. 161-180.

MATEC Web of Conferences **140**, 01034 (2017) DOI: 10.1051/matecconf/201714001034
ICEESI 2017

Potential of Nanocellulose Composite for Electromagnetic Shielding

Nurul Fatihah Nabila Yah[1], *H.A.Rahim*[1,*], *Yeng Seng Lee*[2], *Wee Fwen Hoon*[1], *Mohd Fareq Malek*[3], *Nik Adilah Hanin Zahri*[1], and *Hayati Hasibuan Zainal*[1]

[1]Bioelectromagnetic Research Group (BioEM), School of Computer and Communication Engineering, Universiti Malaysia Perlis, Kampus Pauh Putra, 02600, Arau, Perlis, Malaysia

[2]Faculty of Technology Engineering, Universiti Malaysia Perlis, Sg. Chuchuh Campus, Padang Besar, Perlis, Malaysia

[3]Faculty of Engineering and Information Sciences, University of Wollongong in Dubai, Dubai, United Arab Emirates

Abstract. Nowadays, most people rely on the electronic devices for work, communicating with friends and family, school and personal enjoyment. As a result, more new equipment or devices operates in higher frequency were rapidly developed to accommodate the consumers need. However, the demand of using wireless technology and higher frequency in new devices also brings the need to shield the unwanted electromagnetic signals from those devices for both proper operation and human health concerns. This paper highlights the potential of nanocellulose for electromagnetic shielding using the organic environmental nanocellulose composite materials. In addition, the theory of electromagnetic shielding and recent development of green and organic material in electromagnetic shielding application has also been reviewed in this paper. The use of the natural fibers which is nanocelllose instead of traditional reinforcement materials provides several advantages including the natural fibers are renewable, abundant and low cost. Furthermore, added with other advantages such as lightweight and high electromagnetic shielding ability, nanocellulose has a great potential as an alternative material for electromagnetic shielding application.

1 Introduction

Recent years have shown a remarkable progress in development of wireless technology worldwide. In particular, technology that aims for higher frequency and wider bandwidth for establishing a high-speed communication system were rapidly developed to accommodate consumer need. However, all of these systems result in an increasing electromagnetic pollution or electromagnetic interference (EMI) to the surrounding. It has become a major concern which brings the need to shield the unwanted electromagnetic signals for both proper operation and human health.

Electromagnetic interference (EMI) can be defined as conducted or radiated electromagnetic signal that interferes with the operation of other electronic devices [1]. It can cause malfunction of the devices and even can be harmful to life [2]. To address this problem, studies on good materials for electromagnetic interference shielding is necessary. Electromagnetic interference shielding refers to the reflection and/or absorption of electromagnetic radiation by a material [3].

Due to these circumstances, researches on the new and better materials for electromagnetic interference shielding have gained attention. Metals such as aluminium, copper and steel in forms like sheet metal, metal screen and metal foam are the most widely used materials for EMI shielding [2]. But, there are several limitations in the applicability of metals in shielding application since they are heavy, not easily handled and they suffer from corrosion. Therefore in the last few years researchers have focused their interest in developing new shielding materials.

The present paper reviews the potential of nanocellulose material in electromagnetic shielding application. The second part of the paper discusses the theory and mechanism of electromagnetic shielding. The third part covers the recent development in researches on green and organic material as alternative material in electromagnetic application. Next the fourth part encompasses nanocellulose composites, its application and potential in electromagnetic shielding field. Lastly the challenges and potential of this research is concluded.

* Corresponding author: haslizarahim@unimap.edu.my

2 Theory of Shielding

Three mechanisms of shielding have been reported which are absorption, reflection and multiple-reflection. For the EMI shielding by reflection, the material must have mobile charge carriers which interact with the electromagnetic wave [43]. Meanwhile, absorption is depends on the thickness of the materials. Shielding by absorption is significant when the shield have electric or magnetic dipoles. The third shielding which is multiple-reflection refers to internal reflection within the shielding materials. The loss due to multiple-reflection can be ignored if the shield is thicker than skin depth.

Figure 1 shows the electromagnetic wave attenuation through a shielding material with different mechanism of shielding.

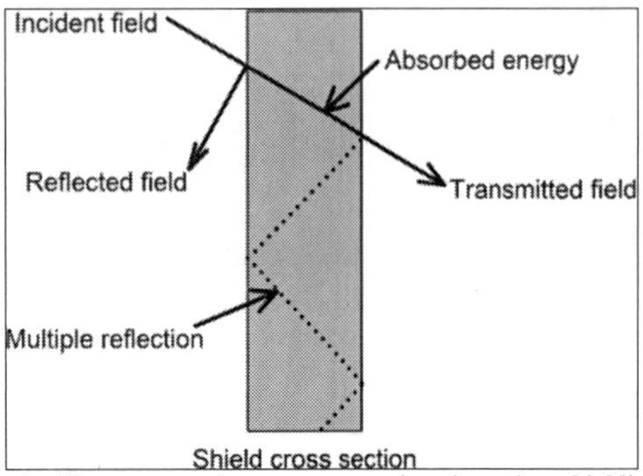

Fig. 1, Electromagnetic wave attenuation through a shielding material [5]

Electromagnetic shielding effectiveness (EMI SE) of a material is the ratio of transmitted power (P_t) to incident power (P_i), expressed in decibels (dB) as in Eq. (1): [5]

$$SE_{total}\,(dB) = -10\log(P_t\,/P_i) \quad (1)$$

When an electromagnetic wave reaches a shielding material, two waves will be created at the external surface; a reflected wave (E_R) and a transmitted wave(E_{I-R}). As the transmitted wave travels through the shield, the strength of the wave exponentially decreases due to absorption. Once the wave reaches the second surface, a portion of the wave will be transmitted (E_T) and a portion will be reflected into the shield (E_{MR}). If the shield is thicker than the skin depth the reflected wave will be absorbed by the shielding material, thus can be ignored. The contribution of reflection, absorption and multiple-reflection to the total shielding effectiveness is defined in Eq. (2) : [4-6]

$$SE_{total}\,(dB) = SE_R + SE_A + SE_{MR} \quad (2)$$

Using S parameter, the transmittance (T), reflectance (R), and absorbance (A) through the shielding material is obtained; T = $S^2{}_{12} = S^2{}_{21}$, R = $S^2{}_{11} = S^2{}_{22}$, A = 1 – R – T. If multiple-reflection is ignored, the effective absorbance (A_{eff}) inside the shielding material can be defined as;

$$A_{eff} = (1 - R - T)/(1 - R) \quad (3)$$

Therefore (SE_R) and (SE_A) become:

$$SE_R = -10\log(1 - R) \quad (4)$$

$$SE_A = -10\log(1 - A_{eff}) = -10\log(T/1 - R) \quad (5)$$

Based on shielding theory, shielding effectiveness (SE) is as follow:

$$SE_A\,(dB) = 20d\,(\frac{\mu_r\omega\sigma}{2})^{1/2}\log(e) \quad (6)$$

$$SE_R\,(dB) = 10log\,(\frac{\sigma}{16\mu_r\omega\varepsilon_0}) \quad (7)$$

$$SE_{total}\,(dB) = 20d\,(\frac{\mu_r\omega\sigma}{2})^{1/2}\log(e) + 10log\,(\frac{\sigma}{16\mu_r\omega\varepsilon_0})$$

Where d is the sample thickness (cm), e is 2.718, μ_r is relative magnetic permeability, $\omega = 2\pi f$ is the angular frequency, $\sigma = 2\pi f\,\varepsilon^*$ is electrical conductivity (S/cm) and ε_0 is the dielectric constant of vacuum.

3 Green or Organic Material in Electromagnetic Shielding Application

For environmental awareness, green or organic materials are the materials of choice for shielding material among a few researchers. In [7], the authors discussed the reflection loss performance of hexagonal base pyramid microwave absorber using different agricultural waste material. Three sample of agricultural waste are used that is rice husk, rice straw and banana leaves. Microwave absorbers based on dielectric properties of those materials measured with free space measurement technique are designed using Computer Simulation Technology Microwave Studio (CST MWS). The simulation shows that many parameters can affect the performance of an absorber.

Liyana et al. [8] fabricated microwave absorbers using sugar cane bagasse material. They investigated the potential of sugar cane bagasse as an alternative material for EMI shielding. High carbon contain in agricultural waste particularly sugar cane bagasse increase its ability or potential to be used as absorber material. Besides that, research and experimentation also shows that corn stover could be used as the material for microwave absorber [9]. The preliminary results prove that the performance of the organic absorber is comparable to the commercially available absorber.

Table 1 shows a comparison of reflection loss performance for different type of material as microwave absorber. Agricultural wastes were chosen by many researchers due to their eco-friendly nature, inexpensive, abundantly available and good microwave absorption properties.

DOI: 10.1051/matecconf/201714001034

Table 1. Reflection loss performance of different types of materials

Researchers	Material	Target frequency	Reflection loss performance
Hassan Nornikman et al., 2011 [10]	Rice husks	7 – 13GHz	Better than -10dB
Liyana et al., 2012 [8]	Sugar cane bagasse	0.1 – 20GHz	Better than -30dB
H. Nornikman et al., 2010 [7]	Rice husks Rice straw Banana leaves	1 – 20GHz	Better than -10dB
Salleh et al., 2011 [11]	Coconut shell-derived	8 – 12GHz	Better than -10dB
Smythe et al., 2014 [9]	Corn stover	5.4 – 18GHz	
Farhany et al., 2012 [12]	Dried banana leaves	0.1 – 20GHz	Average -82.3dB
Noordin et al., 2012 [13]	Oil palm ash	8 -12GHz	Better than -10dB Best at -49dB
Shaaban, Se, Ibrahim, & Ahsan, 2015 [14]	Rubber wood sawdust	1-3GHz	Best at -16.4dB
Kaur, Aul, & Chawla, 2015 [15]	Dried banana leaves and coal	8.2-12GHz	Average -45.2dB

4 Nanocellulose Composites

Cellulose which is an organic compound easily obtained from nature is a structural component of the cell walls of many plants. It is used mainly for making paper and cardboard but recently it has attracted a lot of interest in various industrial applications [16]. This biopolymer has been used in various application and fields due to its availability, eco-friendly nature, sustainability and also flexibility [171816].

Cellulose also has received special attention for development of conducting materials. For example, Jabbour et al. [19] used papermaking techniques to produce graphite/carbon fiber/cellulose fiber composites papers with tuneable conductivity and good mechanical properties. The produced conductive papers are flexible, easy to handle and have an extremely good conductivity. Carbon nanotube (CNT) / cellulose composites material using papermaking process was fabricated in [20]. Higher electric conductivity compared to the usual polymer-based composite materials was achieved without affecting the paper strength.

In recent years, cellulosic fibers in micro and nano scale are gaining attention to replace man-made fibers as reinforcement to make an environment-friendly composite. Researchers worldwide have been focusing on the production of nanocellulose (NC) with crystalline structure because of its exceptional material properties [21].

A few examples of biodegradable polymer–CNF based nanocomposites and their mechanical properties were discussed in [17]. This includes Poly(vinyl alcohol) – CNF, Poly(lactic acid) – CNF, Poly(ethylene oxide) – CNF, Chitosan – CNF, Starch – CNF and Soy protein – CNF. It has been discovered that NC can be applied to energy devices such as paper batteries, supercapacitors and paper displays.

It has been acknowledged that nanocellulose has great potential as reinforcement in nanocomposites. Therefore nanocelluose can be applied to various applications and fields in the upcoming decades. However, its applicability for electromagnetic shielding notably has not been widely explored. Polymer based composites are still commonly used as electromagnetic interference shielding materials [1]. Additionally, it has been discovered that cellulose are used for conducting material which is very efficient in electromagnetic shielding. High carbon content also increases its ability to be used for electromagnetic shielding. Besides, different types of filler could be incorporated into nanocellulose composite in order to get a better and higher performance of electromagnetic shielding. For example graphene, carbon nanotube (CNT), Carbon Black (CB) and other C-based materials [22]. Therefore, nanocellulose composite has a high potential to be develop as an alternative material for electromagnetic shielding that is lightweight and with a wider bandwidth.

Conclusion

The recent development of green material in electromagnetic shielding and nanocellulose composite were reviewed. It has been show that nanocellulose has an exciting potential as reinforcement in nanocomposites and are indeed the future materials. Nanocellulose can be applied to various applications and fields. Further studies and researches on nanocellulose as alternative material to replace current petroleum based material are highly required especially in electromagnetics shielding applications.

Acknowledgement

This research was funded by the grant of Fundamental Research Grant Scheme (No. 9003-00591) under Ministry of Education, Malaysia.

References

1. Thomassin, J. M. *et al.* Polymer/carbon based composites as electromagnetic interference (EMI) shielding materials. *Mater. Sci. Eng. R Reports* **74,** 211–232 (2013).
2. Suchea, M. *et al.* Nanostructured composite layers for electromagnetic shielding in the GHz frequency range. *Appl. Surf. Sci.* **352,** 151–154 (2015).
3. Chung, D. D. . Electromagnetic interference shielding effectiveness of carbon materials. *Carbon N. Y.* **39,** 279–285 (2001).
4. Al-Saleh, M. H. & Sundararaj, U. Electromagnetic interference shielding mechanisms of CNT/polymer composites. *Carbon N. Y.* **47,** 1738–1746 (2009).
5. Bayat, M., Yang, H., Ko, F. K., Michelson, D. & Mei, A. Electromagnetic interference shielding effectiveness of hybrid multifunctional Fe3O4/carbon nanofiber composite. *Polym. (United Kingdom)* **55,** 936–943 (2014).
6. Kuester, S., Barra, G. M. O., Ferreira, J. C., Soares, B. G. & Demarquette, N. R. Electromagnetic interference shielding and electrical properties of nanocomposites based on poly (styrene-b-ethylene-ran-butylene-b-styrene) and carbon nanotubes. *Eur. Polym. J.* **77,** 43–53 (2016).
7. Nornikman, H., Malek, F., Soh, P. J. & Azremi, A. A. H. Reflection loss performance of hexagonal base pyramid microwave absorber using different agricultural waste material. *2010 Loughbrgh. Antennas Propag. Conf. LAPC 2010* 313–316 (2010). doi:10.1109/LAPC.2010.5666029
8. Liyana, Z. *et al.* Investigation of sugar cane bagasse as alternative material for pyramidal microwave absorber design. *IEEE Symp. Wirel. Technol. Appl. ISWTA* 66–70 (2012). doi:10.1109/ISWTA.2012.6373879
9. Smythe, B., Casserly, S. & Arakaki, D. Organic-based microwave frequency absorbers using corn stover. *IEEE Antennas Propag. Soc. AP-S Int. Symp.* 920–921 (2014). doi:10.1109/APS.2014.6904788
10. Nornikman, H. *et al.* Setup and Results of Pyramidal Microwave Absorbers Using Rice Husks. *Prog. Electromagn. Res.* **111,** 141–161 (2011).
11. Salleh, M. K. M. *et al.* Experimental verification of multi-layer coconut shell-derived microwave absorbers. *2011 IEEE Int. RF Microw. Conf. RFM 2011 - Proc.* 115–118 (2011). doi:10.1109/RFM.2011.6168709
12. Farhany, Z. S. *et al.* Potential of dried banana leaves for pyramidal microwave absorber design. *IEEE Symp. Wirel. Technol. Appl. ISWTA* 60–65 (2012). doi:10.1109/ISWTA.2012.6373878
13. Noordin, I. R. M. *et al.* Investigation of oil palm ash microwave absorber for broadband application. *Proc. - 2012 IEEE 8th Int. Colloq. Signal Process. Its Appl. CSPA 2012* 232–235 (2012). doi:10.1109/CSPA.2012.6194724
14. Shaaban, A., Se, S., Ibrahim, I. M. & Ahsan, Q. Preparation of rubber wood sawdust-based activated carbon and its use as a filler of polyurethane matrix composites for microwave absorption. *New Carbon Mater.* **30,** 167–175 (2015).
15. Kaur, R., Aul, G. D. & Chawla, V. Improved Reflection Loss Performance of Dried Banana Leaves Pyramidal Microwave Absorbers by Coal for Application in Anechoic Chambers. **43,** 157–164 (2015).
16. Potulski, D. C., De Muniz, G. I. B., Klock, U. & De Andrade, A. S. Influ??ncia da incorpora????o de celulose microfibrilada nas propriedades de resist??ncia mec??nicas do papel. *Sci. For. Sci.* **40,** 345–351 (2014).
17. Kim, J. H. *et al.* Review of nanocellulose for sustainable future materials. *Int. J. Precis. Eng. Manuf. - Green Technol.* **2,** 197–213 (2015).
18. Pandey, J. K., Takagi, H., Nakagaito, A. N., Saini, D. R. & Ahn, S. H. An overview on the cellulose based conducting composites. *Compos. Part B Eng.* **43,** 2822–2826 (2012).
19. Jabbour, L., Chaussy, D., Eyraud, B. & Beneventi, D. Highly conductive graphite/carbon fiber/cellulose composite papers. *Compos. Sci. Technol.* **72,** 616–623 (2012).
20. Imai, M., Akiyama, K., Tanaka, T. & Sano, E. Highly strong and conductive carbon nanotube/cellulose composite paper. *Compos. Sci. Technol.* **70,** 1564–1570 (2010).
21. Jung, Y. H. *et al.* High-performance green flexible electronics based on biodegradable cellulose nanofibril paper. *Nat. Commun.* **6,** 7170 (2015).
22. Qin, F. & Brosseau, C. A review and analysis of microwave absorption in polymer composites filled with carbonaceous particles APPLIED PHYSICS REVIEWS A review and analysis of

DOI: 10.1051/matecconf/201714001034

microwave absorption in polymer composites filled with carbonaceous particles. **61301,** (2012).

 DOI: 10.1051/matecconf/201714001035

IoT Enabled Low Cost Wearable Device for Location and Information Tracking System

Nur Atika Binti Kamaludin[1*], *Normaliza Omar*[1], *Thennarasan Sabapathy*[1], *Nursabrina Binti Iskandar*[1], *and Muhammad Ramlee Kamarudin*[2]

[1]Biolectromagnetic Research Group, School of Computer and Communication Engineering, Universiti Malaysia Perlis (UniMAP), Malaysia

[2]Centre for Electronic Warfare, Information and Cyber (EWIC), Cranfield Defence and Security, Cranfield University.

Abstract. An internet-of-things (IoT) enabled low cost wearable device for location and information tracking system is developed to monitor and ensure the safety of the children while their parents are away either to work or abroad. This system aims to ensure human safety especially the children and efficiently initiate search and rescue in the event of emergencies. The device will notify the parents on the condition of the surroundings of the children, specifically inside the house and notify the parents if the children are outside the target area by using the application developed in the smartphone. The system uses a microcomputer called Beagle-Bone Black (BBB) as the processor while GPS and LTE technology for wireless communication. Future improvement mainly focuses on reducing the size and power consumption of the device, improving communication module, and optimizing microcomputer's specification.

1 Introduction

Generally, tracking system is the system that is used to track any object or human at any location. Nowadays, this system is very popular to track anything at any time. Recently, there many techniques and technologies can be using to develop and improve this system. The aim of this work is to ensure the safety of the children while their parents are away either to work or abroad. The proposed system will notify the parents on the condition of the surroundings of the children which specifically targeted inside the house and notify the parents if the children are outside the target area.

Nowadays, there are many techniques and technologies can be implanted in tracking system. Various platform has been used for such location tracking system for home users [1]. However, these systems are expensive and has high complexity. Some of the system are also does not support internet-of-things (IoT). The proposed location tracking system in this work uses latest technologies to make it user friendly. It supports IoT and adopts geo-fencing [2] techniques to alert the parents about the location of the children. The development of the proposed system is further explained in following section.

2 Methodology

Methodology discusses about the overall components involved in the proposed system. Mainly these components divided to five parts such as M-WiCom, As shown in Figure 1, the component used it is Beagle-Bone Black also known as M-WiCom for the operation system and smartphone for parents to monitor the position of their children. Physically from that figure, Beagle-Bone Black (BBB) divided into three parts of antenna. There are GPS, LTE and RF module [3]. But in this work, GPS and LTE are only being used. For the GPS antenna will connect with GPS satellite. The function of GPS is to monitor the coordinate or the position of their children. For the LTE antenna is to transmit the information about the children through the smartphone. In this Beagle-Bone Black, there is processor that that control all the system in the Beagle-Bone Black.

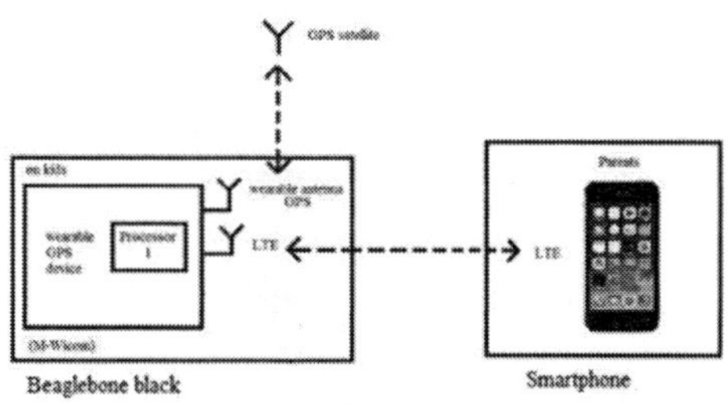

Fig. 1. System concept.

The flowchart in Figure 2 shows the operational flow of the whole system. It starts with setting the current GPS coordinate as origin. Then it specifies radius limit to 100 meters. Next get the latest GPS coordinate will be obtained when the user moves. Whenever the children move, it can read the distance for the movement of the children. The notification will be sent when the distance current GPS coordinate minus the set current GPS

* Corresponding author: atikakamaludin94@gmail.com

DOI: 10.1051/matecconf/201714001035

coordinate as origin more than 100 meters. When the notification on, it will directly send information to the parents through the LTE by using smartphone.

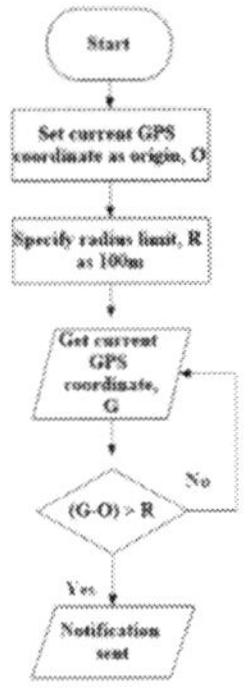

Fig. 2. Flowchart of the system operation.

2.1 Location tracking system using M-WiCom (GPS and LTE enabled)

The Figure 3 shows the M-WiCom. This location tracking system uses M-WiCom which is based on Beagle-Bone Black (BBB) as the processing platform. M-WiCom is selected because it GPS and support with LTE on a single board. Thus it could reduce the complexity of the design where the use of many devices can be avoided.

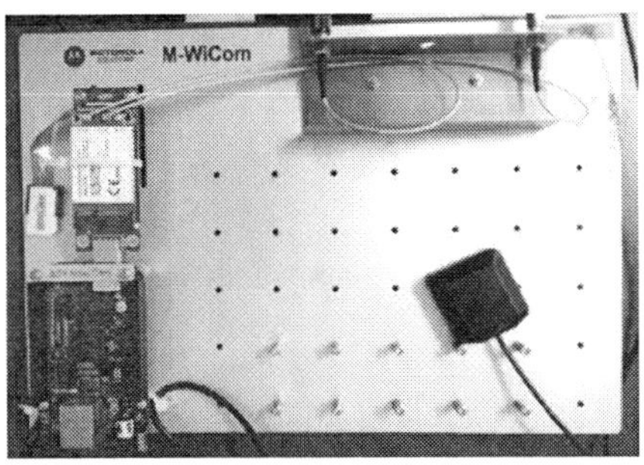

Fig. 3. M-WiCom.

2.1.1 Data Processor

Figure 4 shows the Beagle-bone Black Microcomputer. A microcomputer called Beagle-Bone Black (BBB) with 1 GHz processing speed and 512 MB. The microcomputer can perform similarly to a regular computer. It can boot with various operating systems which are easily available. This microcomputer is used for the following reasons:

1. The board has UART serial interface which is suitable with RF transceiver.
2. It has a USB port which enables it to be connected to a 4G LTE module.
3. The board can perform network programming like regular computer, thus enables it to access external databases.

Fig. 4. Beagle-Bone Black Microcomputer.

On the other hand, modem is needed for the data processor to connect to the internet. For the purpose of providing mobile internet connection, an embedded wireless module is used. The wireless module is Sierra MC7304 as shown in Figure 5, support both 3G/4G LTE and GPS communication. This module is attached to a USB adapter with a SIM card slot which is then connected to the data processor via its USB port.

Fig. 5. Wireless internet with mini PCI-E to USB adapter.

Figure 6 shows the SIM card slot. The SIM card stores data such as user identity, location and phone number, network authorization data, personal security keys, contact lists and stored text messages [2].

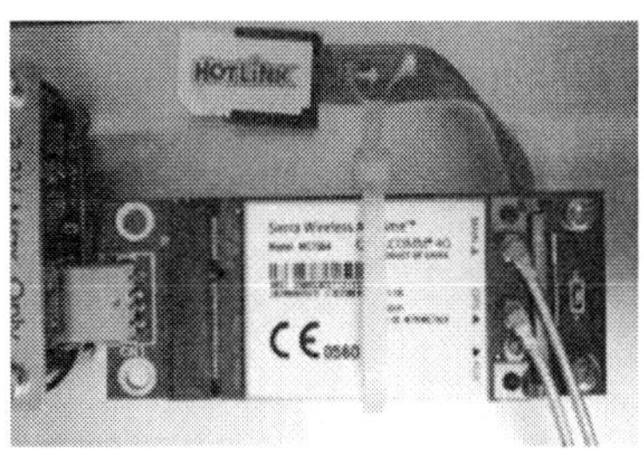

Fig. 6. SIM card slot.

The operating system used in BBB is Ubuntu Linux. Figure 7 shows the block diagram that illustrating the role of microcomputer. The GPS coordinates received by the BBB. Then according to the processed data, the LTE (4G) communication will be established if the notification needed to be sent to the parent's mobile.

Fig. 7. Microcomputer architecture block diagram.

2.1.2 GPS communication device and LTE modem

In this work, the communication device of GPS and LTE are proposed to use. Communication device of GPS is used to monitor the position of the children. Meanwhile, the communication device of LTE is used to send information to the parents. Both techniques are suitable to make some improvement and development of the tracking system.

The GPS is used for tracking the location. There are three ways to target location. Figure 8 shows the ways to target location. The geo-fencing method is selected to give a certain coverage for the user. Geo-fencing is a virtual perimeter and virtual barrier for geographical areas [4]. It could be dynamically generated as in a radius around a store or point location. In this proposed framework, 100 m radius is selected around the center point of a location. Notification will be send to parents when the children move out from the radius value.

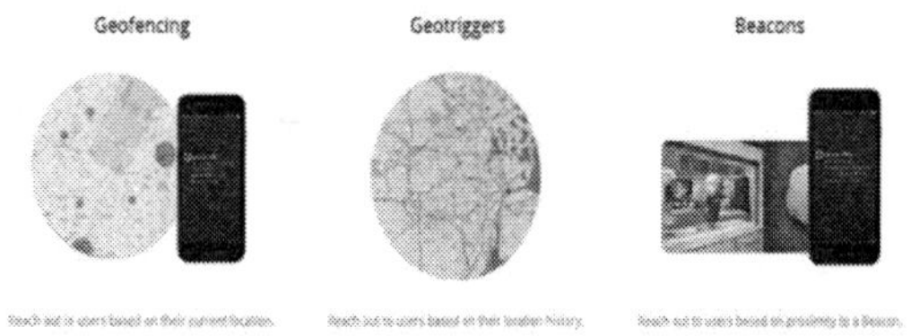

Fig. 8. Ways to target location.

Long Term Evolution or known as LTE is one of the communication devices for this tracking system. This Mobile terminal modem unit interfaces with cellular base stations and send or receives user information data generated by the application unit. So, it interacts with a base station using different cellular interface standard like GSM, LTE or others to send or receive information to distantly located, called server. It also interacts locally with its applications units, like data transfer applications for getting or providing the user application data [5]. LTE is used to send information from M-WiCom to the smartphone.

2.2 Smartphone (Notification to parents)

The use of the smartphone in this system is to receive the notification from the M-WiCom. MIT apps inventor will be developed for the smartphone to receive the notification. MIT Apps Inventor is an innovation beginner's introduction to programming and app creation that transforms the complex language of text-based coding into visual, drag and drop building blocks. The simple graphical interface grants even an inexperienced novice the ability to create a basic, fully functional app within an hour or less. The android apps will be developed using Android. Any information about the position their children can be known through this smartphone. Through that application, the parents can monitor their children's location. Figure 10 shows the application in the smartphone.

Fig.10. Application in the smartphone.

2.3 Software Development

In this section, the stage by stage development of the software part is explained. It start with the software development in BBB by using Linux to Apps development for mobile phone. A detailed explanation is given in the following subsections.

2.3.1 Software development in BBB location tracking system and GPS fencing

Linux software has been used to develop in BBB location tracking system and GPS fencing. To calculate the distance between the two coordinates, equirectangular approximation has been used. The formula below shows how to calculate the distance between two points:
Formula

$$x = \Delta\lambda \cdot \cos \varphi m \quad (1)$$

$$y = \Delta\varphi \quad (2)$$

$$d = R \cdot \sqrt{(x^2 + y^2)} \quad (3)$$

defined:

var x = (λ2-λ1) * Math.cos((φ1+φ2)/2)
var y = (φ2-φ1)
var d = Math.sqrt(x*x + y*y) * R
var R = 6371e3in metres
var φ1 = lat1 in Radians
var φ2 = lat2 in Radians
var Δφ = (lat2-lat1) in Radians
var Δλ = (lon2-lon1) in Radians

2.3.2 Development of ThingSpeak

ThingSpeak is an open source "Internet of Things" application and API to store and retrieve data from things using HTTP over the Internet or via a Local Area Network. With ThingSpeak, the user can create sensor

DOI: 10.1051/matecconf/201714001035

logging applications, location tracking applications, and a social network of things with status updates. Data from the Beagle-Bone Black that contain GPS location will be transmit into ThingSpeak. This ThingSpeak will display into graph that can be easier to analyse.In the ThingSpeak, react name has been used as react PushingBox. PushingBox is a cloud that can send notifications based on API calls. From one request, it can send several notifications like a Push, a Tweet, an Email, message and others. This react can be notification to the parents by using message. From the Figure 13, the initial or origin of coordinate can be put at the latitude and longitude part. Apart from that, it also can set how further we need to notify from the origin coordinate.

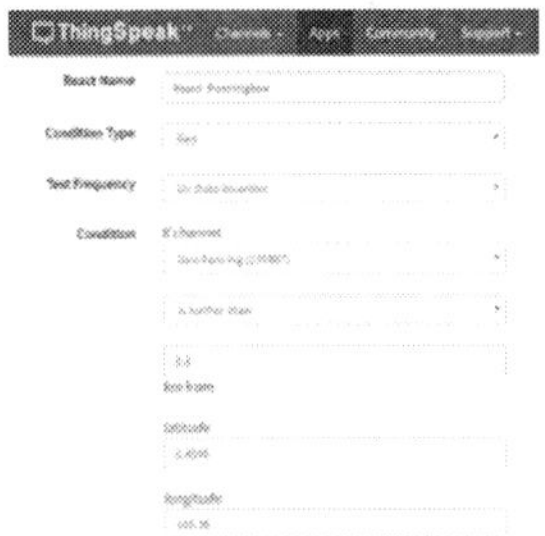

Fig.13. Development for the apps part.

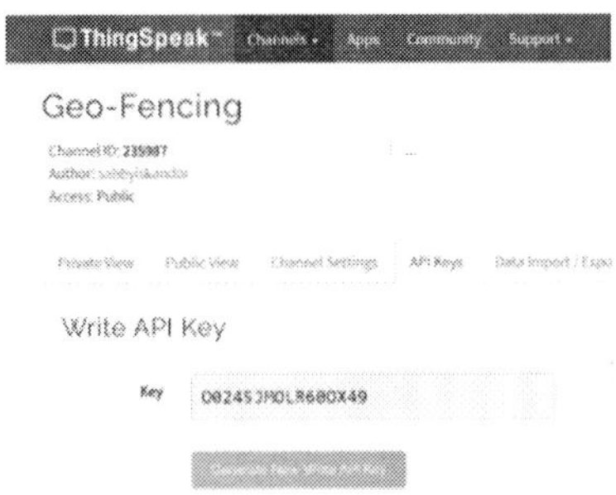

Fig.14. API key.

Figure 14 shows the API key for the ThingSpeak. It also displayed the channel ID. This API key is very important because it will be put on the URL that needed from the apps development by using MIT apps inventor.

2.3.4 Application development in smartphone

This MIT App inventor will divide into two parts, first part for the design layout on the screen and the second part for the block layout.

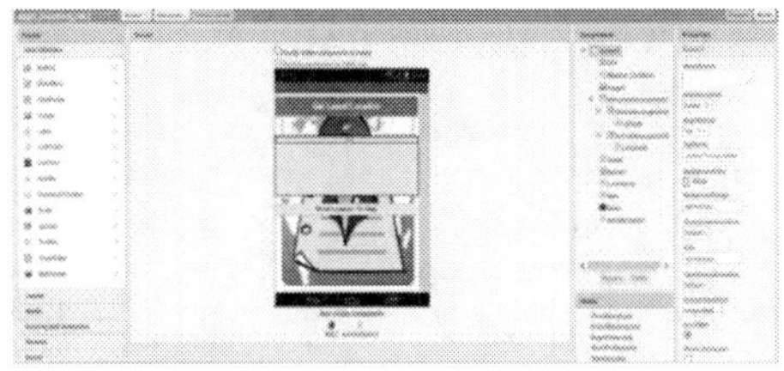

Fig.15. Design layout.

Figure 15 show the design layout for the apps. The design layouts contain the user interface, layout, media, drawing and animation, sensors and social. Each palette can be design depend on the designer desire by referring their own project. Anything can be change on the properties and component part.

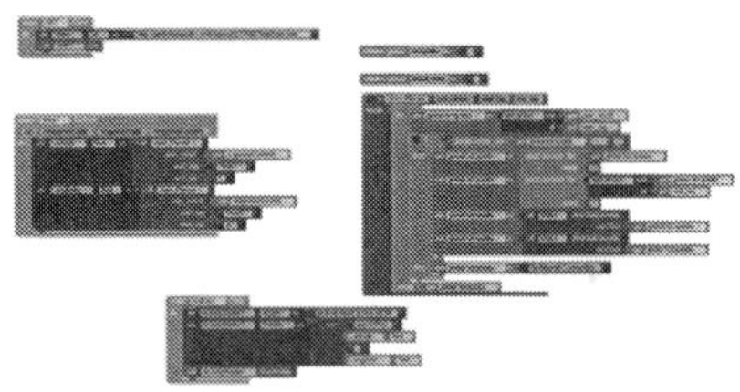

Fig.16. Block layout.

The Figure 16 shows the block layout for the apps design. The block displays the button in the the block form with the different colour. The URL from the web at ThingSpeak will be inside the block. The function for the URL is to call the ThingSpeak web and get the data from that web.

2.3.5 Application development to give notification to parents

Apart from the apps, parents will get the notification by the message with the voice message by using the Newtifry apps. Parents or user need to install the Newtifry from the play store into their smartphone. Newtifry is a push notification service to send server notifications to Android Mobile phones, running android. It uses Google Cloud Messaging APIs. It is intended to replace notifications sent by SMS. SMS cost money to send, and for single user applications this can give you real time notifications for free. Unlike other systems that do the server monitoring for user, Newtifry is just intended to get the message and voice message to the phone and no more. Newtifry is meant to be very simple and just focus on getting the notifications to the phone. User can plug Newtifry into your existing monitoring system to get the notifications going. Users supply the system administrators with a single unique key that identifies at message source, which they can enable or disable at will.

3 Result and Discussion

In this part, the result can be review from the BBB, ThingSpeak, Apps and message notification. The coordinate from the reading of GPS antenna will be sent to the ThingSpeak in a graph form. From the graph at the ThingSpeak, the data can be an analyse it. After that, the data will be sent to the apps in the smartphone and sent the message notification.

3.1 Result monitored in BBB

Result from the BBB will display on the terminal by using Linux software. The terminal will show the value of latitude and longitude at the current position. The Figure 17 shows the terminal from the Linux software.

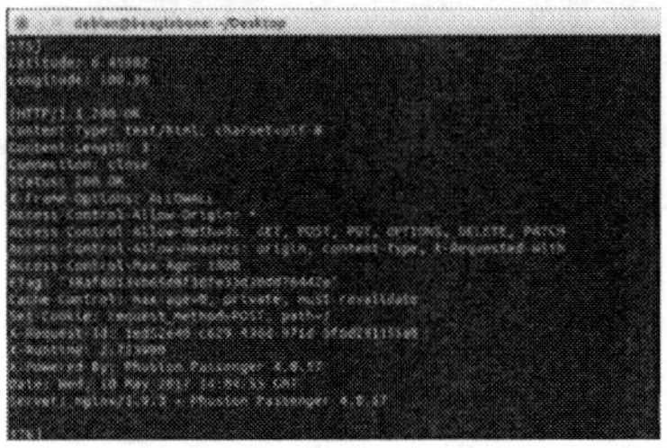

Fig.17. Terminal from the Linux software.

3.2 Result monitored in ThingSpeak

In this ThingSpeak, it will divide the chart into two parts. It contains GPS latitude and GPS longitude. The Figure 18 shows the GPS Latitude set for the origin while the Figure 19 shows the GPS Latitude for latest coordinate. From the graph below, x-axis will display date while the y-axis will display value of latitude. It can see from the graph, each dot with red colour represent the movement of the children. From the graph, it can be easily to see and analyse.

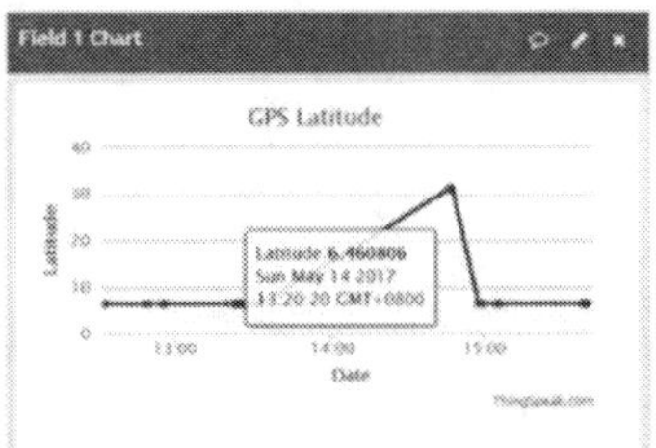

Fig.18. GPS Latitude for the origin.

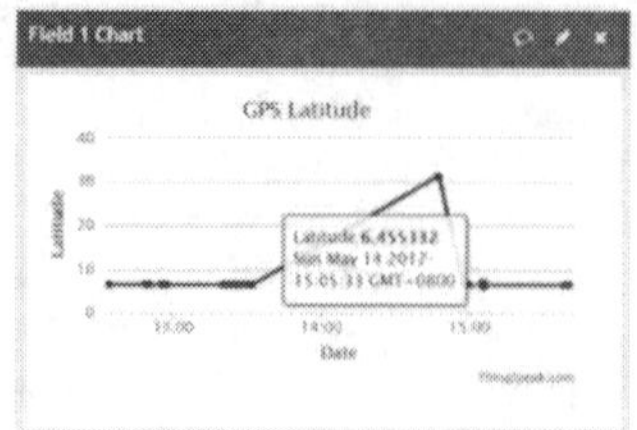

Fig.19. GPS Latitude for the latest coordinate.

The Figure 20 shows the GPS Longitude for the origin while Figure 21 shows the GPS Longitude for the latest coordinate. From the graph below, x-axis will display the date while the y-axis will display value of Longitude. From the graph below, x-axis will display date while the y-axis will display value of Longitude. It can see from the graph, each dot with red colour represent the movement of the children. From the graph, it can be easily to see and analyse.

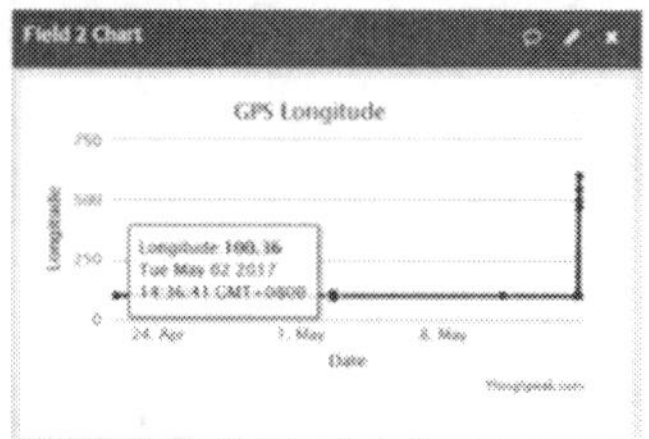

Fig.20. GPS Longitude for the origin.

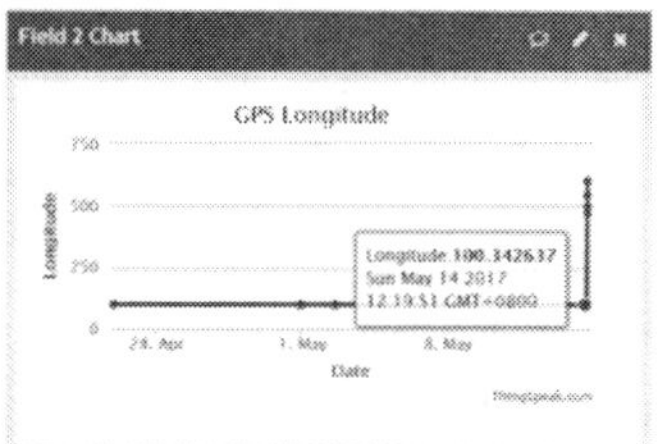

Fig.21. GPS Longitude for the latest coordinate.

The coordinate from the origin to the latest coordinate as shown as the graph above is more than 100 meters. Then, notification message will be sent to the parents and the latest coordinate also will display in the apps.

3.3 Result monitored on Apps

The apps is named as Location tracking system for children. This apps display the GPS latitude and longitude. For the additional part, we show the location on map based on the latest GPS coordinate. This location on map will help the parents share the location to the police, ambulance or any rescue bodies if any accidents happen to their children. The Figure 22 shows the icon apps in the smartphone. This application same with the other apps that already existed in the regular smartphone.

DOI: 10.1051/matecconf/201714001035

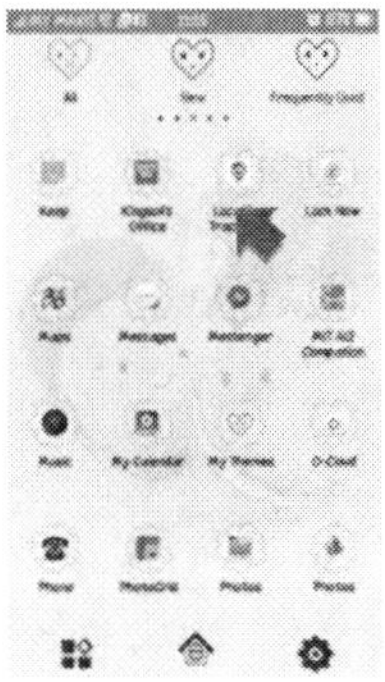

Fig.22. Icon in the smartphone.

The Figure 23 shows the design layout for the apps. After click the icon part, the apps will display as the Figure 23. When the user clicks the button "Get Latest Location", the coordinate will display in the form as latitude and longitude. From the design layout, it will show the latitude value is 6.4544 and the longitude value is 100.36. That coordinate shows the latest location for the children. There is a button "Show Location on Map" at the middle part. After the user knows the latest coordinate for their children, they can click that button "Show Location on Map". This button will display directly to the address in a goggle map exactly like in the Figure 23 by referring the current coordinate.

This location in Google map will help the parents send the location immediately to the police, ambulance or others rescue bodies if anything happens for their children. Figure 24 shows the Location in Google map.

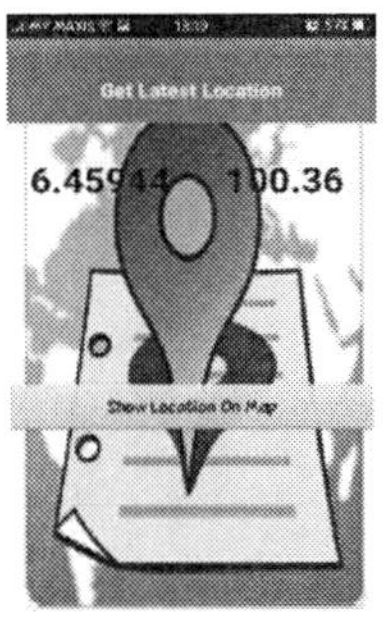

Fig.23. Design apps layout.

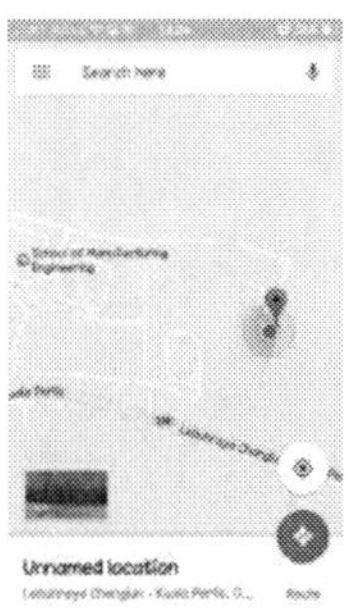

Fig.24. Location in Google map.

3.4 Result monitored on notification message

Apart from the apps, the user or parents will receive the message notification from the Newtifry. The notification will display the date and time as show at the Figure 25. The voice message will produce simultaneously with the message notification. After a click, the source as shown in the Figure 26, the detail message will display. The message will display "Your children is away from your home" as shown at the Figure 26.

Fig.25. Message notification from Newtifry.

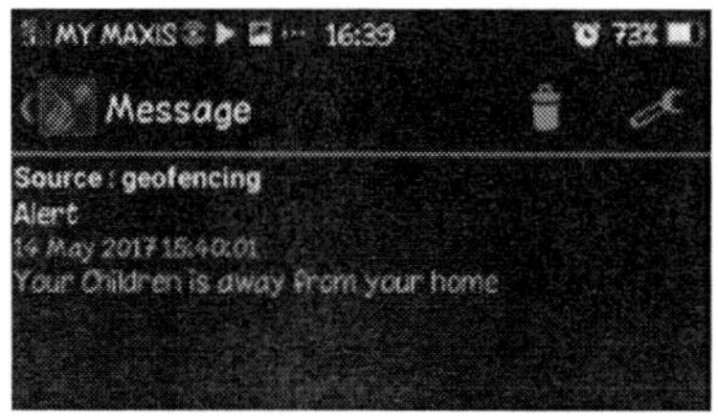

Fig.26. Message details.

4 Conclusion and Future Work

A prototype of a location and information tracking system for children is developed and presented. The GPS antenna is validated to be capable of acquiring coordinate data for the children and them wirelessly to the microcomputer. The microcomputer is able to accept these data and channel them to android apps through the smartphone. Parents as the user of location and information tracking system is able to monitor the movement of their children based on android apps.

References

1. A. B. Mieee, "GPS , GPRS , GIS FOR Tracking System," vol. 1, no. 8, pp. 527–529, 2011.
2. S. Holtmanns, S. P. Rao, and I. Oliver, "User location tracking attacks for LTE networks using the interworking functionality," 2016 IFIP Netw. Conf. (IFIP Networking) Work. IFIP Netw. 2016, pp. 315–322, 2016.

3. A. Nayyar and V. Puri, "A Comprehensive Review of BeagleBone Technology : Smart Board Powered by ARM," vol. 10, no. 4, pp. 95–108, 2016.
4. "Construction Research Congress 2012 © ASCE 2012 818," pp. 818–827, 2012.
5. G. Cardone, A. Cirri, A. Corradi, L. Foschini, R. Ianniello, and R. Montanari, "Crowdsensing in Urban areas for city-scale mass gathering management: Geofencing and activity recognition," IEEE Sens. J., vol. 14, no. 12, pp. 4185–4195, 2014.
6. A. K. Kamaraj Radha M Priyanka M Punitha Asst, "Intelligent Transport System Using Integrated GPS Optimized Reader," pp. 332–336, 2016.

A New Perceptual Mapping Model Using Lifting Wavelet Transform

TahaBasheer Taha[1],Phaklen Ehkan[1], Ruzelita Ngadiran[1]

[1]School of Computer and Communication Engineering, University Malaysia Perlis (UniMAP), Perlis, Malaysia

Abstract. Perceptual mappingapproaches have been widely used in visual information processing in multimedia and internet of things (IOT) applications. Accumulative Lifting Difference (ALD) is proposed in this paper as texture mapping model based on low-complexity lifting wavelet transform, and combined with luminance masking for creating an efficient perceptual mapping model to estimate Just Noticeable Distortion (JND) in digital images. In addition to low complexity operations, experiments results show that the proposed modelcan tolerate much more JND noise than models proposed before.

Keywords: JND; Accumulative Lifting Difference (ALD);Lifting Wavelet Transform (LWT);PerceptualMapping.

1. Introduction

Human Visual System (HVS) cannot perceive visual alerts unless it pass a certain threshold called Just Noticeable Distortion (JND)[1]. Hence, JND models are employed as pre-processing step in implementation of different image processing techniques as enhancement, compression and watermarking to increase the appearance quality of digital images and videos. As consequence, a lot of attempts were adopted to estimate JND in transform domain or in time domain.

An early Discrete Cosine Transform (DCT) based JND masking attempt was proposed by Watson [2], the model estimates the perceptibility of frequency sensitivity, luminance masking and contrast masking, these factors are polled to create what is called Watson distance (D_{wat}). Another DCT based perceptual model is proposed by Niu et.al [3], where the spatial Contrast Sensitivity Function (CSF) , luminance adaptation, and contrast masking are combined to produce the JND estimator.Recently, Fazlali et.al [4] proposed a JND model for estimating maximum watermark embedding strength. DCT is applied on the detailed band in second level of Contourlet Transform (CT) after its partitioning into blocks. Severity of embedding is found by two factors, edge concentration and entropy of each block.

Although an acceptable performance is achieved by DCT transform with JND models, the usage of Discrete Wavelet Transform (DWT) comes with better performance, this due to temporal-spatial characteristics which similar to the Human Vision System behavior[5]. Moreover, In DWT, if the coefficient is modified, only the corresponding region where the coefficient exists is affected [6]. These factors motivated researchers to adopt DWT for creating perceptual masks. Barni et.al [6], proposed a pixel wise JND model where HVS characteristics are estimated by analyzing the imperceptibility of four DWT decompositions. Barni'smodelis utilized in other models, as [7] where the entropy masking is added besides previous factors for JND mask enhancement.Logarithmic calculations used in entropy determination are complex operations which increase the computational overhead, furthermore, and generally speaking, despite of accepted results achieved by analysing perceptual features using traditional DWT, it relies on Fourier transform which involves imaginary number calculations.Also, different filters are used for different pixels' positions [8]. These factors will bound the usage of transform domain models in basic processors found in embedded systems where resources such as CPUspeed, number of logic elements and memoryare limited.

In spatial domain attempts, JND valuesare directly estimated, Chou and Li's [9] proposed a model where JND is found by estimating the background luminance using Weber law and contrast masking effect is determined with maximum signal from four edges. An improvement of Chou's model is proposed by Yang et.al[10] where edge detection is added to avoid over estimation in edges areas. Wu et.al [11]proposed a pixel adaptive JND model to enhance the estimation of texture in previous two attempts, the model characterized by considering the disorder of texture in texture masking and adding it to luminance factor. However, texture areas are yet underestimated and noise tolerance can be significantly enhanced.

The complexity of transform domain and under estimation of spatial domain attempts when creating perceptual masksevokes the need for a perceptual model

DOI: 10.1051/matecconf/201714001036

that combine features of both, simplicity and reliable estimation. In this paper an efficient and high accuracy perceptual mapping model is presented by taking advantage of Lifting Wavelet Transform (LWT) presented by Sweldens [8].LWT is characterized by simplicity, on location processing, and filter similarity. As a consequence, proposed model will ease the usage of texture masking in embedded systems and limited processors applications.

In next section an introduction of the LWT is presented, proposedmethodology explained in section 3, followed by experimental results and comparisons with existing works in section 4. Conclusion written in section 5

2. Lifting Wavelet Transform

Lifting wavelet transform (LWT) is introduced by Sweldens[8] to efficiently construct wavelet coefficient without relying on Fourier transform. It provides low computational process in compare with filter banks used in regular Discrete Wavelet Transform (DWT). Also, LWT has integer to integer computation which makes it suitable to be mapped to hardware design where the floating points are not often recommended. LWT is applied by three major steps [8][12]:

Split

Split step also called the lazy wavelet, where the signal is being separated into smaller subsets. Generally, coefficients are divided into odd and even samples.

Predict

Even samples are used to predict the odd ones, pixel at odd position are predicted by its two neighbors at even positions, then the difference between the predicted value and real odd value is stored in the location of odd samples. Signal after prediction step is the details band. Predict step can be written as (1)

$$Y2n+1=Xo-\text{PREDICT}(Xe) \quad \ldots(1)$$

Update

When the average of each two pixels in a signal is calculated, the overall structure of decomposed signal is obtained with half energy [13]. However, due to the non-linearity changing in image pixels, even samples interposed the odd ones cannot be taken directly as they need to be "Updated" with the differences computed in predict steps (2) . Produced signal is the approximation band.

$$Y2n=Xe+\text{UPDATE}(Y2n+1) \quad \ldots(2)$$

Inverse of lifting scheme is performed by reversing the order and exchanging the sign of predict and update steps [12]. Figures 1, 2 show LWT decomposition and reconstruction operations respectively.

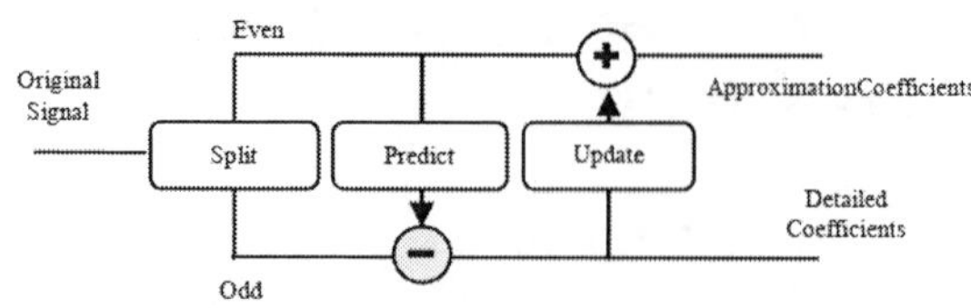

Fig.1. Decomposition of lifting wavelet transform

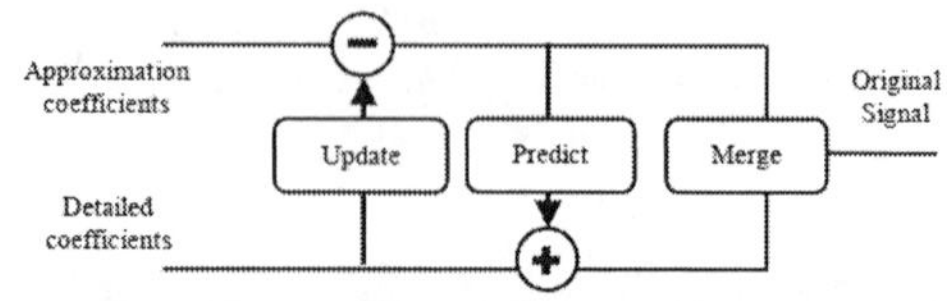

Fig.2. Reconstruction of lifting wavelet transform

The core idea beyond lifting scheme's signal analysis is to divide the original signal into approximation and details bands. Higher Coefficients values in details band refer to higher nonlinear changes in pixels' values, this changes is the region where texture areas found. Proposed method exploits this property to create ALD texture masking model.

3. Proposed Method

To simulate human visual system, Luminance and Texture are employed and combined for creating the proposed perceptual masking model.

3.1 Luminance Masking:

A modified version of luminance equation listed in [14] is considered in this scheme, it states that, according to Chou's law[9], if the gray level of a certain pixel refers to very dark areas and very bright areas, as zero or 255, then the masking value can be 20. For example, pixel can be changed from zero to 20 and the differencewill not be observed by the human eye. And when the gray level is in middle intensity (127)it can tolerate up to three values. For better appearance, tolerance for both very bright areas (255) and very dark areas (0) are set as five. Equation of luminance sensitivity is given as follows:

$$\text{LMask}(I,J)=\begin{cases}\text{floor}\left(\left(\text{Im1}(I,J)-128\right)/32\right)+1) & \text{if } S1(I,J)>127\\ \text{floor}\ (128-(\text{Im1}(I,J))/32+1) & \text{Otherwise}\end{cases} \quad \ldots(3)$$

WhereLMaskis luminance mask, Im(I,J) is the luminance of pixel at (I,J).

3.2 Texture Masking:

As defined in [15], according to statistical approach, "texture is a quantitative measure of the arrangement of intensities in a region". So, frequent change in image

DOI: 10.1051/matecconf/201714001036

intensities produce more textiles area. In fact, the measurement of that disparity is already given by the detailed band equation of lifting scheme (Equation 1), presented by the difference between the actual value and expected value of pixel's intensity.This feature will be exploited to design the proposed texture scheme by dividing the LWT detailed subband (D2) into 5*5 blocks then subtracting neighboured pixels in each row of the block from each other, summarize the absolute of the differences of that subtraction and accumulate the results into single value. The result isAccumulative Lifting Differences (ALD) which is the proposed texture mappingmodel.(Figure 3).

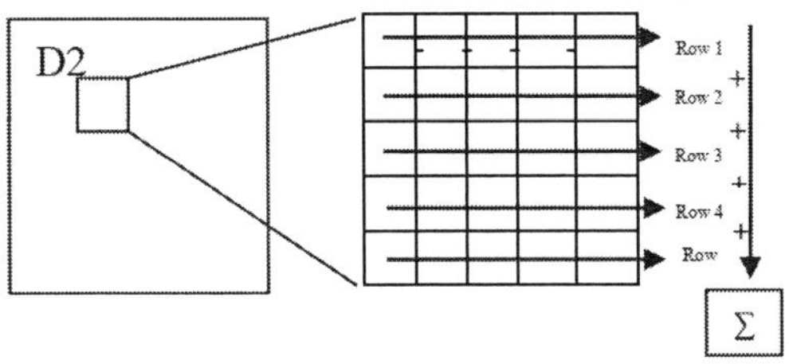

Fig.3.ALD block calculation

Details band (D2) shown in figure 3, has been elected because its coefficients have the difference between approximate value and real value which refer totexture variation. ALD equation is given in (4).

$$ALD\,(I,J)=\sum_{I-2J-2}^{I+2J+2}\sum |D2(J,J-1)-D2(I,J)| \quad ...(4)$$

Where I,J are centre pixels of each 5*5 block. Final mask (T) is the combination of ALD and LMask(5)

$$T(I\,,J) = L(I,J) + ALD(I,J) \quad ...(5)$$

After LWT decomposition, each block in spatial domain will have the same T value for the centre coefficient in S1 band.

4. Results and Discussion:

To evaluate the performance of proposed model, an image with texture and smooth areas is chosen where a set of distributed pixels have been elected for testing.Each pixel indicated by arrow and number in figure 4. Luminance mask values and texture intensity calculated using proposed ALD model is shown in table 1.

For luminance masking, tolerance (mapping value) is increased as the pixel value is close to dark and bright areas. For texture model, ALD values on smoothsky areasare 31, 25, while in mountain surface where high texture intensity can be observed, higher ALD values are obtained, as 495,244. Figure 5 shows the normalized texture mask for Mountain image, where the brighter areas refer to more tolerance to modification.

Fig.4. Original Mountain image

Table 1: Luminance toleration and ALD values for different Blocks

Block	Location of centre pixel (Y , X)	Luminance value	Tolerance of luminance value	ALD Value
1	46,46	139	5	31
2	46,256	168	9	495
3	46,466	147	6	25
4	256,46	88	9	123
5	256,256	140	5	244
6	256,466	72	11	239
7	466,46	102	7	391
8	466,256	65	11	204
9	466,466	109	11	109

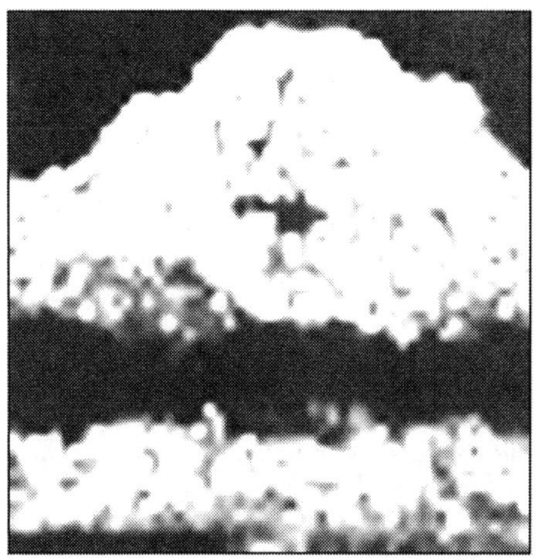

Fig.5.ALD Texture Mask for Mountain image

DOI: 10.1051/matecconf/201714001036

For objective evaluation, Mean Structure Similarity Index (SSIM)[16],a metric that used to predict observed quality of images, is utilized in this paper.Comparisons and evaluation can be established by fixing SSIM value and measure the amount of noise can be injected. Mean Square Error (MSE), is the metric that used for evaluation purpose, where (by fixing SSIM)higher MSE value refers to more tolerance to noise. JND noise guided equation is:

$$fnoise\ (x,y) = f(x,y) + srT(x,y) \qquad \ldots(6)$$

Where s is the random noise as +1 or -1 values, and r is severity of noise.Table 2 shows MSE obtained by applying proposed maskon different standard images with constant SSIM value.

Table 2 : MSE for different images with fixed SSIM value

Image (512*512)	SSIM	MSE
Airplane	0.973	36.42
Babon	0.973	99.39
Boat	0.973	28.11
Cameraman	0.973	102.47
Lena	0.973	26.78

Proposed model is compared with three other JND estimation models, Chou's andLi[9], Yang et.al [10] and Wu et.al [11] models.Baboon image (512*512) is chosen for the comparison between models. Figure 6 shows the noise contaminated images and JND masks for the proposed and compared models. For Chou's and Li's model (a) (b), MSE is 33.35, in Yang et.al (c) (d) MSE is 38.09, while in Wu et.al model (e) (f) MSE is 54.59. In proposed model (g) (h), MSE is 99.39 for the same SSIM value (0.9738). Results shows that the proposed model has better estimation for texture areas and tolerate more noise than other models.

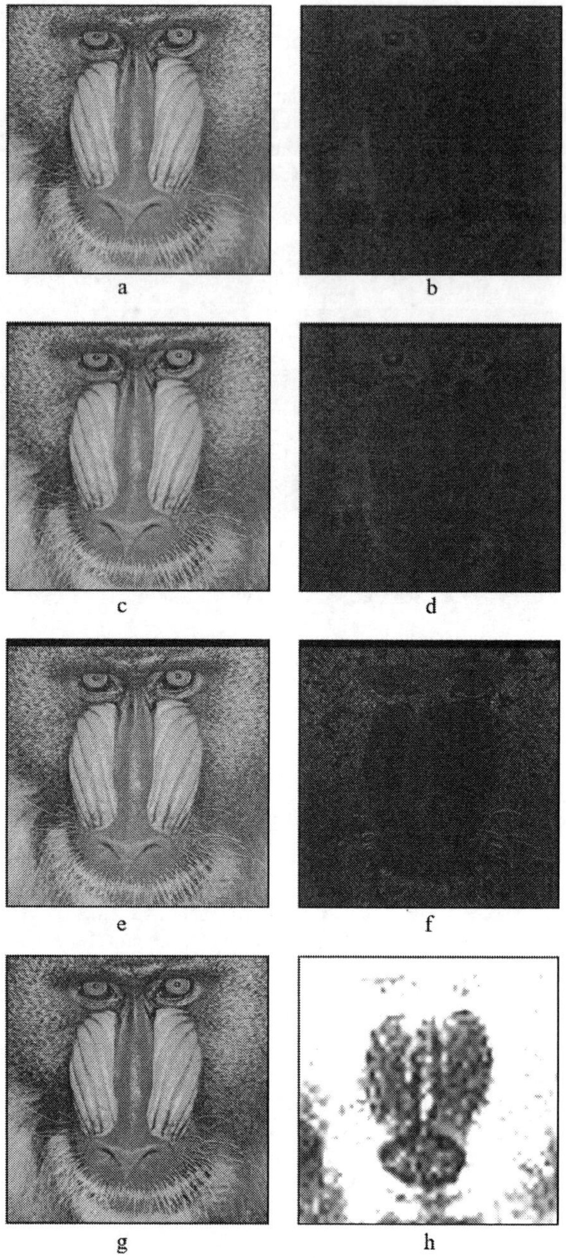

Fig.6. Baboon Image with injected noise (Left), andnoise mask (Right) for different models.

5. Conclusion

Accumulative Lifting Differences (ALD), a new texture mapping model based on low complexity lifting wavelet transform (LWT) is proposed in this paper. ALD is combined with luminance masking to create a novel perceptual model that helps to create JND mask for digital images. Proposed model combines the simplicity of time domain models and the performance of transform domain models. Experimental results shows that the proposed model has efficient estimations of texture regions andoutperforms compared models in amount of noise can be injected without affecting perceptual appearance.

References

1. Wu, J., Qi, F., & Shi, G. (2010, March). An improved model of pixel adaptive just-noticeable difference estimation. In Acoustics Speech and Signal Processing (ICASSP), 2010 IEEE International Conference on (pp. 2454-2457). IEEE.
2. Watson, A. B. (1993, December). DCTune: A technique for visual optimization of DCT quantization matrices for individual images. In Sid International Symposium Digest of Technical Papers (Vol. 24, pp. 946-946). SOCIETY FOR INFORMATION DISPLAY.
3. Niu, Y., Kyan, M., Krishnan, S., & Zhang, Q. (2011). A combined just noticeable distortion model–guided image watermarking. Signal, Image and Video Processing, 5(4), 517-526.
4. Fazlali, H. R., Samavi, S., Karimi, N., &Shirani, S. (2016). Adaptive blind image watermarking using edge pixel concentration. Multimedia Tools and Applications, 1-16.
5. Li, W., Zhang, Y., & Yang, C. (2013). A survey of JND models in digital image watermarking. In Proceedings of the 2012 International Conference on Information Technology and Software Engineering (pp. 765-774). Springer Berlin Heidelberg.
6. Barni, M., Bartolini, F., &Piva, A. (2001). Improved wavelet-based watermarking through pixel-wise masking. IEEE transactions on image processing, 10(5), 783-791.
7. Akhbari, B., &Ghaemmaghami, S. (2005, November). Watermarking of Still Images in Wavelet Domain based on Entropy Masking Model. In TENCON 2005-2005 IEEE Region 10 Conference (pp. 1-6). IEEE.
8. Sweldens, W. (1995, September). Lifting scheme: a new philosophy in biorthogonal wavelet constructions. In SPIE's 1995 International Symposium on Optical Science, Engineering, and Instrumentation (pp. 68-79). International Society for Optics and Photonics.
9. Chou, C. H., & Li, Y. C. (1995). A perceptually tuned subband image coder based on the measure of just-noticeable-distortion profile. IEEE Transactions on circuits and systems for video technology, 5(6), 467-476.
10. Yang, X. K., Ling, W. S., Lu, Z. K., Ong, E. P., & Yao, S. S. (2005). Just noticeable distortion model and its applications in video coding. Signal Processing: Image Communication, 20(7), 662-680.
11. Wu, J., Qi, F., & Shi, G. (2010, March). An improved model of pixel adaptive just-noticeable difference estimation. In Acoustics Speech and Signal Processing (ICASSP), 2010 IEEE International Conference on (pp. 2454-2457). IEEE.
12. Gholipour, M. (2011, June). Design and implementation of lifting based integer wavelet transform for image compression applications. In International Conference on Digital Information and Communication Technology and Its Applications (pp. 161-172). Springer Berlin Heidelberg.
13. Jensen, A., & la Cour-Harbo, A. (2001). The Discrete Wavelet Transform via Lifting. In Ripples in Mathematics (pp. 11-24). Springer Berlin Heidelberg.
14. Qi, H., Zheng, D., & Zhao, J. (2008). Human visual system based adaptive digital image watermarking. Signal Processing, 88(1), 174-188.
15. Shapiro, L., & Stockman, G. C. (2001). Computer vision. 2001. ed: Prentice Hall.
16. Wang, Z., Bovik, A. C., Sheikh, H. R., &Simoncelli, E. P. (2004). Image quality assessment: from error visibility to structural similarity. Image Processing, IEEE Transactions on, 13(4), 600-612

AUTHOR INDEX

AUTHOR INDEX